Springer
Tokyo
Berlin
Heidelberg
New York
Hong Kong
London
Milan
Paris

Hideki Takayasu (Ed.)

The Application of Econophysics

Proceedings of the Second Nikkei Econophysics Symposium

With 137 Figures

Springer

Hideki Takayasu
Senior Researcher
Sony Computer Science Laboratories, Inc.
3-14-13 Higashi-Gotanda, Shinagawa-ku
Tokyo 141-0022, Japan

ISBN 4-431-14028-X Springer-Verlag Tokyo Berlin Heidelberg New York

Library of Congress Cataloging-in-Publication Data

Nikkei Econophysics Symposium (2nd : 2002 : Tokyo, Japan)
The application of econophysics : proceedings of the Second Nikkei Econophysics Symposium / Hideki Takayasu (ed.)
p. cm.
Includes bibliographical references.
ISBN 4-431-14028-X
1. Finance--Statistical methods--Congresses. 2. Finance--Mathematical models--Congresses. 3. Statistical physics--Congresses. I. Takayasu, Hideki, 1958- II. Title.

HG176.5.N55 2002
330'.01'5195--dc22

2003065192

Printed on acid-free paper

Typesetting: Authors

SPIN: 10929506

Preface

The global economy is changing very rapidly, and new approaches to rebuilding economies are attracting a great deal of attention. Econophysics is an exciting new academic field in which economic phenomena are analyzed in a scientific manner using data from the financial markets and corporations.

This book is the proceedings of the second Nikkei symposium, "Application of Econophysics," held in Tokyo, November 12–14, 2002. In the first Nikkei symposium on econophysics, held in November 2000 under the title "Empirical Science of Financial Fluctuations," we established empirical laws based on financial data. Many basic discoveries were reported in the proceedings of the same title published in 2001 by Springer-Verlag.

As expected from the title of the second Nikkei symposium, the research trend in that symposium was clearly directed to application. Predictability of markets, investment strategies, and risk hedge methods were the hot topics. Also, there was remarkable progress in basic studies on market characteristics, novel data analysis methods, and theoretical models.

The symposium was supported by the Financial Markets Department, Bank of Japan; the Economic and Social Research Institute, Cabinet Office, Government of Japan; the Japan Center for Economic Research; the Physical Society of Japan; and the Japan Association for Evolutionary Economics. On behalf of all participants, I would like to thank those supporters, as well as the following companies, without whose financial support the symposium would not have been possible: The Bank of Tokyo-Mitsubishi, Ltd.; Hitachi, Ltd. Financial Technology Consulting Center; Sony Computer Science Laboratories, Inc.; Kozo Keikaku Engineering, Inc.; and the World Gold Council.

As the chief organizer, I am grateful for the cooperation of the organizers, H.E. Stanley (Boston University), T. Shimizu (Bank of Japan), and Y. Kuratsu (RP Tech). I would also like to thank the keynote speaker of the symposium, D. Sornette. Special appreciation is extended to the members of the conference secretariat, as represented by K. Suzuki, for their kindness and efficiency, and to the staff of Springer-Verlag Tokyo, for editorial support. Finally, I would like to thank all the authors for their contributions to this volume.

H. Takayasu
Tokyo 2003

Contents

I-2 Predictability

I-3 New Methods

I-4 Various Markets

I-5 Models

II. Other Topics

II-1 Income Distribution

I. Market Properties

I-1 Basic Statistics

Economic Fluctuations and Statistical Physics: The Puzzle of Large Fluctuations

H. E. Stanley[1] Xavier Gabaix[2], Parameswaran Gopikrishnan[1], and Vasiliki Plerou[1]

[1] Center for Polymer Studies, and Department of Physics, Boston University, Boston, MA 02215 USA
[2] Department of Economics, MIT, Cambridge, MA 02142 and National Bureau of Economic Research, Cambridge, MA 02138

Abstract. We present an overview of recent research joining ideas of economic theory and statistical physics to try to better understand puzzles regarding economic fluctuations. One of these puzzles is how to describe outliers, phenomena that lie outside of patterns of statistical regularity. We review evidence consistent with the possibility that such outliers may not exist. This possibility is supported by recent analysis by Plerou et al. of a database containing the sale price of each trade of every stock. Further, the data support the picture of economic fluctuations, due to Plerou et al., in which a financial market alternates between being in an "equilibrium phase" where market behavior is split roughly equally between buying and selling, and an "out-of-equilibrium phase" where the market is mainly either buying or selling.

1 Introduction

Collaborations between economists and physicists seem to some observers to be making progress in answering significant questions. In particular, these collaborations have the potential to change the paradigm for understanding economic fluctuations. Until relatively recently, theories of economic fluctuations invoked the label of "outlier" (bubbles and crashes) to describe fluctuations that do not agree with existing theory. These outliers are of interest, as they correspond to extremely large and unpredictable changes of sufficient magnitude to wreak havoc.

The paradigm of "statistical regularity plus outliers" does not exist in the physical sciences. Indeed, if events occur that do not conform to predictions of the appropriate theory, then that theory is immediately relegated to the dust bin and new theories are sought. An example are the "outliers" that led to the demise of classical mechanics, eventually replaced by the theory of relativity.

Traditional economic theory does not predict outliers, but recent analysis of truly huge quantities of empirical data suggests that classic theories not only fail for a few outliers, but that there occur similar outliers of every possible size. In fact, if one analyzes only a small data set (say 10^4 data points), then outliers appear to occur as "rare events." However, when orders of magnitude more data (10^8 data points) are analyzed, one finds orders of magnitude more outliers—so ignoring them is not a responsible option, and studying their properties becomes a realistic goal. One finds that the statistical properties of these "outliers" are identical to the statistical properties of everyday fluctuations. For example, a histogram giving the number of fluctuations of a given magnitude x for fluctuations ranging in magnitude from everyday fluctuations to extremely rare fluctuations ("financial earthquakes") that occur with a probability of only 10^{-8} is a perfect straight line in a double-log plot.

An analogy with earthquake research is perhaps not entirely inappropriate. If one studies limited data sets, a paradigm arises in which there are everyday (unnoticeable except by sensitive seismometer) "tremors," punctuated from time to time by rare events ("earthquakes"). Thanks to the empirical work, we now know that the partition of shocks into "tremors" and "earthquakes" is not valid. Rather, if one examines enough data, one sees that the shocks occur for all possible magnitudes. The law named after Gutenberg and Richter refers to a statistical formula that gives all the data from the smallest tremors to the "big ones." This law is that the histogram giving the number of shocks of a given size is a straight line in a log-log plot [1–3]—there are no outliers..

Thus, an inappropriate paradigm can arise when a limited quantity of data are considered in which data are partitioned into everyday events (often describable by one statistical law) and rare events which, since they are not described by the law are terms outliers. Has an inappropriate paradigm arisen in economic research? In economic research, there are fluctuations in stock prices, number of shares trading hands, and total number of fluctuations. Recent empirical studies calculating histograms for all three quantities are linear on log-log plots (albeit with different slopes). In mathematical language, the occurrence probability of such quantity's fluctuations appear to be described by a power law.

In economics, neither the existence of power laws nor the exact exponents has any accepted theoretical basis. Professionally, empirical laws such as the aforementioned power laws are called "stylized facts," a term that to my ear always sounds dismissive. Accordingly, some theoretical understanding is urgently needed or else these laws will continue to be largely irrelevant. Of course facts, even facts without any interpretation, may have practical value. For example, the Gutenberg-Richter law enables one to calculate the risk of a shock (tremor or earthquake) of a given magnitude, and hence informs the building codes of Los Angeles and Tokyo. Similarly, the empirical laws governing economic fluctuations enable one to calculate the risk of an economic shock of a given magnitude.

The lack of a coherent theory is unfortunate, especially in economics where facts without theoretical foundation is considered a deplorable situation. Accordingly, my collaborators and I have been seeking to develop a theoretical framework within which to interpret these new empirical facts, and recently some progress is beginning to occur [4,5]. This work is potentially significant since it provides a theoretical framework within which to interpret the new empirical laws. Specifically, the model fulfills these requirements for such a basic "microscopic" model of the stock market. It is founded on realistic features of the stock market, and reflects the view that market participants have of the functioning of the market, as well as the main determinants of their trading behavior.

2 First Discovery of Scaling and Universality

That at least *some* economic phenomena are described by power law tails has been recognized for over 100 years since Pareto investigated the statistical character of the wealth of individuals by modeling them using the *scale-invariant* distribution

$$f(x) \sim x^{-\alpha}, \tag{1}$$

where $f(x)$ denotes the number of people having income x or greater than x, and α is an exponent that Pareto estimated to be 1.5 [6,7]. Pareto noticed that his result was *universal* in the sense that it applied to nations[1] *"as different as those of England, of Ireland, of Germany, of the Italian cities, and even of Peru"*. A physicist would say that the universality class of the scaling law (1) includes all the aforementioned countries as well as Italian cities, since by definition two systems belong to the same universality class if they are characterized by the same exponents.

In the century following Pareto's discovery, the twin concepts of scaling and universality have proved to be important in a number of scientific fields [9–11]. A striking example was the elucidation of the puzzling behavior of systems near their critical points. Over the past few decades it has come to be appreciated that the scale-free nature of fluctuations near critical points also characterizes a huge number of diverse systems also characterized by strong fluctuations. This set of systems includes examples that at first sight are as far removed from physics as is economics. For example, consider the percolation problem, which in its simplest form consists of placing pixels on a fraction p of randomly-chosen plaquettes of a computer screen. A remarkable fact is that the largest connected component of pixels magically spans the screen at a threshold value p_c. This purely geometrical problem has nothing to do, at first sight, with critical point phenomena. Nonetheless, the fluctuations that occur near $p = p_c$ are scale free and functions describing various aspects of the incipient spanning cluster that appears at $p = p_c$ are described by power laws characterized by exponent values that are universal in the sense that they are independent of the details of the computer screen's

[1] However, the exponent for income does not seem stable across time [8].

lattice (square, triangle, honeycomb). Nowadays, the concepts of scaling and universality provide the conceptual framework for understanding the geometric problem of percolation.

It is becoming clear that almost any system comprised of a large number of interacting units has the potential of displaying power law behavior. Since economic systems are in fact comprised of a large number of interacting units has the potential of displaying power law behavior, it is perhaps not unreasonable to examine economic phenomena within the conceptual framework of scaling and universality [9–20] . We will discuss this topic in detail below.

3 Inverse Cubic Law of Stock and Commodity Price Fluctuations

So having embarked on a path guided by these two theoretical concepts, what does one do? Initially, critical phenomena research—guided by the Pareto principles of scaling and universality—was focused finding which systems display scaling phenomena, and on discovering the actual values of the relevant exponents. This initial empirical phase of critical phenomena research proved vital, for only by carefully obtaining empirical values of exponents such as α could scientists learn which systems have the same exponents (and hence belong to the same *universality class*). The fashion in which physical systems partition into disjoint universality classes proved essential to later theoretical developments such as the renormalization group [11]—which offered some insight into the reasons why scaling and universality seem to hold; ultimately it led to a better understanding of the critical point.

Similarly, our group's initial research in economics—guided by the Pareto principles—has largely been concerned with establishing which systems display scaling phenomena, and with measuring the numerical values of the exponents with sufficient accuracy that one can begin to identify universality classes if they exist. Economics systems differ from often-studied physical systems in that the number of subunits are considerably smaller in contrast to macroscopic samples in physical systems that contain a huge number of interacting subunits, as many as Avogadro's number 6×10^{23}. In contrast, in an economic system, one initial work was limited to analyzing time series comprising of order of magnitude 10^3 terms, and nowadays with high frequency data the standard, one may have 10^8 terms. Scaling laws of the form of (1) are found that hold over a range of a factor of $\approx 10^6$ on the x-axis [21–25]. Moreover, these scaling laws appear to be universal in that they, like the Pareto scaling law, hold for different countries [26], for cities[27–32] , other social organizations [33–35], and even for bird populations [36].

Recent attempts to make models that reproduce the empirical scaling relationships suggest that significant progress on understanding firm growth may be well underway [37,38,40,39,41–43], leading to the hope of ultimately developing a clear and coherent "theory of the firm." One utility of the recent empirical work is that now any acceptable theory must respect the fact that power laws hold over typically six orders of magnitude; as Axtell put the matter rather graphically: *"the power law distribution is an unambiguous target that any empirically accurate theory of the firm must hit"* [21].

With this background on power laws and scale invariance in geometry and in economics, we turn now to the well-studied problem of finance fluctuations, where a consistent set of empirical facts is beginning to emerge. One fact that has been confirmed by numerous, mostly independent, studies is that stock price fluctuations are characterized by a scale-invariant cumulative distribution function of the power law form (1) with $\alpha \approx 3$ [44–46]. This result is also universal, in the sense that this inverse cubic law exponent is within the error bars of results for different segments of the economy, different time periods, and different countries—and is the same for stock averages as different as the S&P and the Hang Seng [47].

This "inverse cubic law" disagrees with the classic work of Ref. [9] on price fluctuations of cotton, which appear to display scale free behavior ("no outliers") but with much fatter tails characterized by $\alpha \approx 1.7$; this work is of interest because if $\alpha < 2$, then the distribution is of the Lévy form. To understand this discrepancy, Matia and collaborators have wondered if the reason for the fatter tails of cotton is that cotton is a commodity, and commodities exist in limited supply so that when a commodity is needed one must sometimes pay exorbitant prices (e.g., electricity in California). Accordingly, they analyzed a large number of commodities, but they found that these commodities have tails described not by $\alpha < 2$ but rather by $\alpha \approx 3$ [48,49].

Newcomers to the field of scale invariance often ask why a power law does not extend "forever" as it would for a mathematical power law of the form $f(x) = x^{-\alpha}$. This legitimate concern is put to rest by by reflecting on the fact that power laws for natural phenomena are not equalities, but rather are asymptotic relations of the form $f(x) \sim x^{-\alpha}$. Here the tilde denotes *asymptotic equality.* Thus $f(x)$ is not "approximately equal to" a power law so the notation $f(x) \approx x^{-\alpha}$ is inappropriate. Similarly, $f(x)$ is not proportional to a power law, so the notation $f(x) \propto x^{-\alpha}$ is also inappropriate. Rather, asymptotic equality means that $f(x)$ becomes increasingly like a power law as $x \to \infty$. Moreover, crossovers abound in financial data, such as the crossover from power law behavior to simple Gaussian behavior as the time horizon Δt over which fluctuations are calculated increases beyond about a year (i.e., the power law behavior holds for time horizons up to a month or even a year, but for horizons exceeding a year there is a distinct crossover to Guassian behavior. Such crossovers are characteristic also of other scale-free phenomena in the physical sciences [10,11], where the Yule distribution often proves quite useful.

For reasons of this sort, standard statistical fits to data are inappropriate, and often give distinctly erroneous values of the exponent α. Rather, one reliable way of estimating the exponent α is to form successive slopes of pairs of points on a log-log plot, since these successive slopes will be monotonic and converge to the true asymptotic exponent α. One finds that successive slopes for the empirical data converge rapidly to a value $\alpha \approx 3$ while successive slopes for the model diverge. While it is clear that a simple three-factor model [60] cannot generate power law behavior, it is less clear why the empirical data analyzed appear at first glance to be well approximated by the model. The first fact is that the region of linearity of the data is not so large as in typical modern studies because the total quantity of data analyzed is not that large, since only a low-frequency time series comprising daily data is used. Only 28,094 records are analyzed [60] (not 4×10^7 as in recent studies [46,47]) and the model simulations are presented for limited sample size. The second fact is that when one superposes a curved line (the model) on a straight line (the data), the untrained eye is easily tempted to find agreement where none exists—and closer inspection of Figs. 2–5 of Ref. [60] reveals actually a rather poor agreement between model and data due to the pronounced downward curvature of the model's predictions [61].

4 Other Scale-Invariant Quantities Describing Economic Fluctuations

Other quantities characterizing stock movements (such as the volatility, share volume traded, and number of trades) also display a range of power law behavior over a range of typically $\approx 10^2$ [62–65]. The exponents characterizing the power law decays are different for different quantities; it is tempting to conjecture that in finance there may exist a set of relations among the power law exponents found, just as there exist relations among the exponents characterizing different quantities near the critical point. Finally, it is well-known that while the autocorrelation function of price returns decays rapidly, the autocorrelation function of the absolute values of price returns is power-law correlated in time (see [62] and extensive earlier work cited therein).

Consider, for example, the volatility. There are several possible definitions of this quantity, all of which seem to give the same scale invariant properties. But why care about volatility at all? On the cover of the 15 May 2000 issue of Forbes magazine is a large photograph of Henk Paulson, CEO of Goldman Sachs, and the headline quotation *"Volatility is Our Friend."* Why is this the case? Because it is known that volatility clusters.....i.e., there are time correlations in this quantity. Our group has attempted to quantify these correlations, and found evidence of power law behavior [62,66–68]. If we plot an economic earthquake such as Black Monday (19 October 1987) on which date most worldwide stock indices dropped 30-50 percent, and then plot and compare the volatility (the absolute value of the fluctuations), we see a big peak in the volatility curve on Black Monday. But even prior to Black Monday the value of the volatility on our graph seems to be particularly unstable;

there is some precursor to Black Monday evident in its behavior. One can imagine a computer program that would monitor volatility, not necessarily for the entire market but certainly for an individual stock, and the volatility calculation would need to be updated in real time.

The distribution of volatility fluctuations has also been the object of extensive study. It was at one time believed by many that the volatility follows a log-normal distribution—i.e., the number of times the volatility has a certain value follows not a Gaussian but a log-normal distribution, i.e., one has $e^{-(\log x)^2}$ not e^{-x^2}. But until our group's work, no one had studied *all* the data: *every* trade [62]. Our doing it meant we could study relatively rare events, those occurring much less frequently than everyday events. What we find is that the log-normal part of the curve—the middle—though true for the middle, does not describe the tails. The huge volatilities in the tails are described by a different exponent μ. We also see that volatility clusters—i.e., that volatility is correlated in time.

5 Cross-correlations among Fluctuations of Different Stocks

Another capability of such a software package could be the ability to determine how the fluctuations of one stock price correlate with those of another. This question of cross-correlation is one we have been studying [69–76]. To quantify cross-correlations,we draw a circle corresponding to the stock price x and draw a second circle corresponding to the stock price x, say, five minutes later. If we make the difference in the radii proportional to G, the stock price change, then we can think of the market as thousands of circles, each growing and shrinking—a kind of pulsation that is a function of time. The key is that these correlations change in time. Car sales by Ford and GM may be anti-correlated during some time periods and positively correlated during others.

The standard approach to this problem is to calculate, by brute force, a huge square matrix that has as many rows as there are companies in the database. Each element of the matrix is the correlation between the price change of company i and the price change of company j, but to find a genuine correlation we have to be able to distinguish between correlations from coincidences. In order to do that we draw on something developed by Wigner in his work in nuclear physics—random matrix theory. Random matrix theory compares the matrix calculated by brute force from stock market data with a random matrix that also has 1000 rows and 1000 columns—but with every number generated randomly. Somewhere hidden in the huge matrix calculated by brute force from stock market data are the true correlations. To uncover them, we first diagonalize the matrix in order to determine its eigenvalues, and then make a histogram that gives the number of times each given eigenvalue is found. The histogram curve of a random matrix, unlike this one from real data, can be predicted exactly. For a random matrix there is never an eigenvalue > 2.0. The histogram of the empirical stock price data, on the other hand, contains a significant number of eigenvalues > 2.0 . Some are as big as 5.0. These eigenvalues of necessity must correspond to genuine correlations.

The eigenvalue of a matrix has a corresponding eigenvector—a column matrix of 1000 elements—each element of which is a different weight from each of the 1000 stocks. So we can look at the column vectors that correspond to these deviating, genuinely-correlated eigenvalues and ask: what kind of stocks entered into each of these eigenvectors? What we found, fortunately, has implications for portfolios. If we restart the graph at 2.0—removing the distortions of the random values—and look at the 20 eigenvalues > 2.0 , we see that the stocks that make up most of the weights in the corresponding eigenvectors are almost entirely transportation stocks in the first case, almost entirely paper in the second, almost entirely pharmaceuticals in the third, and so on. In other words, the market *automatically* partitions itself into separate business sectors [75–77]. Thus a physicist who know nothing about the stock market can mathematically partition the economy into separate business sectors!

The sectors and the quantitative degree to which each constituent firm conforms to the sector can be monitored and updated as a function of time, e.g., every 15 minutes. Firms that belong to the same business sector can be monitored in a kind of rainbow spectrum. The "good" firms sticking to the business sector are assigned to the "violet" end of the spectrum, and the "bad" firms deviating from the sector are assigned to the "red." When a firm first starts to move to the red end of the spectrum start to deviate, this alerts the trader to consider action.

6 Equilbrium vs. Out-of-Equilibrium Market Phases

Before concluding, we ask what sort of understanding could eventually develop if one takes seriously the power laws that appear to characterize finance fluctuations. It is tempting to speculate that there might be analogies between finance and known physical processes displaying similar scale-invariant fluctuations. One initially promising analogy was with turbulence: In turbulence, one adds energy at a large scale and this energy is dissipated at smaller and smaller scales in a scale-invariant fashion. Similarly, if external news is added at a large scale, then this news is dissipated by traders at smaller and smaller scales in a scale-invariant fashion. Despite some initial claims [78], these similarities are not borne out by quantitative analysis—although one finds non-Gaussian statistics, and intermittency, for both turbulence fluctuations and stock price fluctuations, the time evolution of the second moment and the shape of the probability density functions are different for turbulence and for stock market dynamics [79,80].

More recent work pursues a rather different analogy, phase transitions in spin systems. It is not new to say that the set of all firm fluctuations is like a set of subunit fluctuations in a physics system such as a spin glass. Each fluctuation can be up or down, or any magnitude, and fluctuations interact with one another via interactions that are certainly long-range and of both signs. Further, the interactions change with time. A given subunit fluctuation is influenced (a) by other fluctuations (so the exchange interactions among spins is somewhat like the "herd effect"), and (b) by forces external to the system (so the external field is somewhat like "news" which plays a role in determining the sign and magnitude of fluctuations).

If this speculative analogy were to hold even approximately, then a first step should perhaps be to seek to identify the analogs for the price fluctuation problem of field and temperature in the magnetic problem. Stock prices respond to demand, just as the magnetization of an interacting spin system responds to the magnetic field. Periods with large number of market participants buying the stock imply mainly positive changes in price, analogous to a magnetic field causing spins in a magnet to align. Recent work [81] quantifies the relations between price change and demand fluctuations, and finds results reminiscent of phase transitions in spin systems, where the divergent behavior of the response function at the critical point (zero magnetic field) leads to large fluctuations [10]. More precisely, buying and selling behavior in complex financial markets are driven by demand, which can be quantified by the imbalance in the number of shares transacted by buyers and sellers over a time interval Δt.

If demand is the analog of magnetic field, then what is the analog of temperature? To answer this question, Plerou et al. [50,51] analyze the probability distribution of demand, conditioned on its local noise intensity Σ, and find the surprising existence of a critical threshold Σ_c separating two market phases. Their findings for the financial market problem are identical to what is known to occur in all phase transition phenomena, wherein the behavior of a system undergoes a qualitative change at a critical threshold K_c of some control parameter K. Plerou et al interpret these two market phases as corresponding to two distinct conditions of the financial market: (a) The "$\Sigma < \Sigma_c$ market phase", where the distribution of demand is single peaked with the most probable value being zero, they interpret to be the market *equilibrium phase*, since the price of the stock is such that the probability of a transaction being buyer initiated is equal to the probability of a transaction being seller initiated, and (b) the "$\Sigma > \Sigma_c$ market phase", where the distribution of demand is bimodal, they interpret to be the *out-of-equilibrium phase*, since the price of the stock is such that there is an excess of either buyers or of sellers and there is a non-zero net demand for the stock.

It should be possible to design a software package that could be on every trader's desk allowing instant access to data on any firm in which time is partitioned into two different phases: equilibrium and out-of-equilibrium.

Qualitatively and informally many people use those terms in reference to the stock market, but in this case we would be actually *quantifying* the extent to which the market is in or out of equilibrium. If we graph the price-change of a particular stock as a function of time for a sequence of 15-minute intervals and use two different symbols for data points when the market is in equilibrium and for those for when it is out of equilibrium, we notice that in general a stock price is not changing when the market is in equilibrium and is changing when the market is out of equilibrium. This could be useful in that it could be an indicator of the relative stability of an individual stock. When the market is out of equilibrium, the probability that a stock price is going to change is higher than when the market is in equilibrium.

7 Discussion

Since the evidence for an analogy between stock price fluctuations and magnetization fluctuations near a critical point is backed up by quantitative analysis of finance data, it is legitimate to demand a theoretical reason for this analogy. To this end, we discuss briefly one possible theoretical understanding for the origin of scaling and universality in economic systems. As mentioned above, economic systems consist of interacting units just as critical point systems consist of interacting units. Two units are correlated in what might seem a hopelessly complex fashion—consider, e.g., two spins on a lattice, which are correlated regardless of how far apart they are. The correlation between two given spins on a finite lattice can be partitioned into the set of all possible topologically linear paths connecting these two spins—indeed this is the starting point of one of the solutions of the two-dimensional Ising model (see Appendix B of [10]). Since correlations decay exponentially along a one-dimensional path, the correlation between two spins would at first glance seem to decay exponentially. Now it is a mathematical fact that the total number of such paths grows exponentially with the distance between the two spins—to be very precise, the number of paths is given by a function which is a product of an exponential and a power law. The constant of the exponential *decay* depends on temperature while the constant for the exponential *growth* depends only on geometric properties of the system [10]. Hence by tuning temperature it is possible to achieve a threshold temperature where these two "warring exponentials" just balance each other, and a previously negligible power law factor that enters into the expression for the number of paths will dominate. Thus power law scale invariance emerges as a result of canceling exponentials, and universality emerges from the fact that the interaction paths depend not on the interactions but rather on the connectivity. Similarly, in economics, two units are correlated through a myriad of different correlation paths; "everything depends on everything else" is the adage expressing the intuitive fact that when one firm changes, it influences other firms. A more careful discussion of this argument is presented, not for the economy but for the critical phenomena problem, in Ref. [11].

8 Summary

In summary, physicists are finding this emerging field fascinating. For a long time, physicists did relatively little in economics. A major reason for this is that, until recently, the amount of data routinely recorded concerning financial transactions was insufficient to be useful to physicists. That fact is no longer true. Now every trade is recorded, along with bid-ask quotes for every trade, and these data are made available.

Part of the reason for the invention of the neologism "econophysics" (in the tradition of the neologisms "biophysics," "astrophysics," "geophysics"...) was to enable our physics students to persuade the departmental administrators that their dissertation research topics actually belonged in the physics department. The neologism seems to have caught on, and there are now several conferences each year with the word "econophysics" in the title.

Finally, a word of humility with respect to our esteemed economics colleagues is perhaps not inappropriate. Physicists may care passionately if there are analogies between physics systems they understand (like critical point phenomena) and economics systems they do not understand. But why should anyone else care? One reason is that scientific understanding of earthquakes moved ahead after it was recognized [1,2] that extremely rare events—previously regarded as statistical outliers requiring for their interpretation a theory quite distinct from the theories that explain everyday shocks—in fact possess the identical statistical properties as everyday events; e.g., all earthquakes fall on the same straight line on an appropriate log-log plot. Since economic phenomena possess the analogous property, the challenge is to develop a coherent understanding of financial fluctuations that incorporates not only everyday fluctuations but also those extremely rare "financial earthquakes".

Acknowledgments We thank NSF for financial support and we thank our collaborators: L. A. N. Amaral, S. V. Buldyrev, D. Canning, P. Cizeau, S. Havlin, Y. Lee, Y. Liu, P. Maass, R. N. Mantegna, K. Matia, M. Meyer, B. Rosenow, M. A. Salinger, and M. H. R. Stanley. This work was supported by the NSF Economics Program, grant SES-0215823, and by the NIH, grant P41RR13622.

References

1. B. Gutenberg and C. F. Richter, *Seismicity of the Earth and Associated Phenomenon, 2nd Edition* (Princeton University Press, Princeton, 1954)
2. D. L. Turcotte, *Fractals and Chaos in Geology and Geophysics* (Cambridge University Press, Cambridge, 1992)

3. J. B. Rundle, D. L. Turcotte, and W. Klein, *Reduction and Predictability of Natural Disasters* (Addison-Wesley, Reading MA, 1996).
4. X. Gabaix, P. Gopikrishnan, V. Plerou, and H. E. Stanley, "A Theory of Power-Law Distributions in Financial Market Fluctuations," Nature **423**, 267–270 (2003).
5. X. Gabaix, P. Gopikrishnan, V. Plerou, and H. E. Stanley, "A theory of large fluctuations in stock market activity" *Quarterly Journal of Economics* (to be submitted).
6. V. Pareto, *Cours d'Economie Politique* (Lausanne, Paris, 1897)
7. S. Solomon and P. Richmond, "Stable power laws in variable economies; Lotka-Volterra implies Pareto-Zipf" Eur Phys J B **27** (2002) 257-261.
8. Feenberg, D. and J. Poterba, "Income Inequality and the Incomes of Very High-Income Taxpayers: Evidence from Tax Returns," in *Tax Policy and the Economy*, ed. J. Poterba (Cambridge, MA: MIT Press, 1993), 145-77.
9. B. B. Mandelbrot, "The Variation of Certain Speculative Prices," *J. Business* **36** (1963) 394–419.
10. H. E. Stanley, *Introduction to Phase Transitions and Critical Phenomena* (Oxford University Press, Oxford, 1971).
11. H. E. Stanley, "Scaling, Universality, and Renormalization: Three Pillars of Modern Critical Phenomena," *Rev. Mod. Phys.* **71** (1999) S358–S366.
12. H. Takayasu, ed., *Empirical Science of Financial Fluctuations: The Advent of Econophysics* (Springer, Berlin, 2002).
13. R. N. Mantegna and H. E. Stanley, *An Introduction to Econophysics: Correlations and Complexity in Finance* (Cambridge University Press, Cambridge, 2000)
14. J.-P. Bouchaud, "Power Laws in Economics and Finance: Some Ideas from Physics," *Quantitative Finance* **1** (2001) 105–112.
15. J. P. Bouchaud and M. Potters, *Theory of Financial Risk* (Cambridge University Press, Cambridge, 2000).
16. H. Levy, M. Levy and S. Solomon, *Microscopic Simulation of Financial Markets* (Academic Press, New York, 2000)
17. B. M. Roehner, *Hidden Collective Factors in Speculative Trading* (Springer, Berlin, 2001).
18. B. M. Roehner, *Patterns of Speculation* (Cambridge Univ Press, Cambridge, 2002).
19. W. Paul and A. Baschnagel, *Stochastic processes from physics to finance* (Springer, Berlin, 1999).
20. J. Voit, *The Statistical Mechanics of Financial Markets* (Springer, Berlin, 2001).
21. R. L. Axtell, "Zipf Distribution of US Firm Sizes" *Science* **293** (2001) 1818–1821.
22. M. H. R. Stanley, S. V. Buldyrev, S. Havlin, R. Mantegna, M. A. Salinger, and H. E. Stanley, "Zipf Plots and the Size Distribution of Firms," *Econ. Lett.* **49** (1996) 453–457
23. M. H. R. Stanley, L. A. N. Amaral, S. V. Buldyrev, S. Havlin, H. Leschhorn, P. Maass, M. A. Salinger, and H. E. Stanley, "Scaling Behavior in the Growth of Companies," *Nature* **379** (1996) 804–806.
24. L. A. N. Amaral, S. V. Buldyrev, S. Havlin, H. Leschhorn, P. Maass, M. A. Salinger, H. E. Stanley, and M. H. R. Stanley, "Scaling Behavior in Economics: I. Empirical Results for Company Growth," J. Phys. I France **7** (1997) 621–633.

25. S. V. Buldyrev, L. A. N. Amaral, S. Havlin, H. Leschhorn, P. Maass, M. A. Salinger, H. E. Stanley, and M. H. R. Stanley, "Scaling Behavior in Economics: II. Modeling of Company Growth," J. Phys. I France **7** (1997) 635–650.
26. H. Takayasu and K. Okuyama, "Country Dependence on Company Size Distributions and a Numerical Model Based on Competition and Cooperation," *Fractals* **6** (1998) 67–79.
27. D. Champernowne, "A Model of Income Distribution," *Economic Journal* **83** (1953) 318–351.
28. M. Levy, and S. Solomon, "Dynamical Explanation for the Emergence of Power Law in a Stock Market Model," International Journal of Modern Physics C **7** (1996) 65–72.
29. D. H. Zanette and S. C. Manrubia , "Role of Intermittency in Urban Development: A Model of Large-Scale City Formation," *Phys. Rev. Lett.* **79** (1997) 523–526
30. M. Marsili and Y. C. Zhang , "Interacting Individuals Leading to Zipf's Law," *Phys. Rev. Lett.* **80** (1998) 2741–2744.
31. X. Gabaix, "Zipf's Law for Cities: An Explanation," *Quarterly Journal of Economics* **114** (1999) 739–767.
32. X. Gabaix and Y. Ioannides, "The Evolution of the City Size Distribution," in *Handbook of Urban and Regional Economics, Vol. 4*, edited by V. Henderson and J. Thisse (North Holland, Amsterdam, 2004).
33. V. Plerou, L. A. N. Amaral, P. Gopikrishnan, M. Meyer, and H. E. Stanley, "Similarities between the Growth Dynamics of University Research and of Competitive Economic Activities," *Nature* **400** (1999) 433–437.
34. D. Canning, L. A. N. Amaral, Y. Lee, M. Meyer, and H. E. Stanley, "A Power Law for Scaling the Volatility of GDP Growth Rates with Country Size," *Econ. Lett.* **60** (1998) 335–341.
35. Y. Lee, L. A. N. Amaral, D. Canning, M. Meyer, and H. E. Stanley, "Universal features in the growth dynamics of complex organizations" Phys. Rev. Letters **81** (1998) 3275-3278.
36. T. Keitt and H. E. Stanley, "Scaling in the Dynamics of North American Breeding-Bird Populations," Nature **393** (1998) 257.
37. L. A. N. Amaral, S. V. Buldyrev, S. Havlin, M. A. Salinger, and H. E. Stanley, "Power Law Scaling for a System of Interacting Units with Complex Internal Structure," *Phys. Rev. Lett.* **80** (1998) 1385–1388.
38. J. Sutton, "The Variance of Firm Growth Rates: The Scaling Puzzle," *Physica A* **312** (2002) 577.
39. F. Cecconi, M. Marsili, J. R. Banavar, and A. Maritan, "Diffusion, Peer Pressure, and Tailed Distributions," *Phys. Rev. Lett.* **89** (2002) 088102.
40. X. Gabaix "Power Laws and the Origins of the Macroeconomic Fluctuations", *MIT mimeo* (2001).
41. M. Wyart and J.-P. Bouchaud, "Statistical Models for Company Growth," cond-mat/0210479 (October 2002).
42. G. De Fabritiis, F. Pammolli, and M. Riccaboni, "On Size and Growth of Business Firms," Physica A **324** (2003) 38–44.
43. G. Bottazzi and A. Secchi, "Explaining the Distribution of Firm Growth Rates" (preprint).
44. T. Lux, "The Stable Paretian Hypothesis and the Frequency of Large Returns: An Examination of Major German Stocks," *Appl. Finan. Econ.* **6** (1996) 463–475.

45. P. Gopikrishnan, M. Meyer, L. A. N. Amaral, and H. E. Stanley, "Inverse Cubic Law for the Distribution of Stock Price Variations," *Eur. Phys. J. B* **3** (1998) 139–140.
46. V. Plerou, P. Gopikrishnan, L. A. N. Amaral, M. Meyer, and H. E. Stanley, "Scaling of the Distribution of Price Fluctuations of Individual Companies," *Phys. Rev. E* **60** (1999) 6519–6529.
47. P. Gopikrishnan, V. Plerou, L. A. N. Amaral, M. Meyer, and H. E. Stanley, "Scaling of the Distributions of Fluctuations of Financial Market Indices," *Phys. Rev. E* **60** (1999) 5305–5316.
48. K. Matia, L. A. N. Amaral, S. Goodwin, and H. E. Stanley, "Non-Lévy Distribution of Commodity Price Fluctuations" *Phys. Rev. E: Rapid Communications* **66** (2002) 045103. cond-mat/0202028.
49. K. Matia, Y. Ashkenazy, and H. E. Stanley, "Multifractal Properties of Price Fluctuations of Stocks and Commodities," *Europhys. Lett.* **61** (2003) 422–428.
50. V. Plerou, P. Gopikrishnan, and H. E. Stanley, "Two-Phase Behaviour of Financial Markets," *Nature* **420** (2002) xxx–xxx.
51. V. Plerou, P. Gopikrishnan, and H. E. Stanley, "Symmetry Breaking in Stock Demand", Phys. Rev. E (submitted) cond-mat/0111349.
52. R. N. Mantegna and H. E. Stanley "Stochastic Process with Ultraslow Convergence to a Gaussian: the Truncated Lévy Flight," *Phys. Rev. Lett.* **73** (1994) 2946–2949.
53. R. N. Mantegna and H. E. Stanley, "Scaling Behavior in the Dynamics of an Economic Index" *Nature* **376** (1995) 46–49.
54. B. Podobnik, P. Ch. Ivanov, Y. Lee, A. Chessa, and H. E. Stanley, "Systems with correlations in the variance: Generating power law tails in probability distributions" Europhysics Letters **50** (2000) 711-717
55. R. N. Mantegna and H. E. Stanley, "Ultra-Slow Convergence to a Gaussian: The Truncated Lévy Flight" in *Lévy Flights and Related Topics in Physics* [Proc. 1994 International Conf. on Lévy Flights], edited by M. F. Shlesinger, G. M. Zaslavsky, and U. Frisch (Springer, Berlin,1995), pp. 300–312.
56. R. N. Mantegna and H. E. Stanley, "Modeling of Financial Data: Comparison of the Truncated Lévy Flight and the ARCH(1) and GARCH(1,1) Processes" [Proc. Int'l IUPAP Conf. on Statistical Physics, Taipei], Physica A **254** (1998) 77–84.
57. B. Podobnik, P. Ch. Ivanov, Y. Lee, and H. E. Stanley, "Scale-invariant Truncated Lévy Process," Europhysics Letters **52** (2000) 491–497.
58. P. Ch. Ivanov, B. Podobnik, Y. Lee, and H. E. Stanley (2001), "Truncated Lévy Process with Scale-Invariant Behavior" [Proc. NATO Advanced Research Workshop on Application of Physics in Economic Modeling, Prague], Physica A **299**, 154–160.
59. J. A. Skjeltorp (2001), "Scaling in the Norwegian Stock Market," Physica **283**, 486–528.
60. B. LeBaron, "Stochastic Volatility as a Simple Generator of Financial Power Laws and Long Memory, *Quantitative Finance* **2** (2001) 621–631.
61. H. E. Stanley and V. Plerou, "Scaling and Universality in Economics: Empirical results and Theoretical Interpretation" Quantitative Finance **1** (2001) 563–567.
62. Y. Liu, P. Gopikrishnan, P. Cizeau, M. Meyer, C.-K. Peng and H. E. Stanley, "The Statistical Properties of the Volatility of Price Fluctuations," *Phys. Rev. E* **60** (1999) 1390–1400.

63. V. Plerou, P. Gopikrishnan, L. A. N. Amaral, X. Gabaix, and H. E. Stanley, "Diffusion and Economic Fluctuations," *Phys. Rev. E (Rapid Communications)* **62** (2000) 3023–3026.
64. P. Gopikrishnan, V. Plerou, X. Gabaix, and H. E. Stanley, "Statistical Properties of Share Volume Traded in Financial Markets," *Phys. Rev. E (Rapid Communications)* **62** 4493–4496
65. V. Plerou, P. Gopikrishnan, X. Gabaix, L. A. N. Amaral, and H. E. Stanley, "Price Fluctuations, Market Activity, and Trading Volume" [Proc. 2000 Santa Fe Econophysics Conference], *Quantitative Finance* **1** (2001) 262–269.
66. Y. Liu, P. Cizeau, M. Meyer, C.-K. Peng, and H. E. Stanley, "Quantification of Correlations in Economic Time Series" Physica A **245** (1997) 437–440.
67. P. Cizeau, Y. Liu, M. Meyer, C.-K. Peng, and H. E. Stanley, "Volatility distribution in the S&P500 Stock Index" Physica A **245** (1997) 441–445.
68. B. Podobnik, K. Matia, A. Chessa, P. Ch. Ivanov, Y. Lee, and H. E. Stanley, "Time Evolution of Stochastic Processes with Correlations in the Variance: Stability in Power-Law Tails of Distributions" Physica A **300** (2001) 300-309.
69. T. Guhr, A. Müller-Groeling, and H. A. Weidenmüller, "Random-Matrix Theories in Quantum Physics: Common Concepts," *Phys. Reports* **299** (1998) 189–425.
70. M. L. Mehta and F. J. Dyson, "Statistical Theory of the Energy Levels of Complex Systems. V," *J. Math. Phys.* **4** (1963) 713–719.
71. F. J. Dyson, "The Distribution of Singular Values of Certain Random Matrices," *Revista Mexicana de Física* **20** (1971) 231.
72. A. M. Sengupta and P. P. Mitra, "Distributions of Singular Values for Some Random Matrices," *Phys. Rev. E* **60** (1999) 3389–3392.
73. L. Laloux, P. Cizeau, J.-P. Bouchaud and M. Potters, "Noise Dressing of Financial Correlation Matrices," *Phys. Rev. Lett.* **83** (1999) 1469–1482.
74. V. Plerou, P. Gopikrishnan, B. Rosenow, L. A. N. Amaral, and H. E. Stanley, "Universal and Nonuniversal Properties of Financial Cross-Correlation Matrices," *Phys. Rev. Lett.* **83** (1999) 1471–1475.
75. P. Gopikrishnan, B. Rosenow, V. Plerou, and H. E. Stanley, "Quantifying and Interpreting Collective Behavior in Financial Markets," *Phys. Rev. E: Rapid Communications* **64** (2001) 035106.
76. V. Plerou, P. Gopikrishnan, B. Rosenow, L. A. N. Amaral, T. Guhr, and H. E. Stanley, "A Random Matrix approach to Financial Cross-Correlations" *Phys. Rev. E* **65** (2002) 066126 cond-mat/0108023.
77. B. Rosenow, V. Plerou, P. Gopikrishnan, and H. E. Stanley, "Portfolio Optimization and the Random Magnet Problem," *Europhys. Lett.* **59** (2002) 500–506 cond-mat/0111537.
78. S. Ghashghaie, W. Breymann, J. Peinke, P. Talkner, and Y. Dodge, "Turbulent Cascades in Foreign Exchange Markets," *Nature* **381** (1996) 767–770.
79. R. N. Mantegna and H. E. Stanley, "Turbulence and Exchange Markets," *Nature* **383** (1996) 587–588.
80. R. N. Mantegna and H. E. Stanley, "Stock Market Dynamics and Turbulence: Parallel Analysis of Fluctuation Phenomena" [Proc. International Conference on Pattern Formation in Fluids and Materials], *Physica A* **239** (1997) 255–266.
81. V. Plerou, P. Gopikrishnan, X. Gabaix, and H. E. Stanley, "Quantifying Stock Price Response to Demand Fluctuations," *Phys. Rev. E* **66** (2002) 027104 cond-mat/0106657.

Triangular Arbitrage in the Foreign Exchange Market

Yukihiro Aiba[1], Naomichi Hatano[2], Hideki Takayasu[3], Kouhei Marumo[4] and Tokiko Shimizu[5]

[1] Department of Physics, Aoyama Gakuin University, Chitosedai 6-16-1, Setagaya, Tokyo 157-8572, Japan aiba@phys.aoyama.ac.jp
[2] Institute of Industrial Science, University of Tokyo, Komaba 4-6-1, Meguro, Tokyo 153-8505, Japan hatano@iis.u-tokyo.ac.jp
[3] Sony CSL, Higashi-Gotanda 3-14-13, Shinagawa, Tokyo 141-0022, Japan takayasu@csl.sony.co.jp
[4] Institute for Monetary and Economic Studies, Bank of Japan, Hongoku-cho Nihonbashi 2-1-1, Chuo, Tokyo 103-8660, Japan kohei.marumo@boj.or.jp
[5] Financial Markets Department, Bank of Japan, Hongoku-cho Nihonbashi 2-1-1, Chuo, Tokyo 103-8660, Japan tokiko.shimizu@boj.or.jp

Summary. In the present article, we review two of our previous works. First, we show that there are in fact triangular arbitrage opportunities in the spot foreign exchange markets, analyzing the time dependence of the yen-dollar rate, the dollar-euro rate and the yen-euro rate. Second, we propose a model of foreign exchange rates with an interaction. The model includes effects of triangular arbitrage transactions as an interaction among three rates. The model explains the actual data of the multiple foreign exchange rates well. Finally, we suggest, on the basis of the model, that triangular arbitrage makes the auto-correlation function of foreign exchange rates negative in a short time scale.

Keywords. Foreign exchange, Triangular arbitrage, Auto correlation, Stochastic model

Introduction

We recently pointed out the existence of the triangular arbitrage opportunity in the foreign exchange market [1, 2]. The triangular arbitrage is a financial activity that takes advantage of the three foreign exchange rates among three currencies [3]. It makes the product of the three foreign exchange rates converge to its average, thereby generating an interaction among the rates.

In order to study effects of the triangular arbitrage on the fluctuations of the exchange rates, we introduced [1] a stochastic model describing the time evolution of the exchange rates with an interaction. The model successfully described the fluctuation of the data of the real market.

We showed further [2] that our model gives an explanation to an interesting feature of the fluctuation of foreign exchange rates. The auto-correlation function of the fluctuation of the foreign exchange rates has been known to be negative in a short time scale [4]. Our model suggests that an important ingredient of the negative auto-correlation is the triangular arbitrage.

The triangular arbitrage as an interaction

The triangular arbitrage is a financial activity that takes advantage of three exchange rates. When a trader exchanges one Japanese yen to some amount of US dollar, exchanges the amount of US dollar to some amount of euro and exchanges the amount of euro back to Japanese yen instantly at time t, the final amount of Japanese yen is given by

$$\mu \equiv \prod_{i=1}^{3} r_i(t), \tag{1}$$

where

$$r_1(t) \equiv \frac{1}{\text{yen-dollar ask } (t)} \tag{2}$$

$$r_2(t) \equiv \frac{1}{\text{dollar-euro ask } (t)} \tag{3}$$

$$r_3(t) \equiv \text{yen-euro bid } (t). \tag{4}$$

If the rate product μ is greater than unity, the trader can make profit through the above transaction. This is the triangular arbitrage transaction. Once there is a triangular arbitrage opportunity, many traders will make the transaction. This makes μ converge to a value less than unity, thereby eliminating the opportunity. Triangular arbitrage opportunities nevertheless appear, because each rate r_i fluctuates strongly.

The probability density function of the rate product μ (Fig. 1) has a sharp peak and fat tails. It means that the fluctuations of the exchange rates have correlation that makes the rate product converge to its average $\langle\mu\rangle \simeq 0.99998$. The average is less than unity because of the spread; the spread is the difference between the ask and the bid prices and is usually of the order of 0.05% of the prices.

For later convenience, we here define the logarithm rate product ν as the logarithm of the product of the three rates:

$$\nu(t) = \ln \prod_{i=1}^{3} r_i(t) = \sum_{i=1}^{3} \ln r_i(t). \tag{5}$$

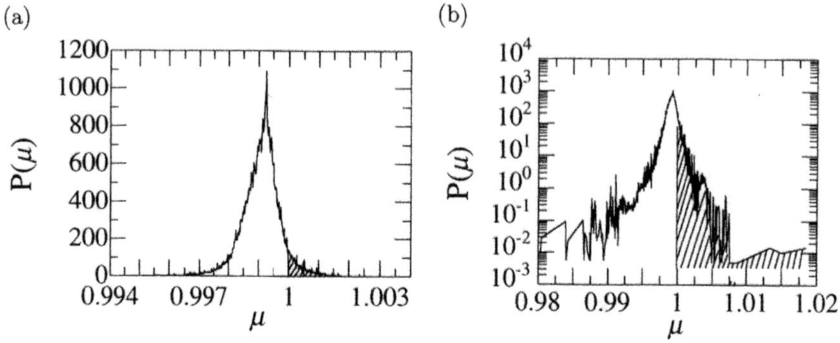

Fig. 1. The probability density function of the rate product μ. (b) is a semi-logarithmic plot of (a). The shaded area represents triangular arbitrage opportunities. The data were taken from January 25 1999 to March 12 1999.

There is a triangular arbitrage opportunity whenever this value is positive.

In one of our previous works [1], we constructed a stochastic model of the time evolution of foreign exchange rates that takes account of the effect of the triangular arbitrage transaction. The basic equation of this model is the time evolution of the logarithm of each rate:

$$\ln r_i(t+T) = \ln r_i(t) + f_i(t) + g(\nu(t)), \quad (i = 1,2,3) \tag{6}$$

where f_i denotes independent fluctuation that obeys a truncated Lévy distribution [5] and g represents an interaction function defined by

$$g(\nu) = -a(\nu - \langle \nu \rangle), \tag{7}$$

where a is a positive constant which specifies the interaction strength and $\langle \nu \rangle$ is the time average of ν. The time-evolution equation of the logarithm rate product ν is given by summing eq. (6) over all i:

$$\nu(t+T) - \langle \nu \rangle = (1-3a)(\nu(t) - \langle \nu \rangle) + \sum_{i=1}^{3} f_i(t). \tag{8}$$

The model equation (8) well describes a fat-tail probability distribution of $\nu(t)$ of the actual market (Fig. 2) [1].

From the physical viewpoint, we can regard the model equation (6) as a one-dimensional random walk of three particles with a restoring force, by making $\ln r_i$ the position of each particle (Fig. 3). The logarithm rate product ν is the summation of $\ln r_i$, hence is proportional to the center of gravity of the three particles. The restoring force $g(\nu)$ makes the center of gravity converge to a certain point $\langle \nu \rangle$. The form of the restoring force (7) is the same as that of the harmonic oscillator.

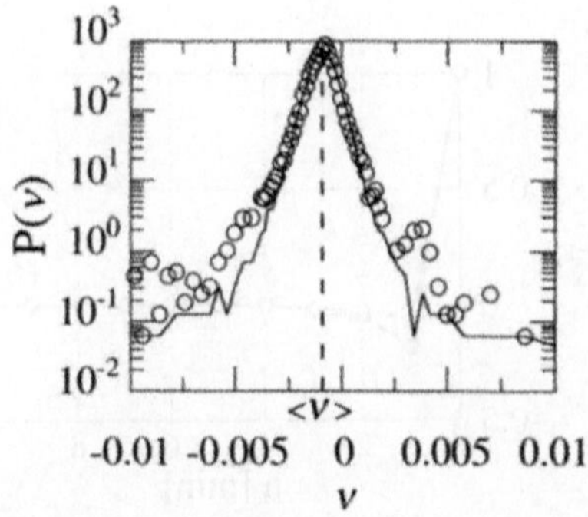

Fig. 2. The probability density function of ν. The circle (o) denotes the real data and the solid line denotes our simulation data. The simulation data fit the real data well.

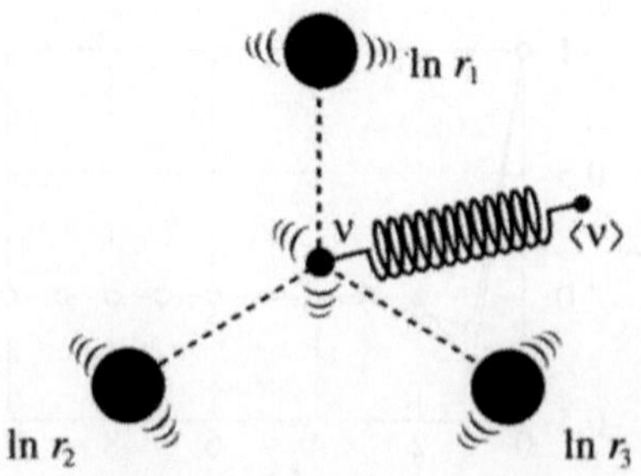

Fig. 3. A schematic image of the model. The three random walker with the restoring force working the center of gravity.

Negative auto-correlation of the foreign exchange rates in a short time scale

In the other of our previous works [2], we pointed out another consequence of the triangular arbitrage, namely the negative auto-correlation of each exchange rate in a short time scale. Let us first show it in the actual data. We analyzed actual tick-by-tick data of the yen-dollar rate, the dollar-euro rate and the yen-euro rate, taken from January 25, 1999 to March 12, 1999 except for the weekends.

The auto-correlation function of the rate fluctuation is defined by the following formula:

$$c_i(n) = \frac{\langle \Delta r_i(t+nT)\Delta r_i(t)\rangle - \langle \Delta r_i(t)\rangle^2}{\langle \Delta r_i(t)^2\rangle - \langle \Delta r_i(t)\rangle^2}, \quad (i=1,2,3; n=0,1,2,...), \qquad (9)$$

where

$$\Delta r_i(t) \equiv \ln \frac{r_i(t+T)}{r_i(t)} \quad (i=1,2,3), \qquad (10)$$

and the angular brackets $\langle ... \rangle$ denote the time average. We fixed the time step T at one minute.

Figure 4 shows that the auto-correlation function of each rate has a negative value for $n=1$. We here claim that the triangular arbitrage is one of the major causes of this negative auto-correlation. In order to see it, we simulated eq. (6) and calculated the auto-correlation function (9). The simulation data (also shown in Fig. 4) are qualitatively consistent with the behavior of the auto-correlation function of the actual data.

Another analysis is possible. Using eq. (6), we can rewrite the auto-correlation function (9) for $n=1$ as:

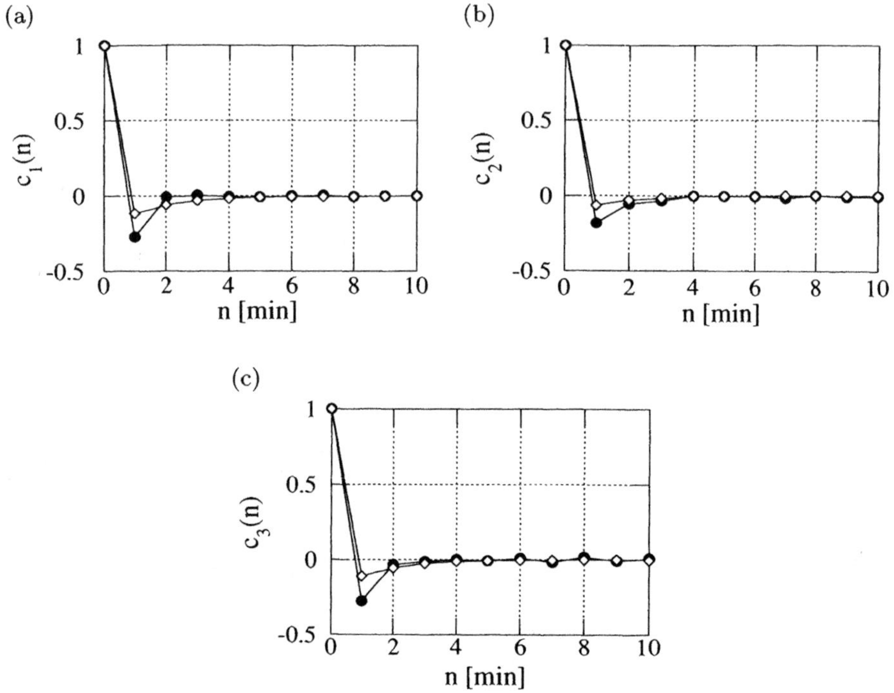

Fig. 4. The auto-correlation function of the rate change of the actual data: (a) $c_1(n)$; (b) $c_2(n)$; (c) $c_3(n)$. The circles ($\bullet$) denote the actual data and the diamonds ($\diamond$) denote the simulation data.

$$c_i(n=1) = \frac{\langle (f_i(t+T)+g(t+T))(f_i(t)+g(t))\rangle - \langle f_i(t)+g(t)\rangle^2}{\langle (f_i(t)+g(t))^2\rangle - \langle f_i(t)+g(t)\rangle^2} \tag{11}$$

$$= -a\frac{\sigma_{f_i}^2 - a(1-3a)\sigma_\nu^2}{\sigma_{f_i}^2 + a^2\sigma_\nu^2}, \tag{12}$$

where ${\sigma_x}^2$ denotes the variance of the variable x. We here used the following relations:

$$\langle f_i(t)\rangle = 0 \quad \text{and} \quad \langle f_i(t+T)f_i(t)\rangle = 0. \tag{13}$$

Note that we have $c_i(n=1) \approx -a < 0$ for small a.

We can estimate σ_ν and σ_{f_i} from the market data. The auto-correlation function for $n=1$ thus-estimated is compared in Table 1 to the one from the actual data and the one from the simulation data.

The value of $c_i(n=1)$ from the actual data is less than those from the simulation data and eq. (12). This may suggest that there are contributions from the triangular arbitrage of other combinations of three rates; for example, the triangular arbitrage among Japanese yen, US dollar and British pound.

Table 1. The value of $c_i(n = 1)$ from the actual data, the simulation data and eq. (12).

Rate	Actual Data	Simulation	Eq. (12)
r_1 (1/yen-dollar ask)	−0.27	−0.12	−0.12
r_2 (1/dollar-euro ask)	−0.18	−0.061	−0.095
r_3 (yen-euro bid)	−0.28	−0.11	−0.13

Conclusions

We first showed that triangular arbitrage opportunities exist in the foreign exchange market. The rate product μ fluctuates around its average. Next, we introduced a model including the interaction caused by the triangular arbitrage transaction. Finally, on the basis of the model, we showed that the triangular arbitrage makes the auto-correlation function of each rate negative for $n = 1$. The comparison with the actual data is good qualitatively, but it also suggests that the triangular arbitrage of various combinations must be considered.

References

1. Y. Aiba, N. Hatano, H. Takayasu, K. Marumo, T. Shimizu, Triangular arbitrage as an interaction among foreign exchange rates, Physica A 310 (2002) 467–479.
2. Y. Aiba, N. Hatano, H. Takayasu, K. Marumo, T. Shimizu, Triangular arbitrage and negative auto-correlation of foreign exchange rates, Physica A to be published.
3. I. Moosa, Triangular arbitrage in the spot and forward foreign exchange markets, Quant. Finance 1 (2001) 387–390.
4. H. Takayasu, M. Takayasu, M. P. Okazaki, K. Marumo and T. Shimizu, in: M.M. Novak (Ed.), Fractal Properties in Economics, Paradigms of Complexity, World Scientific, Singapore, 2000, pp.243–258.
5. J.P. Bouchaud, M. Potters, Theory of financial risks, Cambridge University Press, Cambridge, 2000, pp.34–35.

Time-scale dependence of correlations among foreign currencies

Takayuki Mizuno[1], Shoko Kurihara[1], Misako Takayasu[2], and Hideki Takayasu[3]

[1]Department of Physics, Faculty of Science and Engineering, Chuo University, Kasuga, Bunkyo-ku, Tokyo 112-8551, Japan
[2] Department of Complex Systems, Future University-Hakodate, 116-2 Kameda-Nakano, Hakodate 041-0803, Japan
[3] Sony Computer Science Laboratories, 3-14-13 Higashi-Gotanda, Shinagawa-ku, Tokyo, Japan

Summary. For the purpose of elucidating the correlation among currencies, we analyze daily and high-resolution data of foreign exchange rates. There is strong correlation for pairs of currencies of geographically near countries. We show that there is a time delay of order less than a minute between two currency markets having a strong cross-correlation. The cross-correlation between exchange rates is lower in shorter time scale in any case. As a corollary we notice a kind of contradiction that the direct Yen-Dollar rate significantly differs from the indirect Yen-Dollar rate through Euro in short time scales. This result shows the existence of arbitrage opportunity among currency exchange markets.

Key words. Foreign exchange, cross-correlation, interaction.

1. Introduction

It is widely known that foreign exchange rates sometimes show very violent fluctuations characterized by power law distributions [T. Mizuno][K. Matia]. A very important report appeared recently that arbitrage chances exist by considering three currency markets simultaneously [Y. Aiba]. This result is due to the fact that the distribution of return of rotation transaction of three currencies show a similar fat-tail. As statistical laws among foreign exchange rates in the short time scale are not known sufficiently, we investigate correlations among foreign exchange rates in this paper.

Table.1 The set of Daily data for 25 kinds of foreign exchange we examined.

Base Currency : U.S. Dollar			
Target Currencies :			
Australian Dollar,	Brazilian Real,	British Pound,	Canadian Dollar,
Chilean Peso,	Colombian Peso,	Czech Koruna,	Euro,
Hungarian Forint,	Indonesian Rupiah,	Japanese Yen,	Mexican Peso,
New Zesland Dollar,	Norwegian Kroner,	Peruvian New Sole,	Philippines Peso,
Polish Zloty,	Russian Ruble,	Singapre Dollar,	Slovakian Koruna,
South African Rand,	South Korean Won,	Swedish Krona,	Swiss Franc,
Thai Baht			

Correlations among financial stocks in markets have been actively discussed. The strength of synchronization among stocks was analyzed by using the ultrametric spaces and they discussed about the portfolio of stock [R. N. Mantegna, H. E. Stanley]. The interaction of companies was investigated by analyzing return of many stocks and a directed network of influence among companies was defined [L. Kullmann]. It will be noticed in the present paper that the correlation among foreign exchange rates resembles the stock case.

In the following section we discuss the cross-correlation between exchange rates with no time difference to show the relation between synchronization of exchange rate. Then, we show the maximum correlation between exchange rates is observed with nonzero time shift, namely, the direction of influence is discussed.

2. The correlation among foreign exchange rates

We first analyze a set of daily data provided by Exchange Rate Service [ERS] for 25 exchange rates for about 3 years from January `99 to August `01 as listed in Table 1. We first estimate cross-correlation functions for these exchange rates measured by USD (United States Dollar). The largest correlation value (=0.95) is observed for a pair of CHF(Swiss Franc)/USD and EUR(Euro)/USD. As demonstrated in Fig.1 it is evident that these exchange rates are remarkably synchronized. There are cases with negative correlation values as found for the pair of MXP(Mexican-Peso)/USD and CHF(Swiss Franc)/USD whose correlation value is -0.23. It is known that correlations between geographically closer currencies tend to have larger correlation and there exist key currencies in each area, for example, Euro for West Europe, Yen for Asia, Hungarian Forint for East Europe and Australian Dollar for Oceania [H. Takayasu and M. Takayasu].

Although we can find such large correlations among currencies in daily data, we can expect low correlations in short time scale as it is common that dealers in major banks tend to work with a single foreign exchange market. In order to

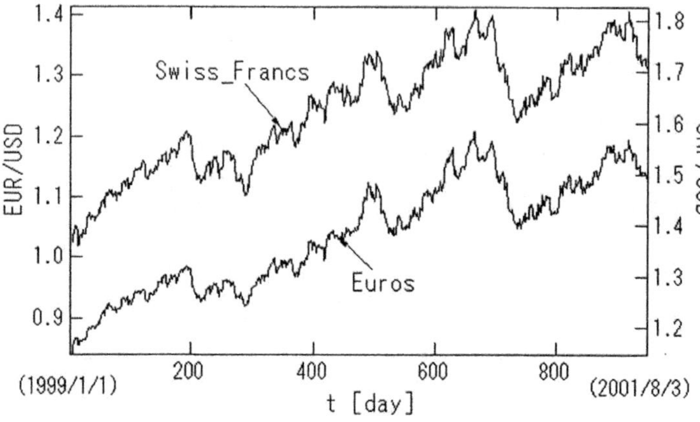

Fig.1 Daily time series of CHF(Swiss-Franc)/Dollar and Euro/Dollar.

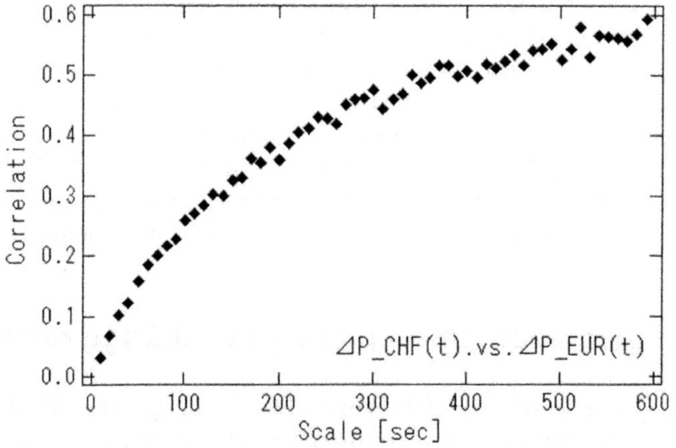

Fig.2 The correlation value as a function of observation time-scale.

clarify this tendency we examine tick-by-tick data provided by Reuters for about 4 months from March `02 to July `02. In Fig.2 we plot the correlation value as a function of coarse-grained time-scale for a pair of CHF/USD and EUR/USD. From this figure we notice that the correlation vanishes if we observe the high-resolution data with the precision of seconds and the correlation value is about 0.5 in the time scale of 5 minutes (300 sec.). From these results it is understood that these two currency markets, CHF/USD and EUR/USD, are working independently in very short time-scale.

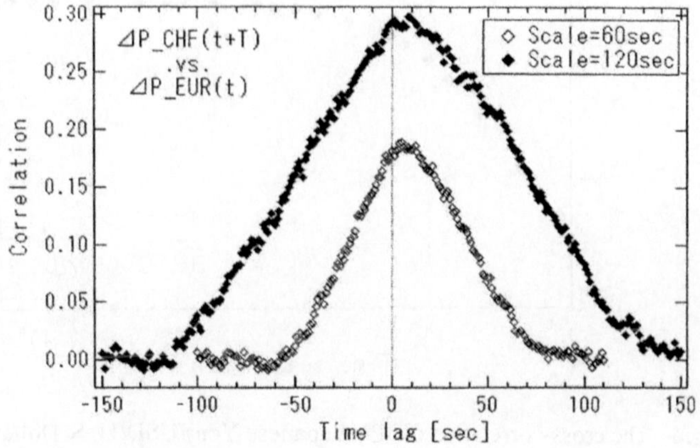

Fig.3 The cross-correlation with a time shift of CHF/USD and EUR/USD.

3. The cross-correlation with a time shift

In order to clarify the nature of short time interaction among currencies we calculate the cross-correlation with a time shift, namely, we observe the correlation of these two markets with a time difference by the following equation,

$$C(dt) = \frac{\langle dp_A(t) \cdot dp_B(t+dt) \rangle - \langle dp_A(t) \rangle \langle dp_B(t+dt) \rangle}{\sigma_A \cdot \sigma_B}, \qquad (1)$$

where, $dp_A(t)$ is the rate change in the market A at time t, $dp_B(t+dt)$ is the rate change in the market B at time $t+dt$, σ is the standard deviation of rate changes in each market. In Fig.3, we show the correlation value between CHF/USD at time $t+dt$ and EUR/USD at time t as a function of time difference dt. Here, we show two plots for different coarse-graining time-scales, 60 sec. and 120 sec.. In both cases it is found that the largest value of correlation is observed around $dt = 10$ seconds, which implies that in an average sense the EUR/USD market is going about 10 second ahead and the CHF/USD market is following it.

4. Currency correlation in short time scale

Here, we discuss about value of currency correlation in each foreign exchange market in a short time scale. From Fig.2, it is noticed that the correlation between foreign exchanges is very low in a short time scale. Namely, each exchange rate is changing rather independently. In order to clarify this property, we analyze the

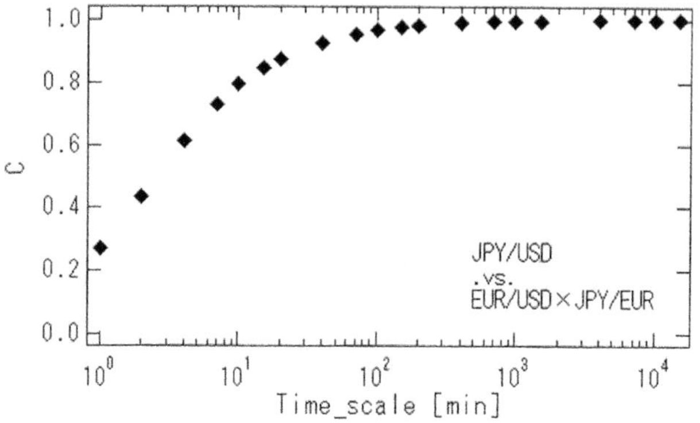

Fig.4 The cross-correlation of JPY(Japanese Yen)/USD(U. S. Dollar) and {EUR(Euro)/USD} × {JPY/EUR}.

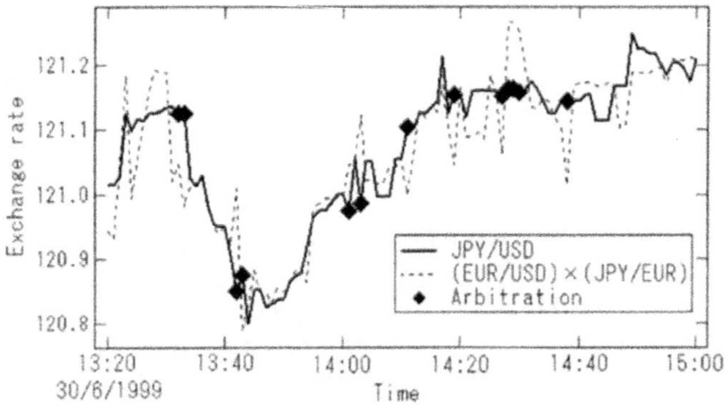

Fig.5 Triangular arbitrage opportunity. A full line is JPY/USD and a dashed line is {EUR(Euro)/USD}×{JPY/EUR}. Triangular arbitrage occurred at the time of ◆.

exchange rate of Yen-Dollar by analyzing a set of tick-by-tick data provided by CQG for about 2 years from February `99 to March `02.

We introduce two definitions of Yen-Dollar rate: One definition of JPY/USD is the usual transaction rate and the other JPY/USD is defined through Euro as {EUR/USD} × {JPY/EUR}. Here, all exchange rates are given by the middle rate (=(Bid rate + Ask rate)/2). In Fig.4 we plot the cross-correlation value of these exchange rates in different time scales. The correlation value is not unity in the time scale less than about 1 hour. This result clearly shows that the value of a currency differs in different markets in the time scale less than an hour. This is a kind of self-contradiction of markets causing the occurrence of triangular arbitrage opportunity as shown in Fig.5 [Y. Aiba].

5. Discussion

We have clarified the detail properties of correlation among foreign exchanges; the short time correlation is generally very small even between the pair of currencies showing large correlation in daily data. Within the time scale of a minute we can observe the direction of influence from one currency market to others. In very short time scale we can find contradiction of exchange rates between JPY/USD and {EUR/USD} × {JPY/EUR}. For example, if you observe carefully the two rates, JPY/USD and {EUR/USD} × {JPY/EUR}, then you can buy Yen cheaper in one market and can sell it with higher rate in the other market even taking into account the effect of the spread (=Ask rate – Bid rate) that is about 0.05%. Although, no time lag is considered regarding the actual transactions, it is now clear how triangular arbitrage opportunity appears.

Acknowledgement

We would like to thank Mr. Hiroyuki Moriya of Oxford Financial Education for providing us with the CQG data, Prof. Tohru Nakano and Prof. Mitsuo Kono for stimulus discussions.

References

Mizuno, T., Kurihara, S., Takayasu, M., Takayasu, H., Analysis of high-resolution foreign exchange data of USD-JPY for 13 years, Physica A, in press.

Matia, K., et al (2002) Non-Levy Distribution of Commodity Price Fluctuations, Phys. Rev. E **66**, 045103.

Aiba, Y., et al (2002) Triangular arbitrage as an interaction among foreign exchange rates, Physica A **310**, 467-479.

Mantegna, R. N., Stanley, H. E., (2000) An Introduction to Econophysics: Correlation and Complexity in Finance, Cambridge University Press, Cambridge.

Kullmann, L., Kertesz, J., Kaski, K., (2002) Time-dependent cross-correlations between different stock returns: A directed network of influence, Phys. Rev. E **66**, 026125.

Exchange Rate Service: http://pacific.commerce.ubc.ca/xr/ .

Takayasu, H., Takayasu, M., (2001) Econophysics-Toward Scientific Reconstruction of Economy, Nikkei, Tokyo. (in Japanese)

Univariate and multivariate statistical aspects of equity volatility

Salvatore Miccichè[1,2], Fabrizio Lillo[1,2], Giovanni Bonanno[1,2], and Rosario N. Mantegna[1,2]

[1] Istituto Nazionale per la Fisica della Materia, Unità di Palermo, Viale delle Scienze, I-90128 Palermo, Italia
[2] Dipartimento di Fisica e Tecnologie Relative, Università di Palermo, Viale delle Scienze, I-90128 Palermo, Italia

Abstract. We discuss univariate and multivariate statistical properties of volatility time series of equities traded in a financial market. Specifically, (i) we introduce a two-region stochastic volatility model able to well describe the unconditional pdf of volatility in a wide range of values and (ii) we quantify the stability of the results of a correlation-based clustering procedure applied to synchronous time evolution of a set of volatility time series.

1 Introduction

Volatility of financial time series is a key variable in the modelling of financial markets. It controls several risk measures associated with the dynamics of price of a financial asset. It also affects the rational price of derivative products. In this paper we discuss univariate and multivariate statistical properties of volatility investigated recently by our research group. Specifically, we discuss the probability density function (pdf) of historical volatility for 100 highly capitalized stocks traded in the US equity markets. We first briefly recall the empirical observation that widespread volatility models such as the Hull and White model [1] and the lognormal model fail in describing the volatility pdf when one asks the unconditional pdf to describe both low and high values of volatility [2]. Starting from these observation we propose here a two-region model of stochastic volatility, which is able to well describe the unconditional pdf of volatility for any values of it.

In this paper we also discuss multivariate properties of volatility time series. The existence of correlation among price returns of different stocks traded in a financial market is a well-known fact [3–5]. Correlation based clustering procedures have been pioneered in the economic literature. Examples are references [6,7]. Recently, a new correlation based clustering procedure has been introduced in the econophysics literature. It has been shown that this correlation-based clustering procedure and some variants of it are able to filter out information which has a direct economic interpretation from the correlation coefficient matrix [8–12]. A similar filtering procedure has also been performed on volatility time series [13]. Here we discuss the problem of the stability associated with the Minimum Spanning Tree (MST) obtained from volatility data. This is done by using two different approaches. One is based on the stability of the time series of the degree (number of links of the stock in the MST) of each stock [13] whereas the second one is based on the detection of the frequency of links conserved during the dynamics of correlation. This last approach has been recently proposed in Ref. [14]. With both approaches, we verify that volatility MSTs are characterized by a small but not negligible level of stability over time intervals as long as several years.

The paper is organized as follows. In Sect. 2 we discuss empirical results about the unconditional probability density function of volatility time series. In Sect. 3 we introduce a new volatility model able to well describe the unconditional pdf of volatility over a wide volatility range. In Sections 4 and 5 we comment on the stability of the MSTs of volatility time series of set of stocks simultaneously traded in a market. In Sect. 6 we briefly draw our conclusions.

2 Empirical Results

The volatility σ of a financial asset is a statistical quantity which needs to be determined starting from market information [15]. It is the standard deviation of asset return (or, almost equivalently, of logarithm price changes of the asset). Different methodologies are used to infer volatility estimation from market data ranging from a direct calculation from past return data (historical volatility) to the computation of the volatility implied in the determination of an option price computed using the Black and Scholes formula [16] or some variants of it. We discuss here some results obtained by our group in the investigation of the statistical properties of volatility for the 100 most capitalized stocks traded in US equity markets during a 4 year time period. The empirical data were taken from the trade and quote (TAQ) database, maintained by the New York Stock Exchange (NYSE). In particular, our data cover the whole period ranging from January 1995 to December 1998 (1011 trading days). This database contains all transactions occurred for each stock traded in the US equity markets. The capitalization considered is the one recorded on August 31, 1998. For each stock and for each trading day we consider the time series of stock price recorded transaction by transaction. Since transactions for different stocks do not happen simultaneously, we divide each trading day (lasting $6^h\ 30'$) into 12 intervals of 1950 seconds each. In correspondence to each interval, we define 12 (intraday) stock's prices proxies $P(k)$ – with $k = 1, \cdots, 12$ defined as the transaction price detected nearest to end of the interval (this is one possible way to deal with high-frequency financial data [17]).

We choose 12 intraday intervals since this value ensures that at least one transaction is observed on average in each interval for all considered stocks in the present study. For each stock we can thus compute a historical daily

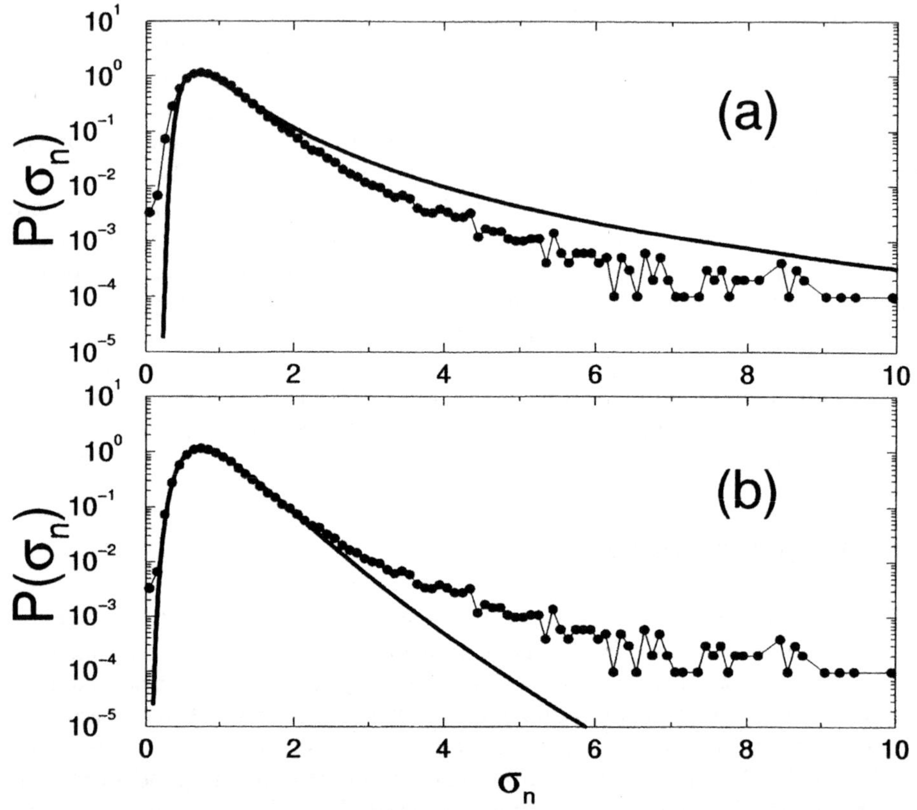

Fig. 1. Best fits of the empirical unconditional pdf of normalized volatility σ_n obtained by investigating 100 stocks traded in US equity markets during the time period January 1995 - December 1998. In both panels, the solid lines are the best fits whereas the solid circles are empirical data. In panel (a) we show the best fitting obtained with the Hull and White pdf of Eq. (2) and in panel (b) we show the best fitting obtained with a lognormal pdf of mean value 0.97 and variance equals to 0.19. In the case of panel (a) the fittings parameters are $2a/\xi^2 + 3 = 3.79$ and $ba/\xi^2 = 0.91$.

volatility as $\sigma(t) = \sqrt{11}\ \mathrm{std}[\ln(P(k+1)/P(k))]$, where std[·] indicates the standard deviation of the argument of the function. Hence, for each stock we have 1011 values of daily volatility. These volatility data have then been analyzed to compute the volatility pdf for each stock. The 100 empirical pdfs we obtain are then fitted with the theoretical pdfs of the considered models. Due to the limited number of records used to estimate the empirical pdfs (1011 records per stock) the results of our fittings are not able to indicate strengths and weaknesses of the two models. For this reason we rescale the volatility value of each stock σ to its mean value $< \sigma >$ and we investigate the pdf of the normalized variable $\sigma_n = \sigma / < \sigma >$ for the ensemble of 100 stocks. In this way we obtain an empirical pdf which is quite accurate being based on the recording of 101,100 events.

In our study [2] we compared our empirical results with two widespread models of volatility. The first model is the Hull and White model [1]. In this model, the variance rate $v \equiv \sigma^2$ is described by the Ito's equation

$$dv = a\ (b - v)\ dt + \xi v dz_v \tag{1}$$

where a and b are parameters controlling the mean reverting nature of the stochastic process, ξ is controlling its diffusive aspects and z_v is a Wiener process. The stochastic process is reverting at a level b at a rate a. The Hull and White model has associated a stationary pdf of the volatility which has the form

$$P(\sigma) = 2\frac{\left(ba/\xi^2\right)^{1+a/\xi^2}}{\Gamma(1 + a/\xi^2)} \frac{\exp(-ba/\xi^2\sigma^2)}{\sigma^{2a/\xi^2+3}} \tag{2}$$

This pdf has a power-law tail for large values of σ. A power-law tail in the empirical volatility pdf has been observed in Ref. [18] for large values of the volatility. Another model is the lognormal model of volatility [19,20]. An Ito's stochastic differential equation associated with a lognormal pdf is

$$d\sigma = a\ (b - \ln \sigma)\ dt + \xi \sigma^{1/2} dz_\sigma \tag{3}$$

where a, b and ξ are control parameters of the model. The two models are characterized by quite different pdfs especially for large values of the volatility where the Hull and White pdf shows a power-law behavior.

The best fittings of our empirical pdf with Eq. (2) and with a lognormal pdf are shown in Fig. 1. Panel (a) shows the best fits obtained with the Hull and White pdf. In this case the volatility low values are only approximately well described by the theoretical pdf. Moreover, for large values of volatility, the best fit overestimates by approximately a factor two empirical results. By inspecting panel (b) we note that the lognormal pdf describes very well low values of volatility in the interval $0 < \sigma_n \leq 2$ but completely fails in describing large values $\sigma_n > 2$. In particular a lognormal pdf underestimates large values of volatility. Both the lognormal model and the Hull and White models fail in describing well the normalized volatility over a relatively wide range of volatility values ($0 < \sigma_n < 10$). This implies that there is still room for the improvement of volatility models down to the basic aspect of well describing the asymptotic pdf of volatility over a realistically wide range.

3 A Stochastic Volatility two-region model

The empirical studies presented so far put forward that volatility pdf is well described by a lognormal distribution for intermediate volatility values and

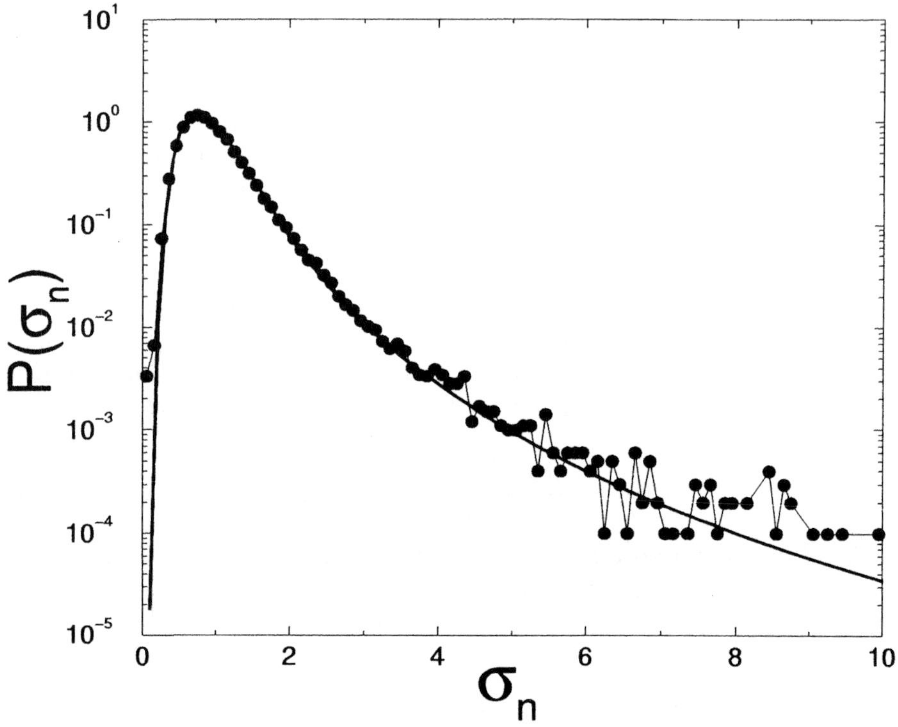

Fig. 2. Empirical unconditional pdf of normalized volatility compared with the pdf predicted by the stochastic model of Eq. (4). Solid line represents the result of a nonlinear best fitting with the pdf of Eq. (5) whereas solid circles are the empirical data. The fitting parameters are $L = 1.7$, $\alpha = 4.78$ and $V_2 = 2.78$.

it is approximated by a power–law for large volatility values. Here we present a two–region model which captures both these features.

Let us consider the following Ito's equation with additive noise:

$$d\sigma = h(\sigma)\; dt + dz_\sigma \tag{4}$$

$$h(\sigma) = \begin{cases} -\,\alpha\; \frac{1}{\sigma} & \sigma > L \\ -2\; V_2\; \frac{1}{\sigma}\; (\log\sigma + \frac{V_1 - V_2}{2V_2^2}) & 0 \le \sigma \le L \end{cases}$$

where α, L, V_1 and V_2 are real constants.

The corresponding Fokker–Planck equation admits as stationary solution the following pdf

$$P(\sigma) = \begin{cases} N_1\; \frac{1}{\sigma^\alpha} & \sigma > L \\ N_2\; e^{-\;(\log\sigma - m)^2/2s} & 0 \le \sigma \le L \end{cases} \tag{5}$$

$$s = \frac{1}{2\,V_2} \qquad m = \frac{1}{2}\frac{V_1 - V_2}{2\,V_2^2}$$

The pdf of Eq. (5) is a lognormal pdf for low volatility values and a power-law pdf for large volatility values. It is worth noting that the lognormal function of Eq. (5) is described by the stochastic differential equation of Eq. (4), which is different from the one of Eq. (3). Lognormal pdfs can be described both by additive and multiplicative stochastic differential equations. In the two-region model of Eq. (4) the description of volatility in terms of an additive stochastic process is more appropriate due to the form of the drift $h(\sigma)$ for $\sigma > L$. The two constants N_1 and N_2 are real constants that can be fixed by imposing that $P(\sigma)$ is continuous and it is normalized to unity. The further requirement that $P(\sigma)$ has continuous first derivative gives a relation between V_1, V_2 and L, which is

$$V_2 = \frac{1 + \alpha + \sqrt{(1+\alpha)^2 - 8V_1 \log L}}{4 \log L} \tag{6}$$

As a result the model is specified if the three parameters α, L and V_1 (or V_2) are given. Parameter α controls the exponent of the power–law tail. Parameter V_1 controls where the lognormal part of the pdf is located, i.e. the value of m. Parameter L controls the turning point between the power–law and the lognormal regimes. Fig. 2 shows a comparison between empirical data (solid circles) and the model of Eq. (4) (solid line). The solid line is obtained with a nonlinear fitting procedure performed at successive steps. We first fit the empirical pdf for large values of σ with a power–law behavior. From this fitting we estimate the exponent of the power–law tail as $\alpha = 4.78$. The second step involves the lognormal region. By fitting the pdf for low values of σ we obtain $s = 0.18$ and $m = 0.31$, which gives $V_2 = 2.78$. By using Eq. (6) we finally find $L = 1.7$ As expected, the proposed model well reproduces the empirical pdf for the entire interval of investigated volatility values. A direct comparison of Fig. 1 and Fig. 2 shows that the proposed two–region model performs better than the models of Eqs (1) and (3) in describing the unconditional pdf.

It is also worth noting that the present model presents a non-trivial persistent form of the autocorrelation function of volatility. This is due to the fact that a stochastic volatility model with a drift $h(\sigma)$ which is inversely proportional to volatility has associated a non exponential decaying autocorrelation function of the random volatility [21].

4 Correlation-based clustering of volatility

We investigate the statistical properties of cross-correlation among volatility time series. In this multivariate investigation we use daily data. To cover a wide time period of the dynamics of correlation-based clustering, we investigate the 12 year time period ranging from January 1987 to April 1999 (3116 trading days). During this time period only 93 of the 100 stocks considered in Sections 2 and 3 have been continuosly traded. For the present investigation, to mantain our set of stocks of constant size we choose to consider here only these 93 stocks.

Starting from the daily price data $P_i(t)$, we compute the historical daily volatility $\sigma_i(t)$ for each stock $i = 1, \cdots, 93$ by using the proxy $\sigma_i(t) = 2\ [\max\{P_i(t)\} - \min\{P_i(t)\}]/[\max\{P_i(t)\} + \min\{P_i(t)\}]$ where $\max\{P_i(t)\}$ and $\min\{P_i(t)\}$ are the highest and lowest price of the day, respectively.

The correlation based clustering procedure introduced in Ref. [8] is based on the computation of the subdominant ultrametric distance [22] associated with a metric distance that one may obtain from the correlation coefficient. The subdominant ultrametric distance can be used to obtain a hierarchical tree and a MST. The selection of the subdominant ultrametric distance for a set of elements whose similarity measure is a metric distance is equivalent to considering the single linkage clustering procedure [23]. Further details about this clustering procedure can be found in [24].

In the present discussion, we wish to summarize the statistical properties of the MST associated to the correlation coefficient matrix of volatility time series. It should be noted that there is an essential difference between price return and volatility probability density functions. In fact the probability density function of price return is an approximately symmetrical function whereas the volatility probability density function is significantly skewed. Bivariate variables whose marginals are very different from Gaussian functions can have linear correlation coefficients which are bounded in a subinterval of $[-1, 1]$ [25]. Since the empirical probability density function of volatility is very different from a Gaussian the use of a robust nonparametric correlation coefficient could more appropriate for quantifying volatility cross-correlation. Indeed our investigations show mixed results. The volatility MSTs obtained starting from a Spearman rank-order correlation coefficient seems to be more stable with respect to the dynamics of the degree of stocks than the ones obtained starting from the linear (or Pearson's) correlation coefficient when we consider the autocorrelation function of the degree time series of the investigated stocks. Conversely no significant difference is noted when we investigate the survival ratio of tree edges [14]. Due to space restrictions, in this paper we only presents results obtained by computing the Spearman nonparametric correlation coefficient between pairs of time series.

The clustering procedure based on the Spearman rank-order correlation coefficient uses the volatility rank time series to evaluate the subdominant ultrametric distance. The time series of the rank value of volatility are obtained by substituting the volatility values with their ranks. Then one evaluates the linear correlation coefficient between each pair of the rank time series [26] and starting from this correlation coefficient matrix one obtains the asso-

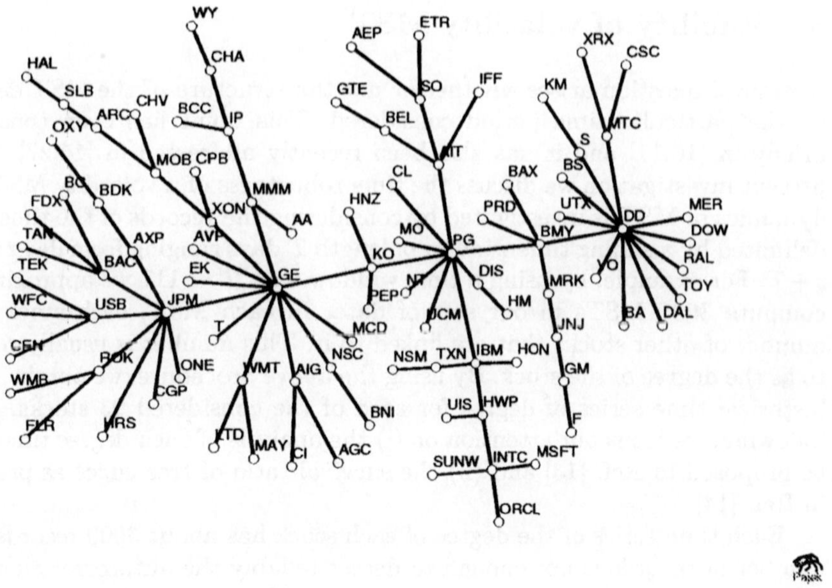

Fig. 3. **Minimum spanning tree obtained by considering the volatility time series of 93 mostly capitalized stocks traded in the US equity market. Each stock is identified by its tick symbol. The correspondence with the company name can be found in any web site of financial information. The volatility correlation among stocks has been evaluated by using the Spearman rank-order correlation coefficient. The MST has been drawn by using the Pajek package for large network analysis http://vlado.fmf.uni-lj.si/pub/networks/pajek/**

ciated MST. An example of the MST obtained starting from the volatility time series and by using the Spearman rank-order correlation coefficient is shown in Fig. 3. This MST is shown for illustrative purposes and it has been computed by using the widest window available in our database ($T = 3116$ trading days). A direct inspection of the MST shows the existence of well characterized clusters. Examples are the cluster of technology stocks (HON, HWP, IBM, INTC, MSFT, NSM, ORCL, SUNW, TXN and UIS) and the cluster of energy stocks (ARC, CHV, CPB, HAL, MOB, SLB, XON). Fig. 3 also shows the existence of stocks that behave as reference stocks for a group of other stocks. Examples are GE (General Electric Co), JPM (JP Morgan Chase & Co) and DD (Du Pont De Nemours Co.).

5 Stability of volatility MST

A natural question arises whether or not the structure of the MST depends on the particular time period considered. This point has been considered briefly in [10,11] and it has also been recently addressed in [14,27]. In the present investigation we discuss the time robustness of a volatility MST. The dynamics of MST is investigated by considering the records of the time series delimited by a sliding time window of length T days ranging from day t to day $t+T$. For example, by using a time window with $T = 117$ we approximately compute 3000 MSTs in our sets of data. In each MST, each stock has a number of other stocks that are linked to it. This number is usually referred to as the degree of the stock. By using the above procedure, we obtain a daily historical time series of degree for each of the considered 93 stocks. In the following, we focus our attention on (i) the analysis of such degree time series as proposed in Ref. [13] and (ii) the survival ratio of tree edges as proposed in Ref. [14].

Each time series of the degree of each stock has about 3000 records. This number of records is not enough to detect reliably the autocorrelation function of the degree time series for each stock. Hence, in Ref. [13] we decided to investigate the properties of the degree time series obtained by joining all the 93 degree time series of each stock. This is done separately for each value of the time window T. From the time series obtained as described above we compute the autocorrelation function.

In Fig. 4 we show the autocorrelation function of the degree time series obtained starting from the time evolution of the volatility MST. The volatility MSTs are obtained starting from the Spearman rank-order correlation coefficient. By inspecting Fig. 4 we note that the autocorrelation function presents two distinct regimes. The first is observed for $\tau < T$, i.e. when the time lag τ is less then the size of the sliding window. In this regime the MSTs obtained at different times t are partially overlapping. The overlapping of the time windows is therefore reflected in the approximate exponential decay of the autocorrelation function. Conversely, when $T > \tau$ no overlap is present in the volatility time series used to compute MSTs. Hence the value of the autocorrelation function provides information about the time stability of the MST in this regime. We note that at $t \approx T$ the values of the autocorrelation function are approximately equal to 0.19, 0.21 and 0.22 for a time window of $T = 117$, $T = 227$ and $T = 357$ trading days respectively. These results indicate that the degree of the stocks in the MSTs of volatility has a low but not negligible level of time stability. In Ref. [13] we have shown that the time stability of the MST computed starting from stock returns is higher than the stability of the MST computed from volatility time series.

In Fig. 5 we show the result of a different investigation of the time stability of volatility MST. Specifically we show the time dynamics of the survival ratio of tree edges. In our analysis we use the definition of the survival ratio ρ introduced in Ref. [14]. The survival ratio, i.e. the fraction of edges found

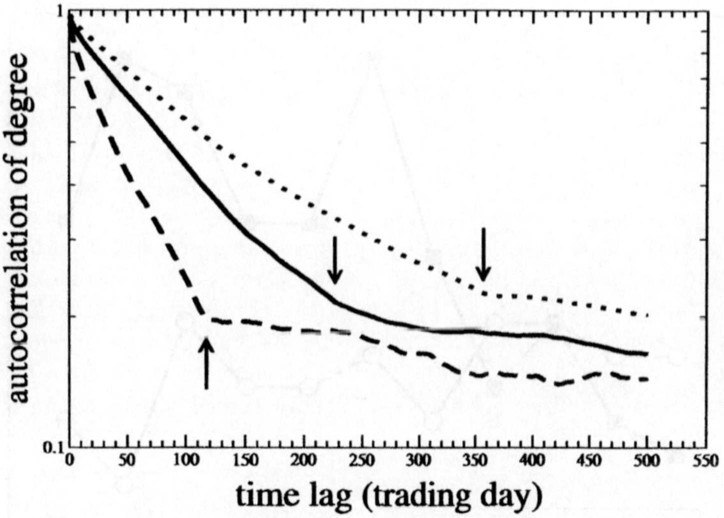

Fig. 4. **Autocorrelation function of the degree of MSTs obtained starting from volatility time series by using a correlation based clustering based on the Spearman rank-order correlation coefficient and the procedure described in the text. A different line-style indicates a different value of the sliding time window. The values of the time windows are $T = 117$ (dashed line), $T = 227$ (solid line) and $T = 357$ (dotted line) trading days. For each line, arrows indicate the value of the time lag corresponding to the time window T. They indicate the point where MSTs start to be computed from non-overlapping time windows.**

common in a pair of successive graphs computed at a time interval Δt, is obtained by considering the MSTs computed each 250 trading days for time windows T=117, and 227. Also in this investigation we are therefore considering results obtained by considering non overlapping time series. From Fig. 5 we note that the survival ratio ρ systematically above 5% and its value increases for the larger value of T. It is also worth noting that Fig. 5 shows a slow dynamics of the survival ratio over the years. This time evolution probably reflects a dynamics of stock volatility cross-correlation.

6 Discussion

Volatility time series of equities are characterized by non trivial statistical properties. The unconditional pdf can be described by using a simple two-region stochastic volatility model. This empirical observation suggests to look

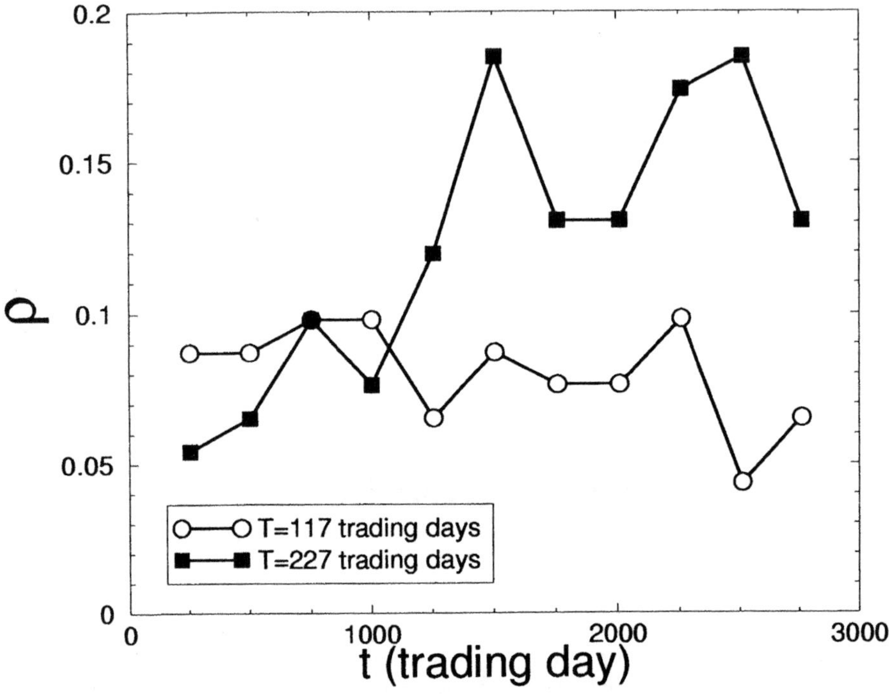

Fig. 5. Survival ratio ρ of the volatility MST edges. Volatility MSTs are computed by using the Spearman correlation coefficient. The time interval Δt between successive MSTs is set to 250 trading days. The time window T is $T = 117$ (circles) and $T = 227$ (squares). With this choice no overlap of the volatility time series occurs during MSTs computation.

for stochastic volatility models which can be characterized in terms of a log-normal pdf for small values of volatility and in terms of a power-law pdf for large volatility values. Our studies also show that a set of economic information can be filtered out from volatility time series of stocks simultaneously traded in a financial market. This information is weak but statistically robust and might play a role in some procedures associated with the estimation of financial risk measures.

Acknowledgements – The authors thank INFM and MIUR for financial support. This work is part of the FIRB project RBNE01CW3M006.

References

1. Hull, J.C., White, A. (1987) The pricing of options on assets with stochastic volatilities. Journal of Finance **XLII** 281-300.

2. Miccichè, S., Bonanno, G., Lillo, F., Mantegna, R.N. (2002) Volatility in Financial Markets: Stochastic Models and Empirical Results. Physica A **314** 756-761
3. Markowitz, H. (1959) Portfolio Selection: Efficient Diversification of Investment, Wiley, NY
4. Elton, E.J., Gruber, M.J. (1995) Modern Portfolio Theory and Investment Analysis. Wiley, NY
5. Campbell, J.W., Lo, A.W., MacKinlay, A.C. (1997) The Econometrics of Financial Markets. Princeton Univ. Press, Princeton
6. Elton, E.J., Gruber, M.J. (1971) Improved forecasting through the design of homogeneous groups. J. Business **44** 432-450
7. Panton, D.B., Parker Lessig, V., Joy, O.M. (1976) Comovement of international equity markets: A taxonomic approach. J. Financial Quant. Anal. **11** 415-432
8. Mantegna, R.N. (1999) Hierarchical Structure in Financial Markets. Eur. Phys. J. B **11** 193-197
9. Kullmann, L., Kertesz, J., Mantegna, R.N. (2000) Identification of clusters of companies in stock indices via Potts super-paramagnetic transitions. Physica A **287** 412-419
10. Bonanno, G., Vandewalle, N., Mantegna, R.N. (2000) Taxonomy of Stock Market Indices. Physical Review **E62** R7615-R7618
11. Bonanno, G., Lillo, F., Mantegna, R.N. (2001) High-frequency Cross-correlation in a Set of Stocks. Quantitative Finance **1** 96-104
12. Bonanno, G., Caldarelli, G., Lillo, F., Mantegna, R.N. (2003) Topology of correlation based minimal spanning trees in real and model markets. http://xxx.lanl.gov/cond-mat/0211546
13. Miccichè, S., Bonanno, G., Lillo, F., Mantegna, R.N. (2003) Degree stability of a minimum spanning tree of price return and volatility, Physica A (in press)
14. Onnela, J.-P., Chakraborti, A., Kaski, K., Kertesz, J. (2002) Dynamic asset trees and portfolio analysis. http://xxx.lanl.gov/cond-mat/0208131 (2002).
15. Hull, J.C. (1997) Options, Futures and Other Derivatives. Prentice Hall, Inc.
16. Black, F., Scholes, M. (1973) The pricing of options and corporate liabilities. Journal of Political Economy **81** 637-654
17. Dacorogna, M.M., Gencay, R., Müller, U.A., Olsen, R.B., Pictet, O.V. (2001) An Introduction to High-Frequency Finance. Academic Press
18. Liu, Y., Gopikrishnan, P., Cizeau, P., Meyer, M., Peng, C.-K., Stanley, H.E., (1999) Statistical properties of the volatility of price fluctuations. Phys. Rev. E **60** 1390-1400
19. Cizeau, P., Liu, Y., Meyer, M., Peng, C.-K., Stanley, H.E. (1997) Volatility distribution in the S&P500 stock index. Physica A **245** 441-445
20. Pasquini, M., Serva, M. (1999) Multiscaling and clustering of volatility. Physica A **269** 140-147
21. Lillo, F., Miccichè, S., Mantegna, R.N. (2003) Long-range correlated stationary Markovian processes. http://xxx.lanl.gov/cond-mat/0203442
22. Rammal, R., Toulouse, G., Virasoro, M.A. (1986) Ultrametricity for physicists. Rev. Mod. Phys. **58** 765-788
23. Mardia, K.V., Kent, J.T., Bibby, J.M. (1979) Multivariate Analysis. Academic Press, San Diego CA
24. Mantegna, R.N., Stanley, H.E. (2000) An Introduction to Econophysics: Correlations and Complexity in Finance. Cambridge Univ. Press, Cambridge, UK

25. Embrechts, P., McNeil, A., Straumann, D. (2002) in Risk Management: Value at Risk and Beyond, ed. M.A.H. Dempster, Cambridge University Press, Cambridge, pp 176-223
26. Press, W.H., Teukolsky, S.A., Veterling, W.T., Flannery, B.P. (1992) Numerical Recipes in Fortran: the art of scientific computing. Cambridge Univ. Press, Cambridge, UK, 2nd Ed.
27. Onnela, J.-P., Chakraborti, A., Kaski, K., Kertesz, J. (2002) Dynamic asset trees and Black Monday. http://xxx.lanl.gov/cond-mat/0212037

Time dependent correlations and response in stock market data and models

J. Kertész[1,2], L. Kullmann [1], A. G. Zawadowski[1], R. Karádi[1] and K. Kaski[2]

[1]Department of Theoretical Physics, Budapest University of Technology and Economics, Budafoki út 8, H-1111, Budapest, Hungary
[2]Laboratory of Computational Engineering, Helsinki University of Technology, P.O.Box 9400, FIN-02015 HUT, Finland

Summary. We study some consequences of the absence of microscopic reversibility on financial processes. We analyze high resolution data and find asymmetric time dependent cross correlation functions indicating dominance of some companies in the price formation procedure. These effects can be summarized in a directed netowrk of influence. Furthermore, we show that in the Lux-Marchesi multi agent market model spontaneous fluctuations decay differently from perturbations caused by external effects. The latter are easily controlled in the model, however, in real data the separation of the internal and the external effects is a highly nontrivial task.

1.1 Introduction

Obviously, there is no reason for maintaining the time reversal symmetry or detailed balance in an economic system. Nevertheless, it is of interest to investigate the time dependent cross correlations between stock returns because they contain information about the way how the prices influence each other, which are the dominant stocks and in what extent this dominance is reflected in the price changes under the conditions of an efficient market. Furthermore, the investigation of the difference between the response to external perturbations and the decay of spontaneous fluctuations sheds light to the nontrivial problem of distinguishing between external and internal market mechanisms.

1.2 Time dependent cross correlations

One of the stylized facts of markets is the exponential decay of the autocorrelation of stock returns with a very short characteristic correlation time, which is in the range of a few minutes (see [1]). This shortness is understood as a signature of market efficiency because it prevents the correlation function from being used for arbitrage purposes. As cross correlations could also be used for arbitrage, one should not expect effects much beyond the above scale and therefore high frequency data are needed for their analysis. We took [2] the Trade and Quote (TAQ) database for $N = 54$ days in the time period from 01.12.1997 to 09.03.1998 which includes tick-by-tick data for about 10000 companies. Since this is quite a short time period we selected only those companies which were traded more than 15000 times reducing this way the number of companies to 195. There are several problems related to the nonsimultaneous trading and the very noisy data. These can be overcome by careful smoothening of the data [2]. The averaging procedure involves an appropriate averaging time parameter Δt which should be larger than the characteristic time between trades but short enough not to smear out the effect. In accordance with this argument we used $\Delta t = 100$ but checked that the results are rather robust when changing it within $50 \leq \Delta t \leq 500$. For the maximum of the time shift we choose $2000s$. This is definitely beyond any reasonable characteristic time for correlations in return values due to market efficiency. In fact, using such large values for the time shift allows us to measure the noise level which the possible effect should be compared with.

For the resulting 195*194/2 correlation functions we measured the maximum value, C_{max}, the position τ_{max} at which time shift this maximum was found, and the ratio, R, of the maximum and strength of the noise, which was defined as the variance of the correlation values for larger time shift values between 600 and $2000s$. We looked at those pairs of companies for which this three values exceeded a prescribed threshold values, which we defined for $\Delta t = 100$ as: $\tau_{max} \geq 100$, $C_{max} \geq 0.04$, $R \geq 6.0$. One example of the measured correlation function can be seen in Fig. 1.1. In this case the company XON (Exxon) – which is a large oil company – "pulls" the ESV (Ensco International) which provides drilling service to oil and gas companies. The effect is rather weak but the large value of R shows that it is significant.

The maximal value of the correlations are quite small, in average less than 0.1, (*e.g.* see Fig. 1.1), although the generally quoted equal time cross correlations have much larger values. The root of this effect lies in the choice of the time difference, Δt. Increasing Δt increases the values of the equal time correlations [4].

The results show that the characteristic time shift is around 100 sec. which is consistent with the effective market hypothesis. A time shift larger than the characteristic time of the decay of the auto-correlations of returns would contradict to the efficient market picture and could be used to arbitrage.

We found that usually one "small" company is influenced by many "large" ones and one "large" company pulls many "small" ones. The more frequently traded companies influence, "pull" the less frequently traded ones. This is not surprising since obviously the more frequently traded companies are more important; it is therefore more likely that they influence a smaller company than the other way around. Although this is the generic situation, there are a few exceptions when a less often traded company "pulls" the other one.

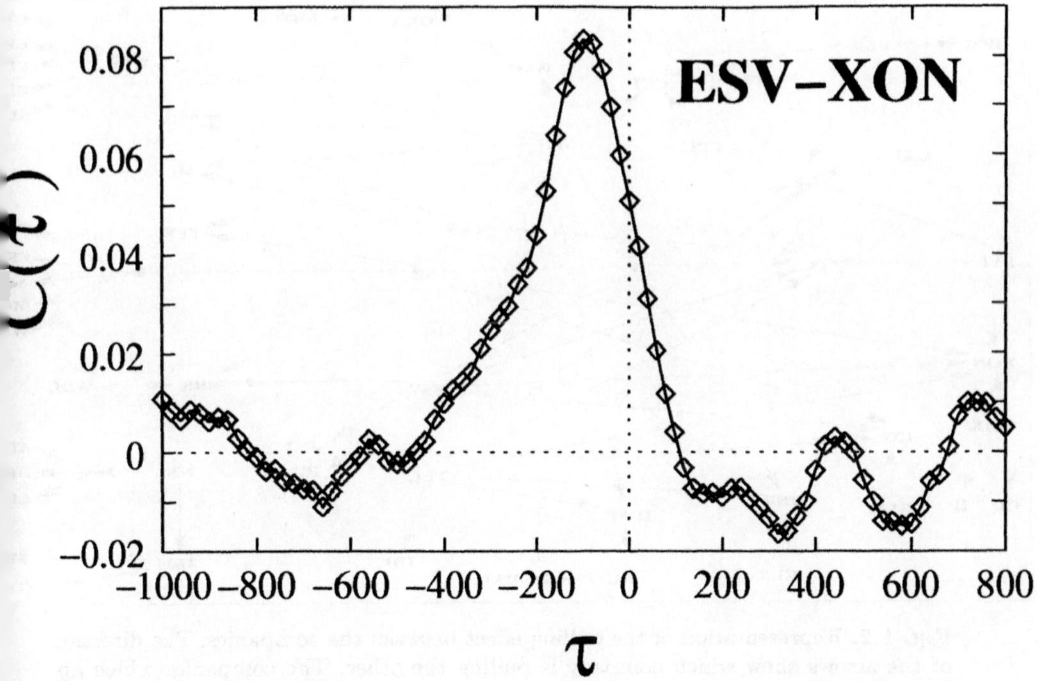

Fig. 1.1. Example for the measured shifted-time correlation function. The two companies are: Ensco International (ESV) and Exxon Corp. (XON). The maximum correlation value is at -100 sec, which means that the return time series of ESV has to be shifted back in order to get the maximal correlation i.e. the price changes happen later in time, ESV is pulled by XON. From [3], with permission from Elsevier.

The relationships can be represented as a graph, see Fig. 1.2, with directed links where there are nodes from which many links go out (meaning that this node is influenced by many others) and there are other nodes where many links go in (these are the big companies influencing the less important ones).

1.3 Fluctuations and response in the Lux-Marchesi model

The Lux-Marchesi model is a single asset – multi agent model [5]. One of the main assumptions is that there exists a fundamental price p_f of stocks (the value of the company and its prospective future growth) around which the real price fluctuates. In this model the agents are let to choose among the three following strategies:

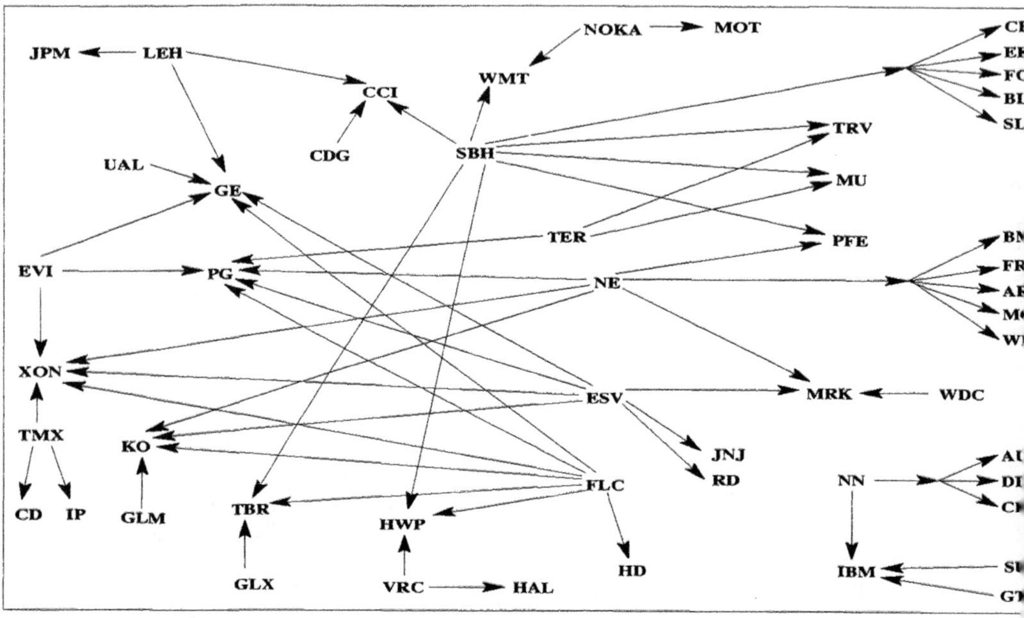

Fig. 1.2. Representation of the pulling effect between the companies. The direction of the arrows show which company is pulling the other. The companies which appear in the figure show the most significant effects. From [3], with permission from Elsevier.

optimists (who buy whatever happens), pessimists (who sell), and fundamentalist (who sell if the market price is above the fundamental price and vice versa). Optimists and pessimists together are called chartists according to the usual terminology.

The dynamics of the model is governed by the rule that agents may switch between the strategies if prospective payoffs are better using another strategy: the bigger the difference between the payoffs the higher the probability that the agent switches to the better strategy (transition probabilities are an exponential function of the profit difference).

The model uses many parameters, some of them of economic origin, some of them determining the size and reaction speed of the market. The price formation is based on the excess demand. In our simulation we used the parameter sets given in the original article of Lux and Marchesi.

The dynamics of the model implies that there are continuous fluctuations around the fundamental price. We examined the decay of these fluctuations using the following simple method: when the price rose to $p = p_f + \Delta p = 10.0 + \Delta p$ we defined this as an important fluctuation and observed the average decay for many runs. Hence we did not try to determine whether the price really sank after reaching $p = p_f + \Delta p$ assuming that (at least at bigger fluctuations) the probability of further rise is much smaller than that of further decline.

We observed exponential decay for the price fluctuations using all parameter sets in accordance with the well known fast decay of the correlation function. Detailed simulations were undertaken for parameter set IV (Fig. 1.3a). Furthermore, in case of relatively large fluctuations we observed that the opinion index and the fraction of chartists significantly differed from their average values and exponential decay was observed for both quantities. The characteristic time of the decay of the price decreased with the size of fluctuation and was in the order of magnitude of 1 time unit (up to fluctuations of 4) (Fig. 1.3b). The opinion index decayed with approximately the same characteristic time. On the other hand the fraction of chartists (which is closely related to volatility) decayed much slower with characteristic times in the order of magnitude of 100 time units.

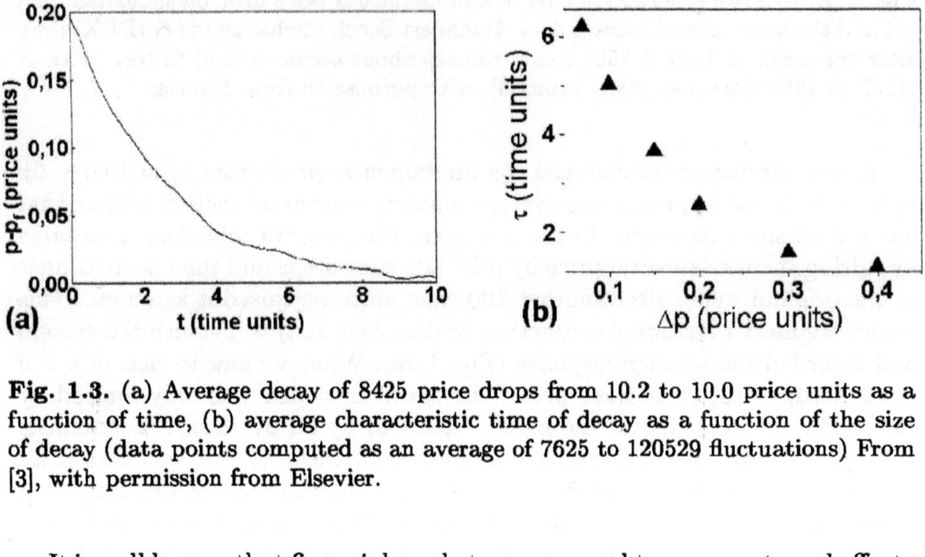

Fig. 1.3. (a) Average decay of 8425 price drops from 10.2 to 10.0 price units as a function of time, (b) average characteristic time of decay as a function of the size of decay (data points computed as an average of 7625 to 120529 fluctuations) From [3], with permission from Elsevier.

It is well known that financial markets are exposed to many external effects from the outside world. This means that the changes in prices are only partly due to the inherent market mechanisms (fluctuations around "equilibrium"), changes can also be caused by news from the outside world (e.g. financial reports; bankruptcy; death of important personalities; outbreak of war; terror attacks, etc.). The analysis of real market data from the point of view of external news is a highly non-trivial task. On the one hand, it is difficult to set an independent level of "importance" of news in our age of information explosion; furthermore, the effects of different news may overlap. On the other hand, the reaction of the market to the news is also hard to tell. The situation is much simpler in an artificial market like the Lux-Marchesi model where we can immediately change the fundamental value of a company or asset (which would be in reality a consequence of the external event).

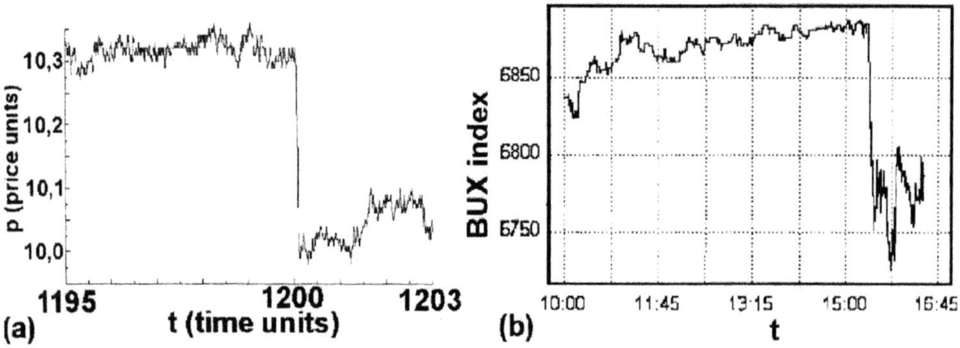

Fig. 1.4. (a) Price evolution in case of a fundamental price drop on simulated data (b) and the same on real market data: Budapest Stock Exchange Index (BUX) drop after the crash of flight AA587 (early rumors about terror attack) in New York at 15:17 on 12th November 2001. From [3], with permission from Elsevier.

In our simulation we changed the fundamental price from $p_f = 10.0 + \Delta p$ to $p_f = 10.0$ and examined the average of many runs as a function of time that has passed since the event. In the computer program we solved the averaging by raising the fundamental price by p for 100 time steps and then decreasing it to the original value, after another 100 time units we raised it again etc. This means we used a rectangular function (Θ-like function) to perturb the system and recorded the average response (Fig. 1.5a). What we saw in case of $p > 0$ is an abrupt drop in market price (the speed of which was only limited by the minimum time step of the model) followed by an overshoot. Economists and traders have long known that a correction exists after a very fast price change(Fig. 1.4). A sad example for this was the reaction of the European stock markets to the terror attack on New York on the 11th September. The prices dropped fast that day (generally losses over 10 % were recorded) but the next day there was already an upward moving trend (a correction after the overreaction of the events).

Another interesting result is that (in case of a price drop) the location of the price minimum in time is independent of the price drop Δp for a wide range of Δp (for parameter set IV the price minimum is located at approximately $t = 0.42 \pm 0.02$ time units up to price drops of 10%). This means that on average one can predict when the minimum of the price occurs (if one knows the parameters of the market) irrespective of how big the fundamental price change Δp is.

Let us define the magnitude M of the overshoot as the difference of the price minimum and the new equilibrium price (which equals the fundamental price after the event). M shows linear dependence on the fundamental price change Δp in a wide range of Δp-s (up to a 10% abrupt fundamental price change, which is already huge on market scales) (Fig. 1.5b).

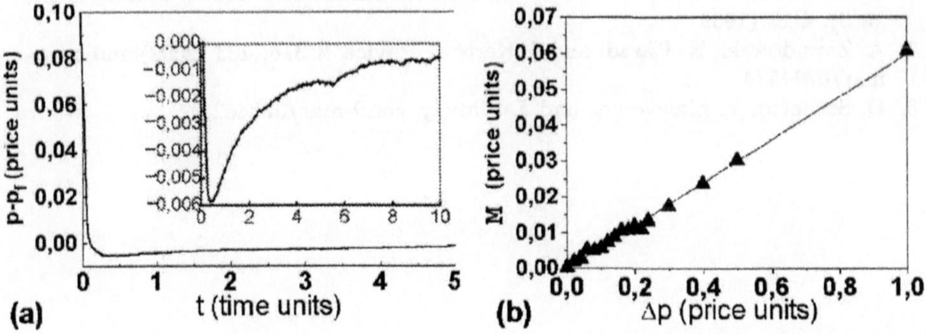

Fig. 1.5. (a) An average of 50000 price drops following an abrupt fundamental price change from 10.1$ to 10.0$; inset shows the overshoot on a different scale (b) average size of overshoot as a function of the fundamental price drop (each data point calculated as an average of 50000 price drops). From [3], with permission from Elsevier.

It has been a long standing problem to sort out the origin of rapid changes in the prices [6]. The study of the difference between the decay of spontaneous fluctuations and externally caused perturbations addresses just this question. For a detailed discussion we refer to [7]. We would like to mention that the empirical study of the response discussed here is under way. After finishing this manuscript we learned about the closely related preprint by Sornette et al [8].

Support of the Academy of Finland, project No. 1169043 (Finnish Centre of Excellence Programme 2000-2005) is acknowledged. Thanks are due T. Lux, J. Török, J.-P. Onnela and A.-L. Barabási for their help. JK is member of the "Center of Applied Mathematics and Computational Physics" of the BUTE.

References

1. R. Mantegna and H.E. Stanley: An Introduction to Econophysics (Cambridge UP, 2000), W. Paul, J. Baschnagel *Stochastic Processes From Physics to Finance*, Springer-Verlag (1999) R. Cont: Quantitative Finance 1, 223 (2001) J. D. Farmer, Computing in Science & Engineering, Nov./Dec. (1999), p. 26-39.

2. L. Kullmann, J. Kertész and K. Kaski, Phys. Rev. E 66, 026125 (2002) and cond-mat/0203256
3. J. Kertész, L. Kullmann, A.G. Zawadowski, R. Karadi and K. Kaski: Correlations and response: Absence of detailed balance on the Stock Market, Physica A (in press)
4. G. Bonanno, F. Lillo, R. N. Mantegna, Quantitative Finance, **1**, 96, (2001)
5. T. Lux, M. Marchesi, Nature 397, 498 (1999) T. Lux, M. Marchesi, International Journal of Theoretical and Applied Finance 3, 675 (2000)
6. D.M. Cutler, J.M. Poterba, L.H. Summers, Journal of Portfolio Management, 15(3), 4-12 (1989)
7. A. Zawadowski, R. Karadi and J. Kertész, Phyica A 316, 403 (2002) and cond-mat/0204574
8. D. Sornette, Y. Malevergne and J.-F.Muzy, cond-mat/0204626

Distributions and Long-Range Correlations in the Trading of US Stocks

Plamen Ch. Ivanov[1], Ainslie Yuen[2], Boris Podobnik[3] and Youngki Lee[4]

[1] Center for Polymer Studies and Department of Physics, Boston University, Boston, MA 02215 plamen@argento.bu.edu
[2] Signal Processing Laboratory, Department of Engineering, Cambridge University, UK aslw2@cam.ac.uk
[3] Faculty of Civil Engineering, University of Rijeka, Rijeka, Croatia bp@phy.hr
[4] Yanbian University of Science and Technology, Beishan St., Yanji City, Jilin Province, China 133000 youngki@ybust.edu.cn

Summary. We analyze the sequence of time intervals between consecutive stock trades of five large companies representing different sectors of the US economy over a period of four years. We show that independent of the industry sector and average level of activity, the series of intertrade times exhibit common statistical features. Specifically, we find that: (i) the tail of the probability density function of intertrade times may be fit by a stretched exponential form; (ii) the probability densities of the intertrade times for all five companies collapse onto a single curve when appropriately rescaled, and (iii) the intertrade times exhibit correlated behaviour over hundreds of trades within a trading day and an even greater degree of correlation over longer time scales.

Investigations of price dynamics of stocks, commodities and financial indices have long been the main focus of economic research [1, 2, 3, 4]. Recent studies however, have turned to the information offered by other aspects of the trading process such as volume of shares traded at each transaction [5] and number of trades in a unit time [6, 7]. Empirical observations suggest a possible relationship between price and trading activity — a significant change in price is usually accompanied by a flurry of trading (Fig. 1).

To understand the dynamics of the trading process on a trade-by-trade level, we consider the series of time intervals between consecutive trades which we call the intertrade times (ITT). While there have been some attempts to model the dynamics of ITT [8, 9, 10, 11] there are only a few empirical studies of high-frequency single-stock recordings of ITT over limited periods of time [12] or of very low-frequency recordings [13].

Here we empirically investigate the statistical and scaling properties of ITT over extended periods of trading. We compare these properties for five prominent companies with large market capitalization representing different sectors

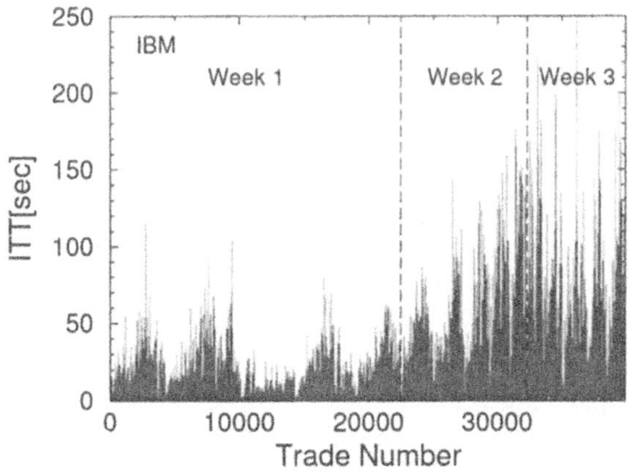

Fig. 1. Intertrade times (ITT) of IBM stock over three weeks of trading representing 4×10^4 trades. Each week consists of five trading days exhibiting an intraday pattern of short intertrade times at the open and close of each day, with longer intertrade times in between.

of the US economy: Morgan J. P. & Co. (Financial), Boeing Co. (Aerospace), General Electric Co. (Conglomerate), International Business Machines Corp. (Technology) and Merck & Co. Inc (Pharmaceutical). We study the time intervals between consecutive stock trades over a period of four years — from Jan. 93 till Dec. 96. All five companies are traded on the New York Stock Exchange, vary substantially in their average market capitalization and exhibit different levels of trading activity with different numbers of trades over the period considered (Table 1).

Company	**Symbol**	**<Mkt Cap.> ($\$10^9$)**	**Number of trades**	**<ITT> (sec)**	α_1	α_2
Morgan JP & Co.	JPM	13.7	401213	58	0.60	0.92
Boeing	BA	19.9	728779	32	0.64	0.97
General Electric	GE	101.7	1374682	17	0.65	0.93
IBM	IBM	45.4	1677319	14	0.67	0.95
Merck & Co.	MRK	56.9	2055443	11	0.69	0.94

Table 1. Descriptive statistics of the five US stocks studied over the period Jan. 93 - Dec. 96. We include only intertrade times occurring during NYSE trading hours from 9.30am until 4pm EST, excluding public holidays and weekends. The period considered covers 1009 trading days. α_1 and α_2 indicate the values of the scaling exponent characterizing correlations in ITT at time scales less than and greater than a trading day.

We first study the probability density function of ITT. Inspection reveals marked differences between companies (Fig. 2). This is not surprising given the variation in trading activity: companies with frequently traded shares have a higher peak of the probability distribution at shorter intertrade times, while more rarely traded companies have tails extended over larger intertrade times.

We find that the tails of the probability distribution of ITT for all five companies are well fit by a stretched exponential distribution

$$P(x,\delta) = k\ e^{-x^{\delta}/l_0}, \tag{1}$$

where k is the normalizing constant, δ is the stretched exponent and l_0 is the characteristic timescale (Fig. 2a).

Since more frequently traded companies have generally smaller values of ITT we choose to rescale the probability distributions by the mean value of ITT over the entire period. After rescaling we find that the curves of all five companies fall onto a single scaled plot — data collapse (Fig. 2c). This common behaviour is also characterized by a stretched exponential form. Such data collapse is reminiscent of a wide class of physical systems with universal scaling properties [14].

Next we investigate the temporal organization of ITT. Empirical observations show that like many other financial time series, the ITT data exhibit nonstationary behaviour and complex variability with a superposed pattern of daily activity (Fig. 1). Patches of inactive trading are often followed by patches of more active trading within a trading day. Similar patterns can be observed on a daily, weekly and even monthly basis, independent of the average level of trading activity for a company. Such observations suggest that there may be a self-similar, fractal organization in the sequence of ITT over a broad range of time scales. To test this hypothesis we apply the detrended fluctuation analysis (DFA) method [15, 16]. The DFA method can accurately quantify fractal features in ITT, as it permits the detection of long-range correlations embedded in nonstationary time series, and avoids the spurious detection of apparent long-range correlations that are an artifact of trends in the data [17].

The DFA method consists of the following steps. We first integrate the ITT series to construct the profile $Y(k) = \sum_{i=1}^{k}(ITT_i - \langle ITT \rangle)$ where $\langle ITT \rangle$ is the series mean. Next, we partition the profile $Y(k)$ into equal windows of size n (scales or number of trades) and fit the local trend in each window with a least-squares polynomial fit. We then detrend the profile $Y(k)$ by subtracting the local trend in each window of size n. For a given window size we calculate the root mean square fluctuation for the detrended profile. This procedure is repeated to obtain the detrended fluctuation function $F(n)$ for different window sizes. A power-law relation $F(n) \sim n^{\alpha}$ indicates the presence of scaling in the ITT series. Thus the fluctuations in the ITT can be characterized by scaling exponent α, a self-similarity parameter that represents the fractal power-law correlation properties of the signal. The scaling or correlation ex-

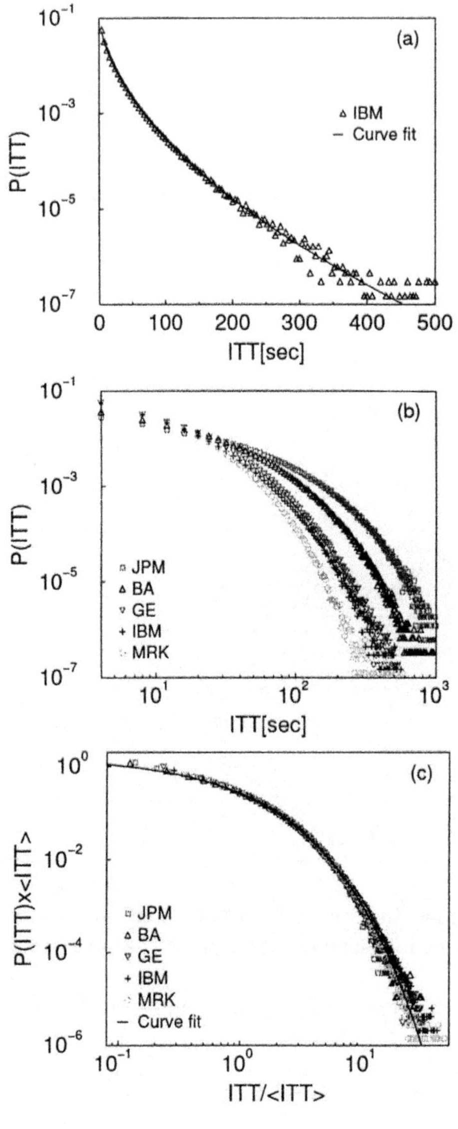

Fig. 2. **(a)** Probability density function of intertrade times (ITT) of IBM stock and a stretched exponential fit (solid line) with parameters $k = 0.16, \delta = 0.54, l_0 = 1.9$ (Eqn. 1). **(b)** Probability density functions of ITT of five US stocks over the period Jan. 93 - Dec. 96 with increasing number of trades from top to bottom (Table 1). **(c)** After rescaling by the mean value of ITT for each company, the density functions collapse onto a single curve suggesting a common functional form. Stretched exponential fit (solid line) with parameters $k = 1.86, \delta = 0.56, l_0 = 0.50$.

ponent α is related to the autocorrelation function exponent γ ($C(n) \sim n^{-\gamma}$ when $0 < \gamma < 1$) and to the power spectrum exponent β ($S(f) \sim 1/f^{\beta}$) by $\alpha = 1 - \gamma/2 = (\beta + 1)/2$ [15]. A value of $\alpha = 0.5$ indicates that there are no correlations and the signal is uncorrelated (white noise). If $\alpha < 0.5$ the signal is said to be *anti-correlated*, meaning that large values are more likely to be followed by small values. If $\alpha > 0.5$ the signal is correlated and exhibits persistent behaviour, meaning that large values are more likely to be followed by large values and small values by small values. The higher the value of α, the stronger the correlations in the signal.

Before performing the DFA analysis we pre-process the data by excluding all outliers in the ITT exceeding ten times the standard deviation of the complete time series for each company. This naturally excludes large ITT values caused by unusual closures inside a trading day as well as data entry errors. This procedure results in the removal of less than 0.06% of data points. In addition, the split transactions that arise when the volume of an order must be matched by several opposing orders often results in a number of transactions having an execution time separated by less than a second. Given the one-second resolution of the data, these transactions result in a number of zero intertrade intervals. Removal of all outliers and all zero intertrade times does not significantly affect the results of the DFA analysis for positively correlated signals [18].

We find that the ITT series for all five companies exhibit long-range power-law correlations over a broad range of time scales from several trades to hundreds of thousands of trades characterized by a correlation exponent $\alpha > 0.5$ (Fig. 3, Table 1). This is consistent with the empirical observation that long segments of high trading activity may be followed by long segments of less active trading (Fig. 1). This scaling behaviour appears independent of average level of trading activity and industry sector. To confirm the presence of such strong persistent behaviour we reshuffle the data. The detrended fluctuation function $F(n)$ for the reshuffled data exhibits white noise behaviour with exponent $\alpha = 0.5$, significantly different from the behaviour of the original ITT series.

For all companies we observe a bump in the scaling curve $F(n)$ due to the periodic daily pattern in trading activity (Fig. 1). The position of the bump corresponds to the average daily number of trades for each company. Previous studies demonstrate that periodic trends superposed on power-law correlated signals do not affect the value of the scaling exponent [17]. For all five companies we observe two scaling regimes, one at short times scales ranging from several trades to a few thousand trades within a trading day with a crossover to a second regime ranging from thousands to hundreds of thousands of trades corresponding to time scales from days to almost a year. In addition we find that the scaling behaviour within a trading day (short time scales) is characterized by a correlation exponent $\alpha_1 = 0.65 \pm 0.03$ (group mean $\pm$ std. dev.) compared with the value of the correlation exponent $\alpha_2 = 0.94 \pm 0.02$ at large time scales. The two scaling regimes in the ITT series may

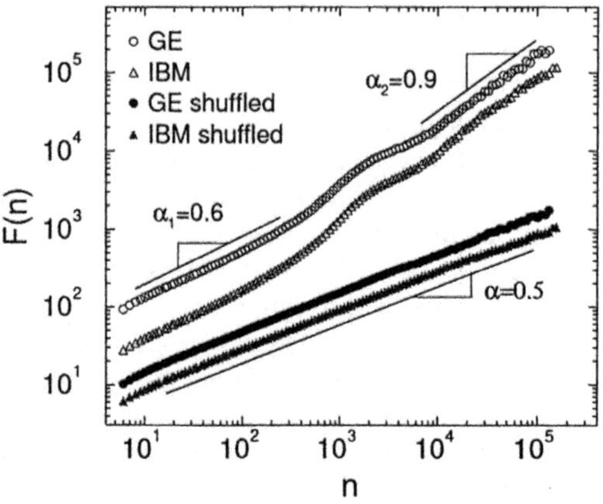

Fig. 3. Detrended fluctuation function $F(n)$ for intertrade time (ITT) series for GE and IBM. Both companies exhibit long-range power-law correlations with a crossover to larger exponent at scales above one trading day (corresponding to $\approx 1.4 \times 10^3$ and $\approx 1.7 \times 10^3$ trades respectively). Shuffled series are characterized by exponent $\alpha = 0.5$ as expected for uncorrelated (white) noise.

be a consequence of the time scales over which news is absorbed. Since trading activity is influenced by information, there will be less coherence between intraday trading as information takes time to disseminate thus leading to a lower value of α. Over time scales greater than a day, more information is available to investors, leading to more coherent and thus more persistent behaviour with a higher value of α.

In summary, we present evidence of common static and dynamic features in the ITT for diverse companies. We find that there is a common form for the probability distribution and common temporal organization in the ITT characterized by long-range scaling behaviour, which are independent of the average level of trading activity, size of company (market capitalization) and industry sector. These common features may relate to previously observed common market structure and common price dynamics [19, 20, 21, 22]. Furthermore, our empirical findings may assist econometric modelling approaches to account for various aspects of the dynamics of ITT.

Acknowledgements We thank T. Lim, K. Hu and S. White for stimulating discussions and the NIH/National Center for Research Resources (P41 RR13622), the Department of Engineering, Cambridge University and King's College, Cambridge for financial support.

References

1. Bachelier L (1900) Theorie de la speculation. Thesis, Sorbonne, Paris. Reprinted in Cootner P (1964) The Random Character of Stock Prices. MIT Press, Cambridge
2. Mandelbrot B (1963) The variation of certain speculative prices. Journal of Business 36:394–419
3. Dothan M (1990) Prices in Financial Markets. Oxford University Press, New York
4. Mantegna R, Stanley HE (2000) An introduction to econophysics. Cambridge University Press, Cambridge, England
5. Gopikrishnan P, Plerou V, Gabaix X, Stanley HE (2000) Statistical Properties of share volume traded in financial markets. Phys. Rev. E 62:4493–4496
6. Plerou V, Gopikrishnan P, Amaral LAN, Gabaix X, Stanley HE (2000) Economic fluctuations and anomalous diffusion. Phys. Rev. E 62:3023–3026
7. Bonanno G, Lillo F, Mantegna R (2000) Dynamics of the number of trades of financial securities. Physica A 280:136–141
8. Engle R, Russell J (1998) Autoregressive conditional duration: A new model for irregularly spaced transaction data. Econometrica 66:1127–1163
9. Gourieroux C, Jasiak J, LeFol G (1999) Intra-day Market Activity. Journal of Financial Markets 2:193–226
10. Grammig J, Maurer K (2000) Non-monotonic hazard functions and the Autoregressive Conditional Duration model. Econometrics Journal 3:16–38
11. Golia S (2001) Long memory effects in ultra-high frequency data. Quaderni di Statistica 3:43-52
12. Raberto M, Scalas E, Mainardi F (2002) Waiting-times and returns in high-frequency financial data: an empirical study. Physica A 312:749-755
13. Sabatelli L, Keating S, Dudley J, Richmond P (2002) Waiting time distributions in financial markets. Eur. Phys. J. B 27:273–275
14. Bunde A, Havlin S (1994) Fractals in Science. Springer, Heidelberg
15. Buldyrev SV, Goldberger AL, Havlin S, Peng CK, Stanley HE, Simons M (1993) Fractal landscapes and molecular evolution: modeling the myosin heavy chain gene family. Biophysics Journal 65:2673–2679
16. Peng CK, Buldyrev SV, Havlin S, Simons, M, Stanley HE, Goldberger AL (1994) Mosaic Organization of DNA nucleotides. Phys. Rev. E 4:1685–1689
17. Hu K, Ivanov PCh, Chen Z, Carpena P, Stanley HE (2001) Effect of trends on detrended fluctuation analysis. Phys. Rev. E 64, 011114
18. Chen Z, Ivanov PCh, Hu K, Stanley HE (2002) Effect of nonstationarities on detrended fluctuation analysis. Phys. Rev. E 65, 041107
19. Lux T (1996) The Stable Paretian Hypothesis and the Frequency of Large Returns: An Examination of Major German Stocks. Appl. Financial Economics 6:463–475
20. Gopikrishnan P, Meyer M, Amaral LAN, Stanley HE (1998) Inverse cubic law for the distribution of stock price variations. Eur. Phys. J. B 3:139–140
21. Liu Y, Gopikrishnan P, Gizeau P, Meyer M, Peng C-K, Stanley HE (1999) Statistical properties of the volatility of price fluctuations. Phys. Rev. E 60:1390–1400
22. Plerou V, Gopikrishnan P, Amaral LAN, Meyer M, Stanley HE (1999) Scaling of the distribution of price fluctuations of individual companies. Phys. Rev. E 60:6519–6529

Time Evolution of Fractal Structure by Price-axis Scaling and Foreign Exchange Intervention Operations

Yoshiaki Kumagai[1]

School of Education, Waseda University, 1-6-1 Nishi-waseda, Shinjuku-ku, Tokyo 169-8050, Japan kumagai@mbg.nifty.com

Summary. We propose a simple method to describe the fractal structure of data, which are irregularly spaced. Using the extreme values determined with a scale, we define a kind of fractal dimension: the fold dimension. This method is based on the same concept as a technical analysis: Kagi chart. In this contribution, we use a high-frequency data set on bid and ask prices of the Dollar/Yen exchange rates, and investigate time evolution of the fractal dimension. Moreover we analyze the effect of foreign exchange intervention operations.

Key words: High frequency data, Technical analysis, Fractal dimension, Exchange rate, Intervention operation

1 Introduction

Transactions in financial markets occur at varying time intervals. Most studies, however, have converted such data into that of equal intervals. High-frequency data are often re-sampled at intervals of 5 or 10-minutes. Moreover, "time", which evolves the price process, cannot be determined uniquely. There are at least four candidates for the proper time scale to use for analyzing market data: physical time, trading time, the number of trade, and cumulated trading volume [1]. We propose another simple method to describe the fractal structure of data, which are fundamentally irregularly spaced.

We expand the definition of extreme values, to be determined with a scale that corresponds the price-axis resolution. Using these extreme values determined with a sale, we can measure a kind of the fractal dimension: the fold dimension. This captures the scaling behavior of the time series without loss of information along with the price-axis. This method is based on the same idea as a Japanese charting technique: Kagi chart. In this contribution, we use a high-frequency data set on bid and ask prices of the Dollar/Yen exchange rate, and investigate time evolution of the fractal dimension. Moreover we analyze the effect of foreign exchange intervention operations.

2 Scaling of fluctuations

The *local maximum determined by a scale C* was defined as follows [2]. Suppose g is a real function defined on $\mathbf{R}^1$. We say that g has a *local maximum determined by a scale C* at point t, if there exist $d_1, d_2 > 0$ such that $g(t-d_1) < g(t) - C, g(t+d_2) < g(t) - C$ and g attains its maximum at t in $[t-d_1, t+d_2]$. The local maximum determined by a scale 0 means the ordinary local maximum. The width C is the accuracy of measurement of extreme values. The*local minimum determined by a scale C* is defined likewise.

Using these extreme values, we define two functions. In this section, suppose that the time series $g(t)$ is defined in a given interval $[0, S]$. Let C denote a vertical spread. In this interval $[0, S]$, we can detect extreme values determined by a scale C. Suppose $g(t)$ attains its extreme values at $t_1^C < t_2^C < t_3^C < \cdots$, ordered increasingly. With $g(t)$ fixed, the number of these maxima and minima in $[0, S]$ is a function of C. Let $m(C)$ denote the number of these extreme values. Next, we introduce a kind of total variation of the time series. Let $R(C)$ denote the total sum of absolute variation between neighboring extreme values:

$$R(C) \equiv \sum_{i=1}^{m(C)-1} |g(t_{i+1}^C) - g(t_i^C)|. \tag{1}$$

$R(0)$ is the total sum of absolute variations : $R(0) = \lim_{n\to\infty} \sum_{k=1}^{n} |g(kS/n) - g((k-1)S/n)|$. Note that $m(C)$ and $R(C)$ are invariants for time scale stretching. The information used is the values and their order in time, not the time themselves. This property is significant to investigate non-equidistant data like high-frequency data in finance. Further properties of $m(C)$ and $R(C)$ are in [2].

Using the function $R(C)$, a kind of fractal dimension was defined [3]. If the relation:

$$R(C)/C \sim C^{-D_f}. \tag{2}$$

holds, we say that the exponent D_f is the *fold dimension* of the time series. The fold dimension D_f is correspond to the latent fractal dimension D_l in self-affine graph. The relation between the local box dimension D_b and the latent fractal dimension D_l is $D_b = 2 - 1/D_l$.

3 Kagi charts and scaling

The charts in the technical analysis of the trend type can be categorized by the x-axis: time series charts and non-time series charts. Rather than having price on the y-axis and time on the x-axis, non-time series charts disregard the passage of time and display price changes on both axes. This category includes Kagi, Renko, Three Lines Break and Point & Figure(P&F) charts.

Kagi charts display a series of connecting vertical lines where the direction and the thickness (or color) of the lines are dependent on the price action. The charts ignore the passage of time. If prices continue to move in the same direction, the vertical line is extended. However, if prices reverse by a "reversal" amount C, a new kagi line is then drawn in a new column. When prices penetrate a previous high or low, the thickness of the kagi line changes.

We now show the algorithm to draw Kagi chart for the price series $p(s), (s = 0, 1, 2 \cdots)$ using the reversal amount C. Lets z^- denotes the candidate for the upward reversal, and z^+ the candidate for the downward reversal. q is an indicator function whether the line extend upwards or downwards at time s. The calculations at time s are as follows.

if $p(s) > p(s-1)$ then
  if $p(s) > z^+$ then
    $z^+ \leftarrow p(s)$
    The line is extended in the same direction (upwards).
  if $q(s-1) < 0$ then
    if $p(s) > z \quad + C$ then
      A short horizontal line is drawn to the next column,
      and a new vertical line is drawn to the closing price.
      $q(s) \leftarrow 1$
      $z^+ \leftarrow p(s)$
if $p(s) < p(s-1)$ then
  if $p(s) < z^-$ then
    $z^- \leftarrow p(s)$
    The line is extended in the same direction (downwards).
  if $q(s-1) > 0$ then
    if $p(s) < z^+ - C$ then
      A short horizontal line is drawn to the next column,
      and a new vertical line is drawn to the closing price.
      $q(s) \leftarrow -1$
      $z^- \leftarrow p(s)$

The reversal amount C is fixed of the size historically used for that particular price range or the size by a constant percentage of the price. An example of Kagi chart is shown in Fig. 1. For more details on these charts, see [5].

Kagi charts and extreme values determined by a scale are based on the same concept of scaling along the y-axis. Local maxima determined by a scale C correspond to the upper ends of the vertical lines in Kagi charts drawn by the reversal amount C. Likewise, local minima determined by a scale C correspond to the upper ends of the vertical lines in Kagi charts drawn by the reversal amount C. Hence the number of the vertical lines in $[0, S]$ is equal to the number of extreme-values $m(C)$. Furthermore, the sum of the vertical lines is equal to the total variation $R(C)$. Making C larger filters out more data. The chartists control the reversal amount C to find a balance that shows enough detail without showing too much. On the other hand, we control C to detect the fractal structure of the time series.

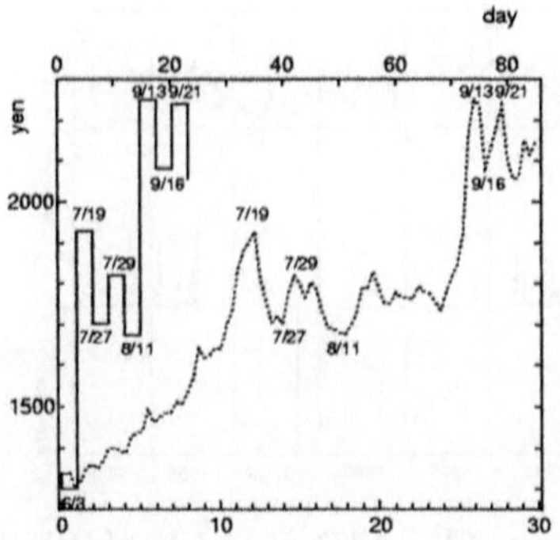

Fig. 1. Kagi chart of NEC in Tokyo Stock Exchange from Jun. 1 to Nov. 30, 1999, drawn by the reversal amount 100 yen. The dotted line represents a bar chart of daily closing prices for the same period (referring the upper axis). The figure does not show the thickness, since it is not related to the scaling of the time series.

4 Data

The data set used is provided by CQG, and contains all quotes (bid and ask) for US Dollar (USD)-Japanese Yen (JPY) exchange rates submitted by the brokers from Jan. 1, 1995 to Nov. 30, 2000 [1]. The records of the quotes do not guarantee the actual execution in this price. The recorded time of submission is given with an accuracy of one minute. The record of the quotes is given in the order of submission.

When using the ordinary methods of fractal analysis, the least division of the time scale of these data is one minute. Therefore, the rest of information from the representative of one-minute interval is not used. On the other hand, by using the fold dimension D_f, we can analyze the data without information losses but time intervals [4]. The efficient slopes of $\log R(C)/C$ with $\log C$ dependent on the scaling regions, since most of the graphs of the empirical time series are not self-affine.

The time evolution of the fractal dimensions: $(2 - 1/D_f)$ and the records of intervention operations by the currency authorities [2] is shown in Fig. 2 and 3. The weekly dimensions are shown in Fig. 2 and the daily ones in Fig. 3. In many cases, the first day of the successive interventions coincides with the local minimum of the dimensions. However, the direction of the cause and effect relationship between the intervention and the change of the dimensions is not clear by this study. In this microscopic scaling region: 0.07-0.17 yen, the price fluctuations show anti-persistence in the ordinary state. When the market becomes volatile, the anti-persistence in this scaling region weakens.

---

[1] The author thanks Hiroyuki Moriya and Sumisho Capital Management for supplying the data.

[2] The statistics of the foreign exchange intervention operations are released by the Ministry of Finance of Japan.

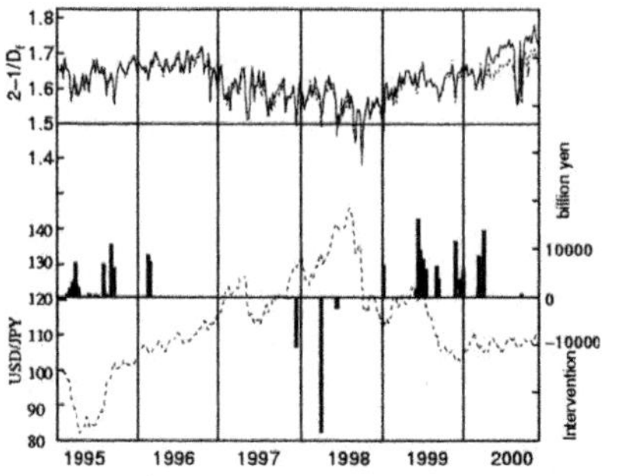

Fig. 2. Evolution of weekly $(2 - 1/D_f)$ of USD/JPY exchange rates and the interventions. The scaling region is 7-17 pips, for bid (dotted line) and ask (solid line). (One pip is 10^{-2} dollar/yen.) The solid bars represent the total amount of the intervention operations of the week. If the operations are selling USD and buying JPY, the amount is denoted by negative. The dashed line shows the exchange rates.

5 Conclusion

In this contribution, we use a simple method to describe the fractal structure of data, which are fundamentally irregularly spaced. This method and Kagi chart are based on the same concept of scaling along the vertical axis. Therefore, this method is the revision of the traders' perception of price series from econophysics. Using the time evolution of the fold dimensions, the fractal structure of foreign exchange market can be described precisely.

Acknowledgements

We are grateful to Soichiro Moridaira and Hiroyuki Moriya for helpful advice on data handling. We also acknowledge the continuing guidance and encouragement of Gyoichi Iwata, Toshio Inaba and Yukitami Tsuji.

References

1. Ghysels E (2000) Some econometric recipes for high-frequency data cooking. Journal of Business and Economic Statistics, 18: 154-163
2. Kumagai Y (2001) Time-space scaling of financial time series. In: Takayasu H (Eds) Empirical science of financial fluctuations. Springer-Verlag Tokyo, pp 250-260
3. Kumagai Y (2002) Fractal structure of financial high frequency data. Fractals, 10: 13-18
4. Kumagai Y (2002) Fractal dimension of high frequency data in foreign exchange market. Journal of the Korean Physical Society, 40: 1100-1104
5. Steve N (1994) Beyond Candlesticks: More Japanese charting techniques revealed. John Wiley & Sons

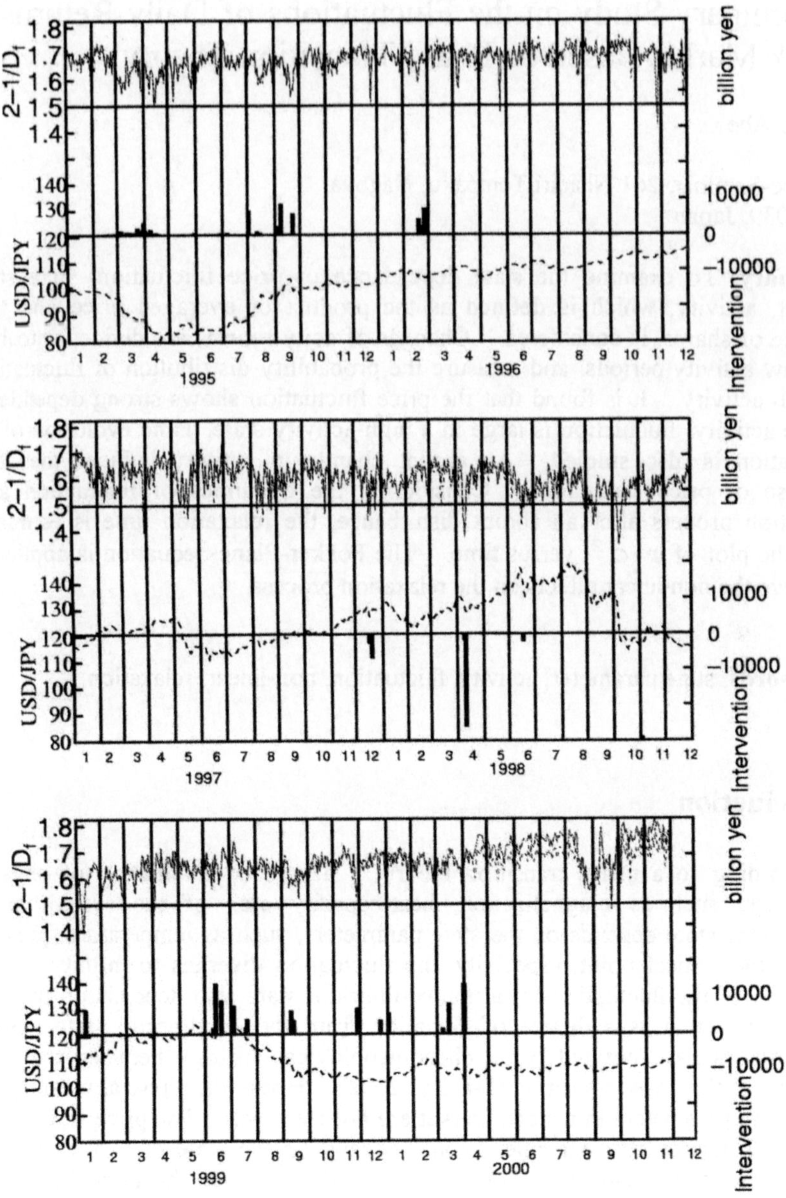

Fig. 3. **Evolution of daily $(2 - 1/D_f)$ of USD/JPY exchange rates and the interventions. The dimensions $(2 - 1/D_f)$, whose scaling region is 7-17 pips, for bid (dotted line) and ask (solid line). The solid bars represent the total amount of the intervention operations of the day. The dashed line shows the exchange rates.**

Preliminary Study on the Fluctuations of Daily Returns in Stock Market Based on Phase Transition Theory

Tetsuo Abe

Science-Seminar, 261 Nishiiri,Tempaku, Nagoya,
468-0039, Japan

Summary. To examine the state dependence of price fluctuations in a stock market, activity, which is defined as the product of averaged price and total volume of shares, is considered. Changes of daily returns are divided into high and low activity periods, and measure the probability distribution of fluctuations in each activity. It is found that the price fluctuation shows strong dependence on the activity; fluctuation is large in a high-activity state. Time evolution of the fluctuation is also studied. An abrupt change in activity induces the steep increase of price fluctuation. Considering the dynamics of fluctuation as a relaxation process after an abrupt disturbance, the relaxation time is estimated from the plot of ln σ^2 versus time. The Fokker-Planck equation is applied to conceive the non-linear effects on the relaxation process.

Key words; state parameter, activity, fluctuation, non-linear, relaxation

Introduction

According to a phase transition theory, a degree of fluctuation of physical properties, such as magnetization, heat-capacity etc., of the system at an equilibrium state depends on the state parameters, such as temperature, pressure etc. At the critical point, especially, the fluctuation diverges to infinity. Time evolution of the fluctuations at a non-equilibrium state also depends on the state parameters, such as a degree of deviation from the equilibrium state. Some investigators have pointed out a phenomenological analogy between the stock market and the phase transition (Stanley, 2002). However, physical meanings of these state parameters in a stock market are not clear yet. The price fluctuations in a stock market, for example, could depend on the state of a stock market. Then what is the state of a stock market, and how does the price fluctuation depend on the state parameters of a stock market?

Probability distribution of fluctuations

It is well known that the empirical probability distribution of a stock price

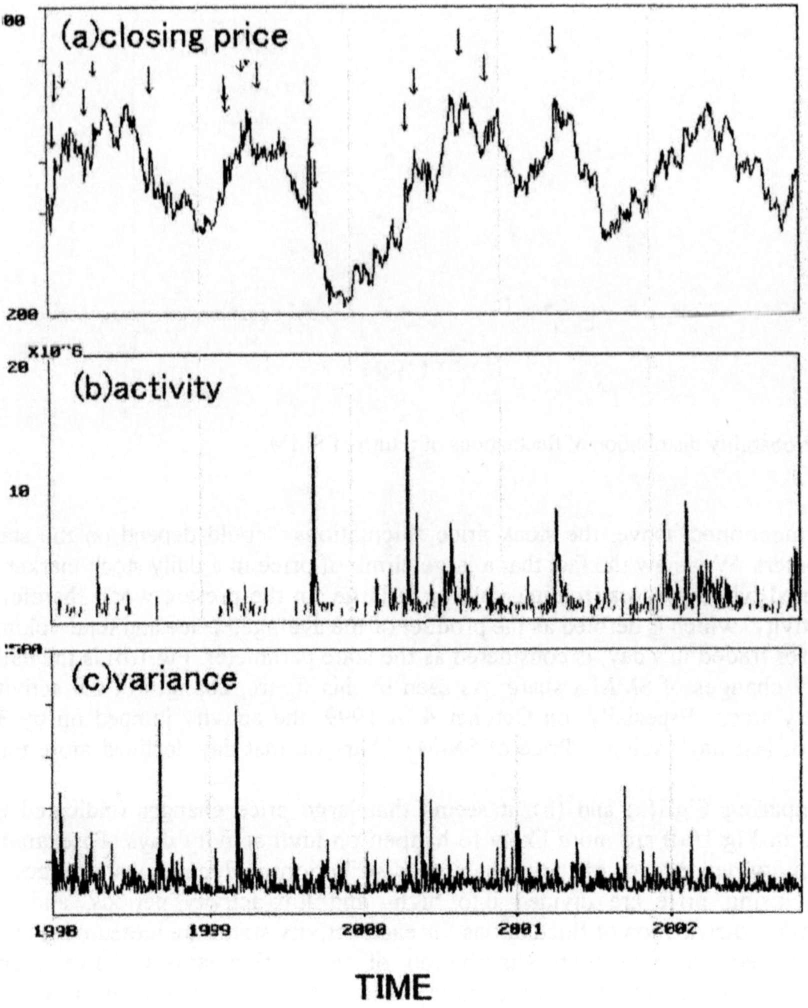

Fig.1 Daily changes of (a) closing price (yen), (b) activity and (c) variance of Sumitomo Metal and Mining in Tokyo stock market for 5 years.

fluctuation is a leptokurtic non-Gaussian distribution, which is characterized by a narrower and larger maximum, and by fatter tails than in Gaussian distribution (Cont, 2001; Mantegna and Stanley, 2000 ; Hisakado and Minakuchi, 2000). Fig.1(a) shows changes of daily closing price of Sumitomo Metal and Mining (SMM) in Tokyo stock market for 5 years (from January 1998 to December 2002). Fig.2 is the probability distribution of daily return fluctuations calculated from Fig.1(a). The leptokurticy is clearly observed.

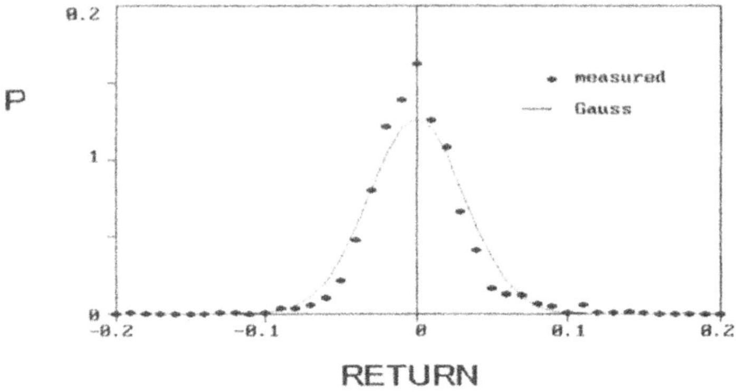

Fig.2 Probability distribution of fluctuations of return of SMM.

As mentioned above, the stock price fluctuations could depend on the state parameters. We know the fact that a large climb of price in a daily stock market is happened when the total trading volume is large. In the present work, therefore, the activity, which is defined as the product of the averaged price and total volume of shares traded in a day, is considered as the state parameter. Fig.1(b) is the daily activity changes of SMM's share. As seen in this figure, changes of the activity are very steep. Especially, on October 4 in 1999, the activity jumped up by 36 times of last day's value. Price of SMM's share on that day declined more than 20%.

Comparing Fig.1(a) and (b), it seems that large price changes (indicated by arrows in Fig.1(a)) are more likely to happen on high activity days. To examine the activity dependence of the probability distribution of fluctuations, changes of daily closing price are divided into high- and low-activity periods, and the probability distribution of fluctuations for each activity states are measured (Fig.3). As expected, the probability distribution of return fluctuations shows strong dependence on the activity; the probability distribution of fluctuation is broader in the high-activity period than in low-activity period. Fig.1(c) shows the changes of variance of SMM. In general, the price fluctuation is large when the activity is high, but there is some exceptional case, for example, at May 17 in 2000, March 1 in 1999 and so on. A large fluctuation does not necessary mean a high activity.

Many papers studied on the probability distribution of stock price fluctuations have been published so far, but most of them have discussed on the character of the fluctuations measured for a long period of marketing time, which includes high and low- activity states. Moreover, as seen later, variance shows a steep increase induced by the abrupt change of the activity. This could be one of the reasons why the empirical probability distribution of stock price fluctuations shows a leptokurticy. It might be difficult to describe such fluctuations by one type of distribution function.

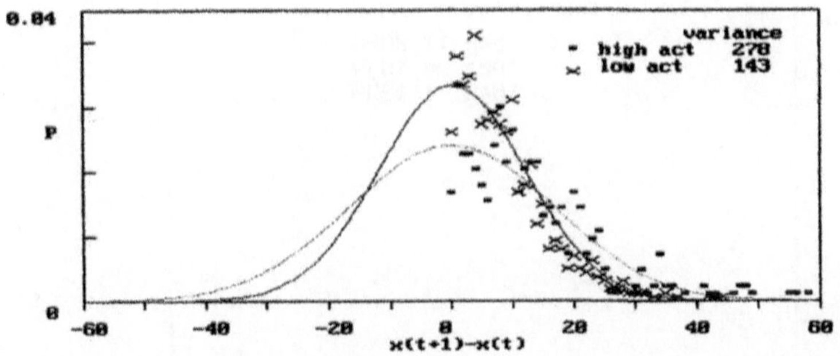

Fig.3. Activity dependence of probability distribution of fluctuations of daily returns.

Another important state parameter in a stock market could be the trend of price changes. In the present work, a dependence of fluctuations on the trend is also examined. There is a tendency that the fluctuation is large at even-trend than at up- or down-trend. The dependence of fluctuations on the trend, however, is not as clear as the one on the activity.

Time evolution of fluctuation

Next, we will discuss about time evolution of fluctuations at non-equilibrium state. As seen in Fig.1, both of activity and variance show steep changes with time. In detail, however, there are several differences between them; after a great increase, the activity generally shows a smooth decay. The variance, on the other hand, decreases with time by the following of several sub-peaks. It seems that an abrupt increase of the activity triggers off the enhancement of unstable fluctuations of price. From such a behavior of the activity and fluctuations, it is conceivable that the dynamics of the price fluctuation is a relaxation process caused by the abrupt disturbance. In the most simple case, the time evolution of variance in a relaxation process after the disturbance at t=0 is given by the following equation;

$$\sigma^2(t) = A\exp(-\tau t) + B \qquad \cdots\cdots\cdots(1)$$

where A and B are constants, respectively, and $1/\tau$ is a so-called "relaxation time", which is the time needed to approach an equilibrium state after the disturbance. From eq(1), the simple linear relationship between ln σ^2 and t is expected. Relaxation time is estimated from the slop of ln σ^2 versus t plot. In general, the actual relaxation process of price fluctuations in a stock market is more complex. Fig.4 shows the changes of ln σ^2 as the function of marketing time at May 1 in 1999, October 4 in 1999 and May 17 in 2000 of SMM,

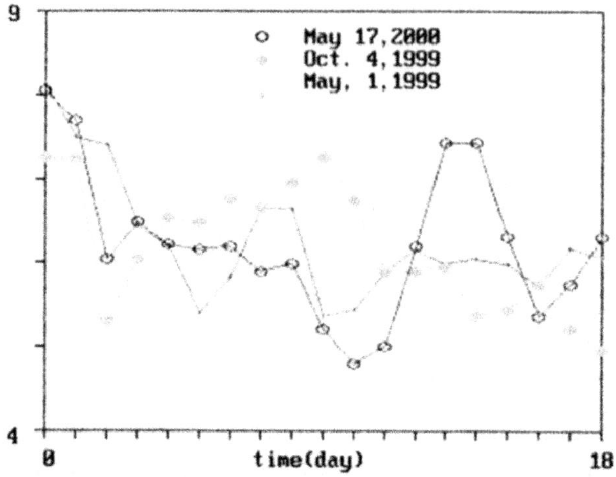

Fig.4. Changes of $\ln\sigma^2$ as a function of marketing time.

respectively. As seen in Fig.4, the first main relaxation process is overlapped with the second sub-process on October 4 in 1999 and May 1 in 1999. This is the very typical case for other relaxation process and makes the analysis difficult. On May 17 in 2000, the relaxation process is relatively smooth; after the fast relaxation process in a first few days, rate of relaxation process becomes slow. The relaxation time for the fast and slow process is estimated to be about 1.3 and 5.4 days, respectively.

Applying the Fokker-Planck equation, the non-linear effect on the relaxation process can be considered (Risken, 1984). When the degree of deviation from an equilibrium state is small, the non-linear Fokker-Planck equation is reasonably linearised (Suzuki, 1976). Assuming the Gaussian distribution for the probability distribution function when the fluctuation is small, we obtain eq(1) for the time evolution of variance. When the degree of deviation from the equilibrium state is large, a non-linear force acting on the system becomes dominant. Time evolution of variance for this case is schematically shown in Fig.5. At first, variance decreases exponentially with time and then the rate of change is slowed down by the non-linear effect. The time when the non-linear effect becomes apparent depends on the state of the system.

Conclusions

The following are the conclusions of the present study on the fluctuations of

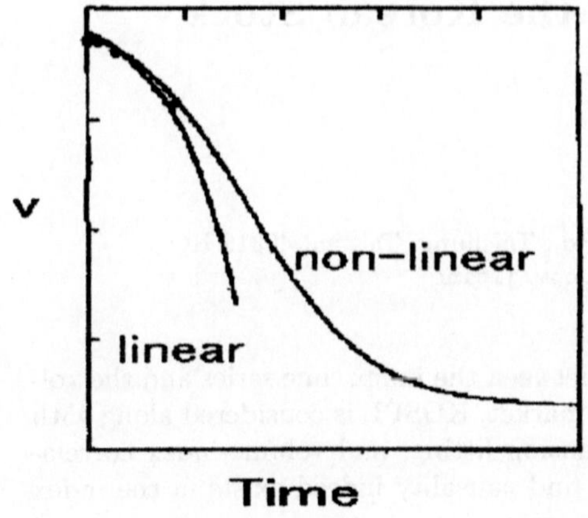

Fig.5. Schematic diagram of time evolution of variance for linear and non-linear case.

daily returns of SMM's share in Tokyo stock market;

(1) The activity (the product of the averaged price and total volume of shares traded per day) is one of the important state parameters when the stock price fluctuation is studied.
(2) The probability distribution of the price fluctuation shows a strong dependence on the activity; in general, the price fluctuation is large when the activity is high,
(3) The dynamics of the stock price fluctuation can be considered as the relaxation process by an abrupt disturbance.
(4) After an abrupt increase, the price fluctuation shows rapid decrease in first few days. The rate of change is gradually slowed down. The averaged relaxation time for each process is estimated to be about 1.3 and 5.4 days, respectively.

References

Cont R.;Quantitative finance,1(2001)223-236
Hisakado M. and Minakuchi K.;Proc. of "Econophysics Application on the Horizon" p.23,2000, Tokyo
Mantegna R.N. and Stanley H.E.;"Econophysics",Cambridge Univ.Press,2000,
Risken H.;"The Fokker-Planck Equation", Springer-Verlag,1984,
Stanley H.E,.:Proceedings of NIKKEI 2nd econophysics ,Tokyo, 2002
Suzuki M; Prog.Theor.Phys.56(1976)77
Suzuki M; Prog.Theor.Phys.56(1976)477

Physical Properties of the Korean Stock Market

Julian Juhi-Lian Ting

No.38, Lane93, Sec.2, Leou-Chuan E. Rd., Taichung, Taiwan, 40312 ROC
jlting@chez.com http://www.chez.com/jlting

Summary The mutual influence between the index time series and the volume time series of the Korean stock market, KOSPI, is considered along with it volatility, first order Landau expansion fitting, and volume-index correlation based on one minute data. We find causality indeed exists in the index time series but not the volume time series.

1.1 Introduction

Nowadays most daily stock exchange information can be obtained from the internet. The data generally contains the opening index, the highest index, the lowest index, the close index, the number of share traded, and the total amount of money involved.

Most analysis use only the close index. For instances, previous researches have shown the distribution of price change has pronounced tail distribution in contrast to Gaussian distribution. Furthermore, the auto-correlation of price change decays exponentially with a characteristic time scale around 5 minute. Johansen and Sornette have considered fitting the Korean stock market for the bubble from 1992 to 1994 using a first order Landau expansion of the index[1]. They showed evidence that market crashes as well as large corrections are preceded by speculative bubbles with two main characteristics: a power law acceleration of the market price decorated with log-periodic oscillations. For most markets the log-frequency $\omega/2\pi$ is close to unity. Emergent markets were seen to have larger fluctuations. They found that the KOSPI bubble has a critical exponent $\alpha = 1.05$ and $\omega = 8.15$. They further extended the expansion up to third order and successfully predicted Nikkei raise in 2000[4].

Are other part of the data simply junk? Should there be other more information that we can extract? Gopikrishnan *et al.*[8] analyzed the statistical properties of number of shares traded for a given time interval from an empirical rule saying that it takes volume to push the index. It is one example of using the other information.

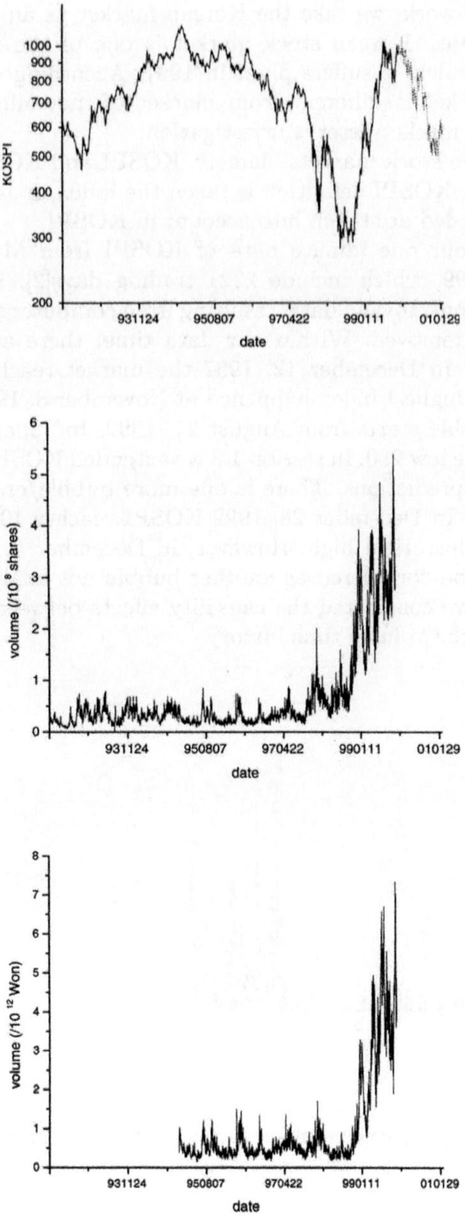

Fig. 1.1. History of KOSPI daily variations and the corresponding volumes. The later part of KOSPI plotted using dotted line is analyzed only in section 1.3. Volume is plotted for the number of share traded or the amount of money involved. The amount of money traded information is rather incomplete. Therefore, only those available part is plotted.

In the present work, we take the Korean market as an example and analyzed its properties. Korean stock market is one of the famous emerging markets. In particular, it suffers plum in 1997. According to previous studies, emerging market is different from market of more liquidity[1]. Therefore,Korean stock market deserve investigation.

Korean has two stock markets, namely, KOSPI and KOSDAK. KOSPI is the major market. KOSPI definition is taken the index in January 4, 1980 as 100. All stocks traded are taken into account in KOSPI.

We analyzed our one minute data of KOSPI from March 20, 1992 to November 29, 1999, which include 2221 trading days[2]. Intra-day time is treated as continuous. Invalid data, resulting from computer system shut down for instance, are removed. Within our data time, there are several events can be identified. In December 12, 1997 the market reached a local minimum 350.68. The highest index happened at November 8, 1994 with an index 1138.75. This bubble starts from August 21, 1992. In June 16, 1998 KOSPI reaches her all time low 280. In section 1.3 we extended KOSPI data up to date in order to make predictions. There is one more bubble/crash can be found in this later data. In December 28, 1999 KOSPI reaches 1028.07, almost recovered previous historical high. However, in December 4, 2000 it dived to 501.73. This can be considered as another bubble not analyzed by previous authors. Finally, we considered the causality effects between the stock index time history and the volume time history.

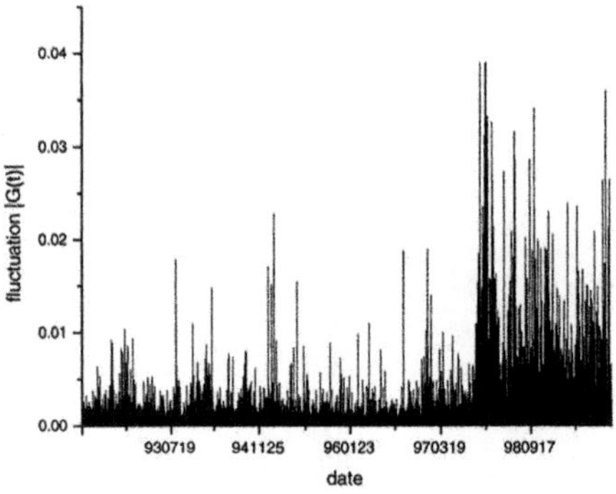

Fig. 1.2. Amplitude of fluctuations, $|G(t)|$, with $\Delta t = 1$ min.

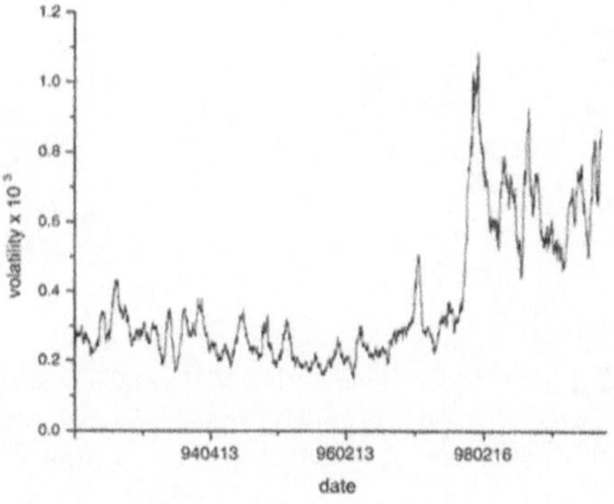

Fig. 1.3. Volatility history over a time window $T = 6600$ min. Dates are labeled at the centers of the average windows.

1.2 Basic Properties

In order to compare Korean stock market with other countries we need to know first some basic properties studied in other markets. The first properties are of course daily index and volume histories. They are plotted in Fig.(1.1).

One of the most studied properties is volatility. It is a found to be a good measure for long time behaviour and risk[3]. It is also the key input parameter for all option pricing models. Many different quantitative definitions of volatility are use in the literature. Following Liu *et al.* we define the price change $G(t)$ as the change in the logarithm of the index,

$$G(t) \equiv \ln Z(t+\Delta t) - \ln Z(t) \cong \frac{Z(t+\Delta t) - Z(t)}{Z(t)}, \qquad (1.1)$$

where Δt is the sampling time interval. The absolute value of $G(t)$ describes the amplitude of the fluctuation, as shown in Fig. (1.2). The large values of $|G(t)|$ correspond to the crashes and big rallies.

We define the volatility as the average of $|G(t)|$ over a time window $T = n \cdot \Delta t$, i.e.,

$$V_T(t) \equiv \frac{1}{n} \sum_{t'=t}^{t+n-1} |G(t')|, \qquad (1.2)$$

in which n is the moving window size. The data is sampled at a frequency $\Delta t = 30$ min. Fig. (1.3) shows the calculated volatility $V_T(t)$ for a large averaging window $T = 6600$ min (about 1 month).

Fig. (1.4) shows the probability density function $P(V_T)$ of the volatility for several values of T with $\Delta t = 30$ min. The 300 min. data is further fitted with Gaussian distribution in Fig. (1.5). The tail is fat. The heel is also high.

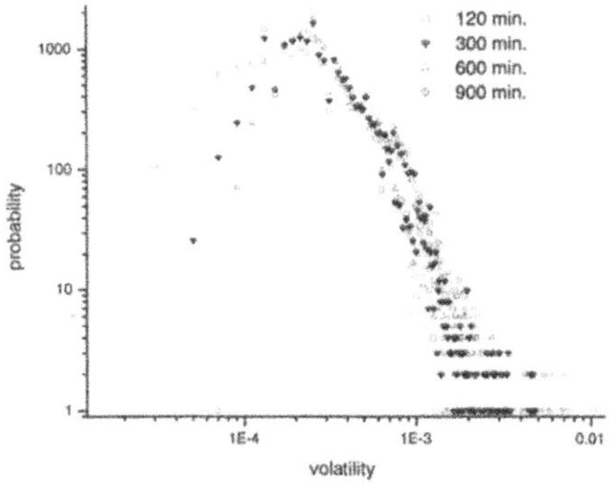

Fig. 1.4. Volatility distribution for various time windows with $\Delta t = 30$ min.

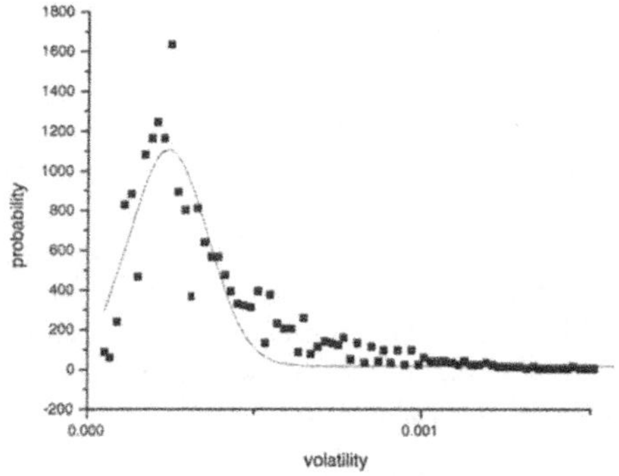

Fig. 1.5. Gaussian fit for the volatility distribution for $T = 300$ min. and $\Delta t = 30$ min.

1.3 prediction of crashes

In this section, we follow Johansen and Sornette to fit the index changes and obtain future trends prediction. We extended our data set up to date. However, our analysis differs from theirs in labeling the date. They convert all dates into decimal numbers. We labeled only by trading dates. The later method might give more accurate prediction, since the trading days changes and actually spread irregularly.

Johansen and Sornette wrote down the first, second, and third order Landau expansion of the index starting or ending at the historical high t_c. The first order expansion read:

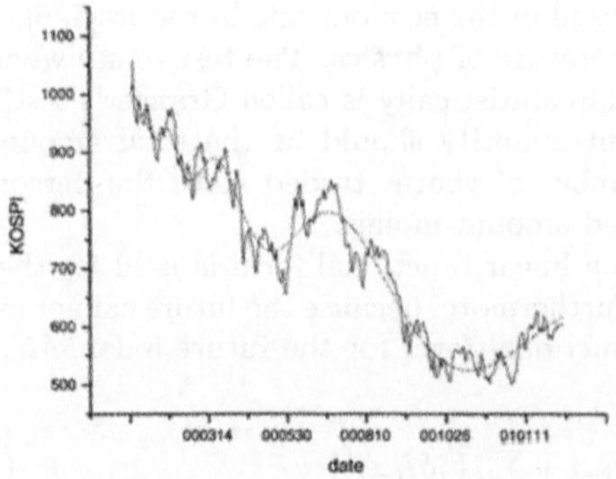

Fig. 1.6. KOSPI fit by first order Landau expansion of the index starting from the recent high in December 28, 1999. Fitted parameters are: $t_c = 1.999$, $\alpha = 0.615$, $\omega = 16.509$, $\phi = 6.012$, $A = 1018.354$, $B = -14.949$, and $C = -3.677$.

$$p_1(t) \approx A + B\tau^{\alpha} + C\tau^{\alpha}\cos[\omega \ln(\tau) + \phi]. \quad (1.3)$$

in which $\tau = |t - t_c|$ is the absolute value of time difference from the critical time of the highest index t_c. The parameter ϕ correspond to the time unit used while the parameter A, B,and C are units determined by the index as well as the historical period. Only t_c, α, and ω are the key parameters: α stands for the power law acceleration, while ω quantifies for the log-periodic oscillation.

Our fitting start from the later high in December 28, 1999 until February 16, 2001 is shown in Fig.(1.6).

The plot shows optimistic for the future. However, our $\omega = 16.509$ is almost double than Johansen and Sornette's, while the critical exponent, $\alpha = 0.615$, is much less than theirs.

1.4 Causality Test

Most people treated stock market variation simply as a time series of the index. We propose that it should be a dual-series consists of the index and the corresponding volume.

Volume is measure of market liquidity while index means the price. However, when people talk about the volume involved, sometimes they mean the number of shares traded sometimes they mean the amount of money involved. Whichever definition used these two time series might be correlated. It is well known, as an empirical rule, that volume changes come before price change. Large trading volume will follow by higher index. This "volume" means the amount of money involved. However, how much ahead? What kind of correlation does they posses? Why is there a lag? This kind of dual series with one influence the other has been analyzed in the econometric literature[5, 6]. The lag is called causality as in the literature of physics. The test to see whether is there any cause-effect relationship statistically is called Granger's test[7].

It appears to us the important quantity should be the total amount of money traded instead of the number of shares traded since the person or institute involved should have fixed amount money.

In the econometrics literature a linear functional form is used for the regression of these two time series. Furthermore, because the future cannot cause the past, we can write down a functional form for the future index, δI_t and the future trading volume, V_t, as

$$\delta I_t = \sum_i a_i V_{t-i} + \sum_j b_j \delta I_{t-j} + u_t \tag{1.4}$$

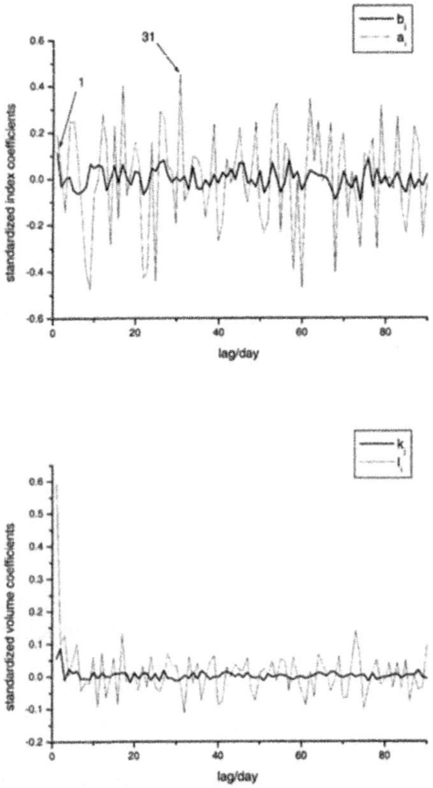

Fig. 1.7. Causality coefficients for index difference and for the corresponding volume.

$$V_t = \sum_i l_i V_{t-i} + \sum_j k_j \delta I_{t-j} + v_t \,. \tag{1.5}$$

The result is plotted in Fig.(1.7).

The figure suggests the index difference time series indeed received slightly higher influence from the previous index difference and the previous volume. However, these effects are pretty small. Furthermore, these peaks will be changed with the truncation of the time series.

1.5 conclusion

We tried to find the correlation between the volume of transactions involved and stock market fluctuations beyond simply considering temporal price series, using Korean stock market as an example in the present work.

From the trading volume history and the volatility history we can find that the Korean stock market has went through fundamental changes after 1997 crash. However, many physical properties are unchanged. In particular the volatility distribution still show fat-tail shape as other markets.

We find the volatility spread relatively uniformly in the past ten years, which can also be found in the stationarity test. The distribution of price change shows fat-heel behaviour.

A slightly higher influence coefficient is found in the index difference fitting with previous index and previous volume. Johansen and Sornette's result suggests us to try regression with log-periodic functional forms in the future. However, it does not seem meaningful to go beyond 90-days for the time lag, since companies will have their seasonal reports by that time.

References

1. Johansen A and Sornette D, *Int. J. Theor. and Applied. Finance*, *cond-mat/9907270.*
2. Daily index can be obtained at http://kse.or.kr/kor/stat/stat_past.htm However, most part of our data are one-minute data.
3. Liu Y, Gopikrishnan P, Cizeau P, Meyer M, Peng C K and Stanley H E, *Phys. Rev. E*, **60**, 1390 (1999).
4. Johansen A and Sornette D, *Int. J. Mod. Phys. C*, **10**, 563 (1999).
5. Gujarati D N, *Basic Econometrics*, 3ed, (McGraw-Hill 1995).
6. Maddala G S, *Introduction to Econometrics*, 2ed, (Macmillan Publishing Co. 1992).
7. Granger G W J, *J. Eco. Dyn. Cont.*, **2**, 329 (1980); ibid **12**, 551 (1988).
8. Gopikrishnan P, Plerou V, Gabaix X, and Stanley H E, *Phys. Rev. E*, **62**, R4493 (2000).

Correlation coefficients between stocks and those distributions of returns in the Tokyo Stock Exchange

Koichi Itoh

Western Hiroshima Prefecture Industrial Research Institute, 2-10-1 Aga, Kure, Hiroshima 737-0004, Japan

Summary. The daily stock price distributions of the Tokyo Stock Exchange from 1990 to 1999 and those correlation coefficients between stocks were calculated. The result showed that correlation the distribution of correlation coefficients can be described Gaussian function, the period, which was decreasing stocks price, the mean value of the correlation coefficients has higher than other period.

Key Words. Cross-Correlation, distribution of returns

1. Introduction

Cross-correlations is great interest for both fundamental and practical reasons. Practical aspects relate to the risk estimation for the optimal portfolio. And many studies about cross-correlation in the New York and European stock Exchange have been investigated [1,2]. The purpose of this paper is to investigate relation between cross-correlation and those price change in the Tokyo Stock Exchange. To investigate, 1370 companies and the period of 10 years from 1990 to 1999 were analyzed.

2. Distribution of returns

Firstly, to calculate distribution of returns, those were defined for $S_i(t)$ which denotes the daily close price of stock i.

$$G_i(t) \equiv \ln S_i(t+\Delta t) - \ln S_i(t) \tag{1}$$

The return distribution for each year from 1990 to 1999 for 1370 stocks and Δt=1day were calculated. Therefore time series have approximately 3 million data points totally. The period from1990 to 1999, the Tokyo Stock Exchange was exposed both high stock price and those suddenly crush.

Figl shows that empirical return distribution for the logarithm of daily price change of Tokyo stock exchange for the 1991,1995 and 1999. The result indicates that the empirical probability function for 1991 was very similar with that of 1995. However, for 1999, it has large fluctuation and fat tail than others.
Fig 2 shows cumulative distribution for returns *P (G>x)* that is the probability of a return lager than or equal to a threshold. The function form of the cumulative distribution has been investigated [3] and it approximately is consistent with a power law .

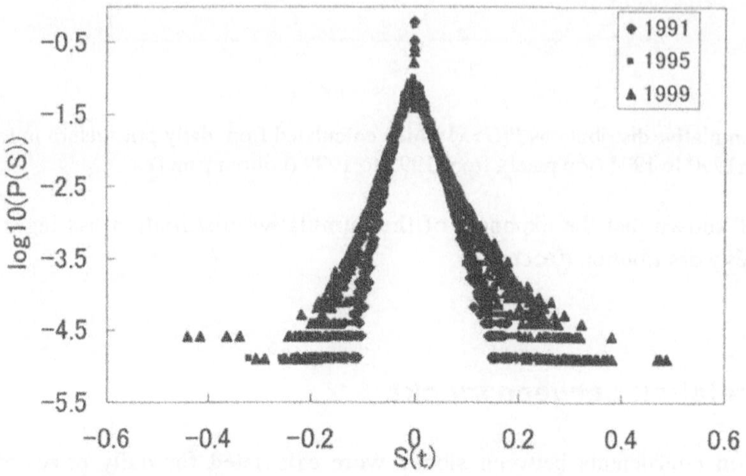

Fig. 1. Empirical return distribution for the logarithm of daily differences in the Tokyo Stock Exchange in the 1991,1995 and 1999.

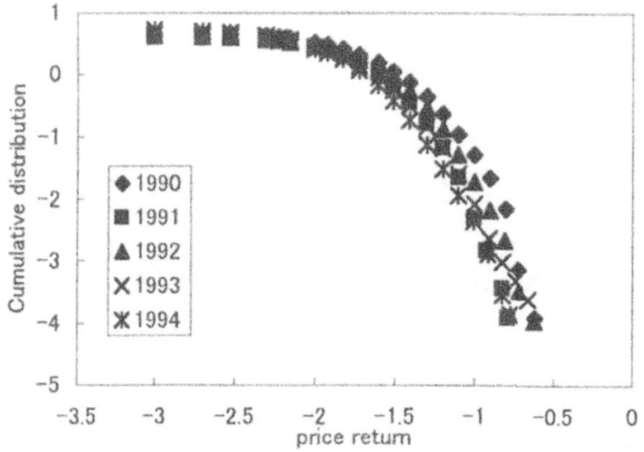

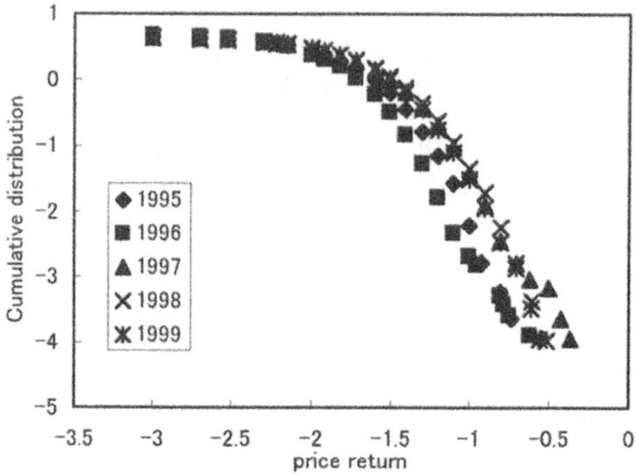

Fig. 2. Cumulative distributions *P(G>x)* which calculated from daily price return in log-log scale from1990 to 1994 (top panel), from 1995 to 1999 (bottom panel) .

It is well known that the exponent of the cumulative distribution has lager than that of Lévy distribution (0< α <2).

3. Correlation between stocks

Correlation coefficients between stocks were calculated for daily price returns from 1990 to 1999 in the Tokyo Stock exchange. Correlation coefficients is defined below [4].

$$C_{ij} = \frac{<S_iS_j> - <S_i><S_j>}{\sqrt{<S_i^2 - <S_i>^2><S_j^2 - <S_j>^2>}} \tag{2}$$

Distributions of correlation coefficients are roughly the same as Gaussan function, therefore they are characterize with mean value (μ) and standard deviation (σ). Table 1 shows that mean value and standard deviation for distributions of Correlation coefficients from 1990 to 1999. The results indicated that the mean value of the distribution is more than zero, however the mean value in 1999 was very small (0.079). The mean value of the 1991 (0.165) is roughly equal to that of 1995 (0.159). It is related to the price return distribution (previously mentioned Sec.2). It is possible that the high mean value in 1991 (0.210) and the standard deviation (0.160) are correspond that the stock price had been downin those period.

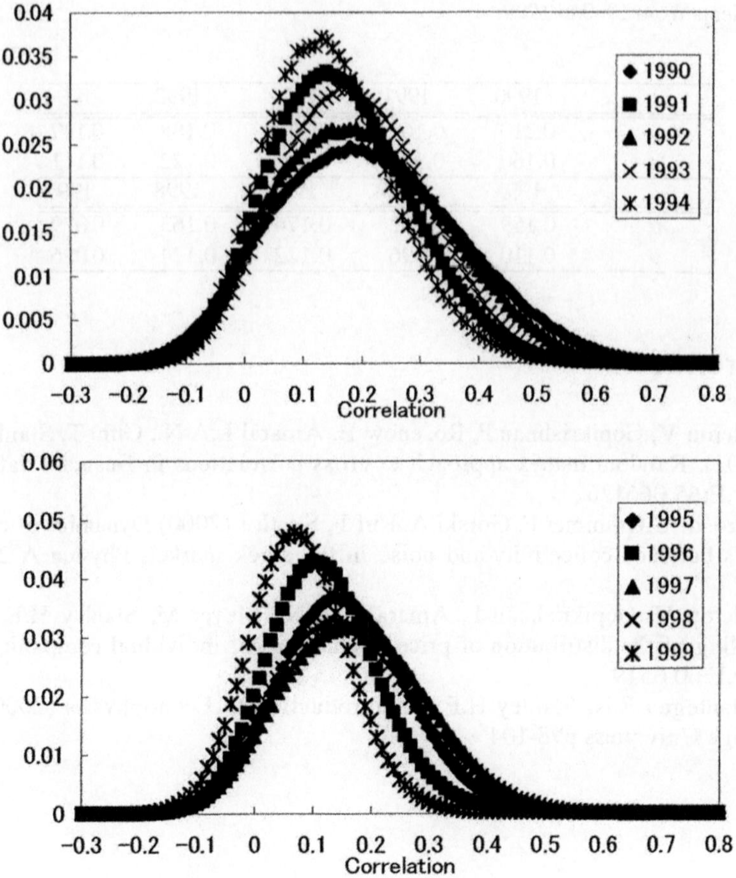

Fig. 3. Distribution of correlation coefficients (C_{ij}) calculated from the daily price return 1990-1994 (top panel) 1995-1999 (bottom pannel)

Table 1. Mean value (μ) and standard deviation (σ) for distributions of correlation coefficients from 1990 to 1999.

	1990	1991	1992	1993	1994
μ	0.210	0.165	0.200	0.188	0.147
σ	0.160	0.120	0.140	0.122	0.111
	1995	1996	1997	1998	1999
μ	0.159	0.122	0.174	0.163	0.079
σ	0.110	0.096	0.122	0.124	0.086

References

1. 1.Plerou V, Gopikrishnan P, Rosenow B, Amaral L.A.N., Guhr T, Stanley H.E. (2002), Random matrix approach to cross correlations in financial data. Phys. Rev.E:65 066126
2. 2.Dro. d. S,Grümmer F, Górski A, Ruf F, Speth J (2000) Dynamics of competitions between collectivity and noise in the stock market. Physica A 287:440-449
3. 3.Plerou V, Gopikrishnan P, Amaral L.A.N., Meyer M, Stanley H.E. (1999), Scaling of the distribution of price fluctuations of individual companies. Phys. Rev.E:60 6519
4. 4.Mantegna R.N, Stanley H.E An Introduction to Econophysics (2000) Cambridge Univ press p98-104

Epochs in Market Sector Index Data - Empirical or Optimistic?

G. Keogh, S. Sharifi, H. Ruskin, M. Crane

School of Computer Applications,
Dublin City University,
Glasnevin, Dublin 9,
Ireland
gkeogh, ssaba, hruskin, mcrane@computing.dcu.ie

Summary. We introduce here the concept of "epochs" in market movements (i.e. periods of co-movements of stocks). These periods in EURO-STOXX market sector data are characterised by linear relationships between price and eigenvalue change. The evidence suggests strong time dependence in the linear model coefficients but residuals are strongly dependent on granularity (i.e. sampling rate) with fit breaking down at rates smaller than five days. Possible reasons for this breakdown are presented together with additional arguments on the relative merits of correlation and variance-covariance matrix eigenanalyses in measuring co-movements of stocks.

1 INTRODUCTION

The leverage effect (i.e. the fact that at-the-money volatilities tend to increase for asset price drops) in financial markets has been much studied over recent years. This increase has been shown to apply for different forecasting horizons, dependent on whether studies focus on the volatility in auto-correlations of actual stock prices or on those of index data. The literature is reticent, however, about the effect of price changes on the combined upward or downward movements of shares and it is this aspect which we address. Specifically, we have developed, for different market sector data from Dow Jones EURO STOXX, a novel approach to the way in which the market recognises and responds to risk. This, implicitly, reflects the volatility, where risk reaction clearly changes as prices rise or fall or crashes occur. The method is based on examining change in the riskiest position, as determined by the maximum eigenvalue from an eigenanalysis of the variance-covariance matrix from day-to-day, and relating this to effects on prices for the various sectors. The change in the largest eigenvalue acts, in some sense, as a barometer of market risk.

Results to date indicate that there are periods in the sector data, for which change in the largest eigenvalue varies linearly with price. As each period appears to end, the relationship changes and a smaller change in the eigenvalue is required to bring about the same change in price. In the limit and during a crash, the slope is still positive (and quite large). This is evidence, we believe, that crash patterns are well-defined, due to common perception of risk change, while upward trends are less predictable. Evidence tends to support an implicit relationship between the instantaneous volatility and the daily changes in the eigenvalues, not least because alternative analysis of the correlation matrix, (for the data normalised with respect to the variance), results in a disappearance of these periods. The above effects vary from sector to sector, being seen to be most pronounced in the technology and telecoms sectors and less so in more defensive sectors.

In summary, evidence of periods in eigenvalue-price correlations is presented together with additional arguments on the relative merits of correlation and variance-covariance matrix analyses in measuring co-movements of stocks.

2 THE BASICS

In order to make money (or more euphemistically "maximise their utility") on stock markets, investors buy and sell assets. As the overall risk associated with a portfolio of stocks can be shown [Elton et al.(2002)] to decrease with the number of assets, more is better when it comes to assets. However, by having more assets, the investor will potentially take on more risk in order to generate higher expected returns. A balance needs to be struck, therefore, between the risk a new asset will add to the portfolio and the expected return. This problem of balance requires a knowledge of the volatility of and correlation between the assets, quantities that only become available (if then!) with time.

In order to measure the correlation between assets, we use the variance-covariance matrix **C** based on a dataset from EURO-STOXX market sector indices. The covariance matrix is updated daily and the individual covariances thus calculated at time T and over a time horizon M are given by (see for example [Litterman & Winkelmann(1998)]):

$$\sigma_{ij}^{T}(M) = \frac{\Sigma_{s=0}^{T} w_{T-s} r_{i,T-s} r_{j,T-s}}{\Sigma_{s=0}^{T} w_{T-s}} \quad (1)$$

where $r_{i,T}$ is the daily return on the ith asset at date T and w_T is the weight applied at date T.

For our purposes, we use asset (i.e. index) price instead of return, $r_{i,T}$ and take unit weights for previous days' data (i.e. $w_{T-s} = 1$ for all s,T) in the calculation of covariances.

As regards volatility measurement, it has been known for some time (see e.g. [Bouchaud & Potters(2000)]) that the eigenstates of the correlation matrix of assets are useful in the estimation of risk in a portfolio made up of those assets. In this paper, we suggest using the *covariance* matrix **C** because, as we will show, this seems to contain more long term information on co-movements of stocks as the covariances retain volatility information which is lost in the normalisation process. Furthermore, the need to use correlation does not apply for indices. For volatility, we suggest the use of day-to-day (or period-to-period) change in the largest eigenvalue of the covariance matrix as a measure of the change in the riskiest position as perceived by the majority of market participants. This, we believe to be novel; up to now, while it has been recognised that the largest eigenvalue (λ^T_{max}) and corresponding eigenvector (z^T_{max}) represented "the market" (with the eigenvector representing in some way the riskiest portfolio), the change in λ^T_{max} over time has not (to our knowledge) been used before as a barometer of the perception of risk in the market.

Specifically, we define the *change in cohesion*, ΔC_{T+1} as being

$$\Delta C_{T+1} = \frac{\lambda^{T+1}_{max}}{\lambda^T_{max}} \tag{2}$$

3 RESULTS AND DISCUSSION

3.1 Epochs: Definition and Possible Causes

We have taken as our data, daily market sector index prices from EUROSTOXX covering 18 different market sectors[1] and, starting with a time horizon of 200 days, have calculated the covariance matrices for successive days as per Equation 1. By calculating the ratio of the change in largest eigenvalue λ_{max} over successive time periods, and plotting this against the price of the particular market sector index, we obtain a view of how the market is tending to organise itself. Referring to Figure 1, which shows the daily change in λ_{max} for the telecom and technology indices, we see evidence for the epochs mentioned above. Central to the idea of the epochs is the notion that "there are many ways for a market to go up but just one to go down". Various authors ([Zumbach et al.(1999)] and others) have identified key events which can have an (occasionally disproportionate) impact on financial markets. Similarly we have identified events which have acted as market breakpoints and can be clearly identified as marking the end of epochs. Some of these are given in Table 1.

It should be emphasised, however, that not all these events appear to be seen by the market as bad *per se*. Often, it seems to be just a pause for breath or a lack of any particular cohesion. Successive epochs seem to be characterised by progressively higher slopes in price change for an increased degree of cohesion, each epoch being brought to an end by a particular event. Finally, it would appear that, as extreme prices are reached, further increases

[1] c.f. www.stoxx.com

Table 1. Key Events as Breakers in Market Cohesion

Date	Event
07/97 - 11/97	Asian Market Crisis
14/09/98	Bad News from South America
23/07/99	Plunge after Market Highs & Greenspan Address
12/12/00	US Supreme Court Judgement Ratifies President Bush
28/03/01	Cut in Federal Reserve Rate
18/09/01	Post September 11th Reaction

in cohesion become impossible and an avalanche will occur with cohesion and price both falling together.

3.2 Epochs: Models and Breakdown

From Figure 1, it will be apparant that cohesion graphs such as those shown, are characterised by periods of simple linear correlation between price and changes in cohesion. Fitting a simple linear model of the form:

$$Y_T = \alpha_T + \beta_T X_T + \epsilon_t \tag{3}$$

(where α_T, β_T are functions of time, Y_T are the daily prices, X_T, the daily change in cohesion and ϵ_t are the errors) we get a good fit ($R^2 \approx 0.45-0.66$) for most market sectors. However, ϵ_t, which should be independently distributed, actually follows a random walk. There is clearly, therefore, time-dependence of order one in ϵ_t with partial auto-correlation function (PACF) at lag one very close to one and a Durbin-Watson[2] (DW) statistic d very much less than 2.

Hence, this implies we should take first differences of Equation 3:

$$\Delta Y_T = \beta + \gamma \Delta X_T + \delta \Delta (T X_T) \tag{4}$$

(where β is small and positive and γ, δ are large and (negative, positive) respectively). Re-examining the modified model we find that the DW statistic is satisfied: $d \approx 2$ so this is now consistent with the null hypothesis of no positive autocorrelation and with no remaining systematic pattern in the residuals (white noise). However, at the *daily* sampling rate, the model fit becomes very poor after taking first differences. Examining different sampling rates (see Table 2), the degradation in R^2 can be clearly seen for the technology market sector index.

This data suggests that the variability (or volatility) increases markedly as the resolution increases and conversely that there is a settling period after which short term effects do not have an impact on subsequent data.

[2] testing for positive autocorrelation in the residuals

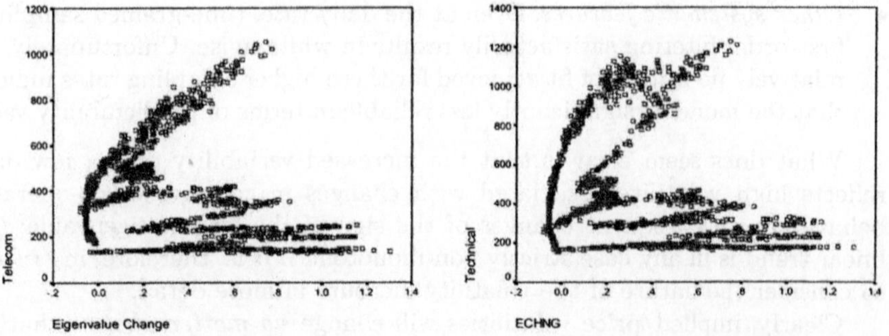

(a) Telecoms Sector (b) Technology Sector

Fig. 1. Daily Change in Cohesion ($\lambda_{max}^{T+1}/\lambda_{max}^{T}$) vs Price for Different Market Sectors

Table 2. Model Fit Statistics for Technology at Different Sampling Rates

Sampling Rate	R^2 Statistic	DW Statistic
Monthly	0.46	1.97
Every 10 Days	0.40	2.05
Every 5 Days	0.40	2.20
Every 2 Days	0.20	2.05

3.3 Possible Reasons for Model Breakdown

As we have seen above, the simple linear regression models break down for fine granularity of sampling. There are several possible reasons for this

- *Day of the week.* It seems possible that market behaviour on certain days of the week, is subject to increased/reduced activity, reflecting start-up, closure and similarly. We rejected this explanation, however, as eigenvalue changes and hence changes in cohesion were highly variable for all same-day data without exception.
- *Statistical leverage or outliers.* This presumes that extreme values occurring at certain time points or reflecting abnormally large localised price fluctuations are peculiar to a high sampling rate. This does not seem to be the case, since even very coarse-grained sampling, (i.e. monthly), retains the epoch skeleton. Furthermore, while epoch turning points are clearly significant, the fit is satisfactory for lower sampling rates, but not for high ones. Similarly, outlying points appear to have little influence on the R^2 or Durbin-Watson statistics for all rates and, in any case, are relatively few in number. It does seem likely, however, that there exists a settling period in the market, a few days in length, during which high variability diminishes the fit.

- *Other systematic features.* Even at the daily rate, (fine-grained sampling), first-order filtering satisfactorily results in white noise. Unfortunately, the relatively poor level of fit achieved for these higher sampling rates indicate that the model is significantly less reliable in terms of predictability value.

What does seem clear is that the increased variability over a few days, reflects high volatility associated with changes in cohesive or co-operative behaviour, with frequent changes of the sign of the largest eigenvalue; (the linear trend is in any case strictly non-monotonic). It is, therefore, instructive to consider the nature of the volatility measure in more detail.

Clearly, implied/price volatilities will change *no matter what* so that the historical basis should be relatively bias free. Equally, λ_{max}^T is an effective measure of volatility, so that the change $\Delta \mathsf{C}_{T+1}$ in the coherence is equivalent to looking at proportional change in volatility. Typically, these high-frequency fluctuations are self-correcting, representing a short-term and not particularly far-reaching lack of cohesive market behaviour, i.e. $\Delta \mathsf{C}_{T+1} \approx 1$. In other words, if coherence represents the change in commonly-perceived risk, deviation from the common perception is sufficient to ensure fluctuating variability over the short-term and corresponding reduction in fit, as observed for our data.

While this position is usually resolved over a slightly longer period, it is the lack of a sustained common perception of risk, (i.e. lack of coherence), which leads to a drop in price. The quantity $\Delta \mathsf{C}_{T+1}$ is clearly large and negative at the turning point of an epoch, corresponding to large change in the maximum eigenvalue, as well as a change of sign, since no change of sign for λ_{max}^T between consecutive time points implies that C is incremented. Unfortunately, since high variability is naturally associated with short-term effects, as mentioned previously, it is non-trivial to determine whether a sustained change can be determined in advance. A number of points are worth noting, however. Firstly, use of the covariance measure exposes the detail of these changes. The epochs, while present to some extent in the correlation matrix information, are far less distinct and thus less useful for our purposes, while dimensionality considerations do not arise for indices. Secondly, we note that outliers are few, so that tolerances can more readily be established on the size of high-frequency fluctuations. (It seems clear that the price change distributions involved do not scale simply, since volatility is dependent on the time interval as usual). It is suggested that coherence provides an intermediate measure which can be directly linked to critical market uncertainty.

4 Conclusions

We demonstrate the existence of epochs in EURO-STOXX market sector data, where the change in the largest eigenvalue of the covariance matrix of daily prices, varies linearly with time. The evidence supports an implicit relationship between instantaneous volatility and the change in the maximum eigenvalue, but volatility patterns are, typically, dependent on the time intervals observed. At the end of an epoch, the relationship changes, with smaller eigenvalue changes leading to the same price change. Epochs are present in all market sectors, but are most strongly defined in the less-defensive sectors, such as technology, telecoms. The epoch end is followed by a return to low coherence in market trading, due to disparate perception of risk. Co-operative behaviour (or strong coherence) in market trading suggests that up-turn patterns are more varied and hence less easy to predict, whereas down-turn patterns are well-defined, due to a common perception of risk change. As a final note, it should be mentioned that we attempted to replicate the results for exchange rate data but as data were limited, partial support only was obtained for results reported here.

5 Acknowledgements

The authors would like to acknowledge Dublin City University and the Institute of Numerical Computation and Analysis (INCA) for financial support for travel and Dr. Michael Mura for his invaluable contributions and insights.

References

[Elton et al.(2002)] Elton, E.J., Gruber, M.J., Brown, S., Goetzmann, W.: Modern Portfolio Theory and Investment Analysis John Wiley and Sons (2002)

[Litterman & Winkelmann(1998)] Litterman, R., Winkelmann, K.: Estimating Covariance Matrices, in *Goldman-Sachs Risk Management Series*, Goldman, Sachs & Co. (1998)

[Bouchaud & Potters(2000)] Bouchaud, J.-P., Potters, M.: Theory of Financial Risks - From Statistical Physics to Risk Management, Cambridge University Press (2000)

[Zumbach et al.(1999)] Zumbach, G.O., Dacorogna, M.M., Olsen, J.L., and Olsen, R.B.: Introducing a Scale of Market Shocks, O & A Research Group Preprint GOZ.1998-10-01, (1999)

I-2 Predictability

Volatility Fingerprints of Large Shocks: Endogenous Versus Exogenous

D. Sornette[1,2] Y. Malevergne[1,3] and J.-F. Muzy[4]

[1] Laboratoire de Physique de la Matière Condensée CNRS UMR 6622, Université de Nice-Sophia Antipolis, 06108 Nice Cedex 2, France
[2] Institute of Geophysics and Planetary Physics and Department of Earth and Space Science, University of California, Los Angeles, California 90095, USA
[3] Institut de Science Financière et d'Assurances - Université Lyon I, 43, Bd du 11 Novembre 1918, 69622 Villeurbanne Cedex, France
[4] Laboratoire Systèmes Physiques de l'Environemment, CNRS UMR 6134, Université de Corse, Quartier Grossetti, 20250 Corte, France
sornette@unice.fr, Yannick.Malevergne@unice.fr and muzy@univ-corse.fr

Summary. Using a parsimonious autoregressive process with long-range memory defined on the logarithm of the volatility, we predict strikingly different response functions of the volatility to external shocks compared with endogenous shocks. These predictions are validated empirically on data from a hierarchy of volatility shocks, on major crashes and finally on major external disturbances such as the events of Sept. 11, 2001 and the coup against Gorbachev on Aug., 19, 1991. Our theory allows us to classify unambiguously these different events in two classes (endogenous and exogenous) with specific signatures for both and characteristic precursors for the endogenous class. It also explains the origin of endogenous shocks as the coherent accumulations of tiny bad news, and thus unify all previous explanations of large crashes including Oct. 1987.

Key words. Multifractal random walk, endogenous, exogenous, jumps, shocks, crises, crashes, volatility persistence

1 Introduction

A market crash occurring simultaneously on most of the stock markets of the world as witnessed in Oct. 1987 would amount to the quasi-instantaneous evaporation of trillions of dollars. Market crashes are the extreme end members of a hierarchy of market shocks, which shake stock markets repeatedly. Among recent events still fresh in memories are the Hong-Kong crash and the turmoil on US markets on oct. 1997, the Russian default in Aug. 1998 and the ensuing market turbulence in western stock markets and the collapse of the "new economy" bubble with the crash of the Nasdaq index in March 2000.

In each case, a lot of work has been carried out to unravel the origin(s) of the crash, so as to understand its causes and develop possible remedies. However, no clear cause can usually be singled out. A case in point is the Oct. 1987 crash, for which many explanations have been proposed but none has been widely accepted unambiguously. These proposed causes include computer trading, derivative securities, illiquidity, trade and budget deficits, over-inflated prices generated by a speculative bubble during the earlier period, the auction system itself, the presence or absence of limits on price movements, regulated margin requirements, off-market and off-hours trading, the presence or absence of floor brokers, the extent of trading in the cash market versus the forward market, the identity of traders (i.e. institutions such as banks or specialized trading firms), the significance of transaction taxes, etc. More rigorous and systematic analyzes on univariate associations and multiple regressions of these various factors conclude that it is not at all clear what caused the crash [2]. The most precise statement, albeit somewhat self-referential, is that the most statistically significant explanatory variable in the October crash can be ascribed to the normal response of each country's stock market to a worldwide market motion [2].

In view of the stalemate reached by the approaches attempting to find a proximal cause of a market shock, several researchers have looked for more fundamental origins and have proposed that a crash may be the climax of an endogenous instability associated with the (rational or irrational) imitative behavior of agents (see for instance [3, 4, 5]). Are there qualifying signatures of such a mechanism? According to [4, 6] for which a crash is a stochastic event associated with the end of a bubble, the detection of such bubble would provide a fingerprint. A large literature has emerged on the empirical detectability of bubbles in financial data and in particular on rational expectation bubbles (see [7, 8] for a survey). Unfortunately, the present evidence for speculative bubbles is fuzzy and unresolved at best, according to the standard economic and econometric literature. Other than the still controversial [9] suggestion that super-exponential price acceleration [10] and log-periodicity may qualify a speculative bubble [4, 6, 11], there are no unambiguous signatures that would allow one to qualify a market shock or a crash as specifically endogenous.

On the other end, standard economic theory holds that the complex trajectory of stock market prices is the faithful reflection of the continuous flow of news that are interpreted and digested by an army of analysts and traders. Accordingly, large shocks should result from really bad surprises. It is a fact that exogenous shocks exist, as epitomized by the recent events of Sept. 11, 2001, and there is no doubt about the existence of utterly exogenous bad news that move stock market prices and create strong bursts of volatility. A case that cannot be refuted is the the market turmoil in Japan following the Kobe earthquake of Jan. 17, 1995 that led to a total cost estimated around $200 billion dollars. Indeed, as long as the science of earthquake prediction is still in its infancy, destructive earthquakes are not endogeneized in advance in stock market prices by rational agents ignorant of seismological processes. One may also argue that the invasion of Koweit by Iraq on Aug. 2, 1990 and the coup against Gorbachev on Aug., 19, 1991 were strong exogenous shocks. However, some could argue that precursory fingerprints of these events were known to some elites, suggesting the possibility the action of these informed agents may have been reflected in part in stock markets prices. Even more difficult is the classification (endogenous versus exogenous) of the hierarchy of volatility bursts that continuously shake stock markets. While it is a common practice to associate the large market moves and strong bursts of volatility with external economic, political or natural events [12], there is not convincing evidence supporting it.

Here, we provide a clear and novel signature allowing us to distinguish between an endogenous and an exogenous origin to a volatility shock. Tests on the Oct. 1987 crash, on a hierarchy of volatility shocks and on a few of the obvious exogenous shocks validate the concept. We first briefly describe the multifractal random walk as an autoregressive process with a long-range memory (decaying as $t^{-1/2}$) defined on the logarithm of the volatility, which captures parsimoniously the long-range correlation and multifractal properties of volatility time series. We then give our prediction for the response of the volatility to external versus internal shocks and present empirical tests.

These results have been communicated recently to the professional finance community through the channel [13]. Related ideas are presented in [14] for more general memory kernels with applications suggested to ecological extinctions in the fossil record (meteorite at the Cretaceous/Tertiary KT boundary versus volcanic eruptions (Deccan traps) versus self-organized critical events), the aviation industry recession (September 11, 2001 versus structural endogenous problems) and to recovery after wars (exogenous wars versus endogenous civil wars). We think that our ideas could also be useful to understand the immune system (external viral or bacterial attack versus internal breakdown) and the learning processes in the brain (external inputs versus internal self-organization and reinforcements). The same concept also allows one to understand the occurrence of foreshocks prior to large earthquakes as conditional triggered seismicity [15].

2 The Multifractal Random Walk model

While returns do not exhibit discernable correlations beyond a time scale of a few minutes in liquid markets, the historical volatility (measured as the standard deviation of price returns or more generally as a positive power of the absolute value of centered price returns) exhibits a long-range dependence characterized by a power law decaying two-point correlation function [16, 17, 18] approximately following a $(t/T)^{-\nu}$ decay rate with an exponent $\nu \approx$

0.2. A variety of models have been proposed to account for these long-range correlations [19, 20, 21, 22, 23].

Not only are returns clustered in bursts of volatility exhibiting long-range dependence, but in addition returns exhibit the property of multifractal scale invariance (or multifractality), according to which moments $m_q \equiv \langle |r_\tau|^q \rangle$ of the returns at time scale τ are found to scale as $m_q \propto \tau^{\zeta_q}$, with the exponent ζ_q being a non-linear function of the moment order q [24, 22].

A flexible and parsimonious model, the so-called multifractal random walk (MRW) [22, 25], explains the multifractal scale invariance in terms of volatility long range dependence. This model is a continuous time limit of a stochastic volatility model where log-volatility correlations decay logarithmically. It possesses a nice stability property related to its scale invariance property: For each time scale $\Delta t \leq T$, the returns at scale Δt, $r_{\Delta t}(t) \equiv \ln[p(t)/p(t-\Delta t)]$, can be described as a stochastic volatility model:

$$r_{\Delta t}(t) = \epsilon(t) \cdot \sigma_{\Delta t}(t) = \epsilon(t) \cdot e^{\omega_{\Delta t}(t)} \ , \tag{1}$$

where $\epsilon(t)$ is a standardized Gaussian white noise independent of $\omega_{\Delta t}(t)$ and $\omega_{\Delta t}(t)$ is a nearly Gaussian process with mean and covariance:

$$\mu_{\Delta t} = \frac{1}{2}\ln(\sigma^2 \Delta t) - C_{\Delta t}(0) \tag{2}$$

$$C_{\Delta t}(\tau) = \mathrm{Cov}[\omega_{\Delta t}(t), \omega_{\Delta t}(t+\tau)] = \lambda^2 \ln\left(\frac{T}{|\tau| + e^{-3/2}\Delta t}\right) . \tag{3}$$

where $\sigma^2 \Delta t$ is the return variance at scale Δt and T represents an "integral" (correlation) time scale. Such logarithmic decay of log-volatility covariance at different time scales as been evidenced empirically in [18, 22]. Typical values for T and λ^2 are respectively 1 year and 0.04. According to the MRW model, the volatility correlation exponent ν is related to λ^2 by $\nu = 4\lambda^2$.

The MRW model can be expressed in a more familiar form, in which the log-volatility $\omega_{\Delta t}(t)$ obeys an auto-regressive equation whose solution reads

$$\omega_{\Delta t}(t) = \mu_{\Delta t} + \int_{-\infty}^{t} d\tau \ \eta(\tau) \ K_{\Delta t}(t-\tau) \ , \tag{4}$$

where $\eta(t)$ denotes a standardized Gaussian white noise and the memory kernel $K_{\Delta t}(\cdot)$ is a causal function, ensuring that the system is not anticipative. The process $\eta(t)$ can be seen as the information flow. Thus $\omega(t)$ represents the response of the market to incoming information up to the date t. At time t, the distribution of $\omega_{\Delta t}(t)$ is Gaussian with mean $\mu_{\Delta t}$ and variance $V_{\Delta t} = \int_0^\infty d\tau \ K_{\Delta t}^2(\tau) = \lambda^2 \ln\left(\frac{Te^{3/2}}{\Delta t}\right)$. Its covariance, which entirely specifies the random process, is given by

$$C_{\Delta t}(\tau) = \int_0^\infty dt \ K_{\Delta t}(t) K_{\Delta t}(t+|\tau|) \ . \tag{5}$$

Performing a Fourier transform, we obtain

$$\hat{K}_{\Delta t}(f)^2 = \hat{C}_{\Delta t}(f) = 2\lambda^2 \ f^{-1} \left[\int_0^{Tf} \frac{\sin(t)}{t} dt + O\left(f \Delta t \ln(f \Delta t)\right) \right] , \qquad (6)$$

which shows that for τ small enough

$$K_{\Delta t}(\tau) \sim K_0 \sqrt{\frac{\lambda^2 T}{\tau}} \qquad \text{for } \Delta t << \tau << T . \qquad (7)$$

This slow power law decay (7) of the memory kernel in (4) ensures the long-range dependence and multifractality of the stochastic volatility process (1). Note that equation (4) for the log-volatility $\omega_{\Delta t}(t)$ takes a form similar to but simpler than the ARFIMA models usually defined on the (linear) volatility σ [20].

3 Linear response to an external shock

Let us assume that new major piece of information $\eta(t) = \omega_0 \, \delta(t)$ impinges on the market at some time (taken without loss of generality to be $t = 0$, since the system is stationary). ω_0 is the amplitude of the external shock. Then, the response of the log-volatility $\omega(t)$, $t > 0$ to this shock is

$$\omega(t) = \mu + \int_{-\infty}^{t} d\tau \ [\omega_0 \cdot \delta(\tau) + \eta(\tau)] \ K(t-\tau) \qquad (8)$$

$$= \mu + \omega_0 \ K(t) + \int_{-\infty}^{t} d\tau \ \eta(\tau) \ K(t-\tau) . \qquad (9)$$

where, for notation convenience, we have omitted the reference to the scale Δt. The expected volatility conditional on this incoming major information is thus

$$\mathrm{E}\left[\sigma^2(t) \middle| \omega_0\right] = \mathrm{E}\left[e^{2w(t)} \middle| \omega_0\right] = e^{2\omega_0 \ K(t)} \ \mathrm{E}\left[e^{2\mu + 2\int_{-\infty}^{t} d\tau \ \eta(\tau) \ K(t-\tau)}\right] \qquad (10)$$

$$= \overline{\sigma^2(t)} e^{2\omega_0 \ K(t)} \simeq \overline{\sigma^2(t)} e^{2\omega_0 \cdot K_0 \sqrt{\frac{\lambda^2 T}{t}}} \quad \text{for } \Delta t << t << T \qquad (11)$$

For time t large enough, the volatility relaxes to its unconditional average value $\overline{\sigma^2(t)} = \sigma^2 \Delta t$, so that the excess volatility $\mathrm{E}[\sigma^2(t) \mid \omega_0] - \overline{\sigma^2(t)}$ due to the external shock decays to zero as

$$\mathrm{E}\left[\sigma^2(t) \middle| \omega_0\right] - \overline{\sigma^2(t)} \propto \frac{1}{\sqrt{t}} . \qquad (12)$$

This universal response of the volatility to an external shock (i.e., independent of the amplitude ω_0 of the shock) is governed by the time-dependence (7) of the memory kernel $K(t)$. This prediction is tested successfully in figure 1.

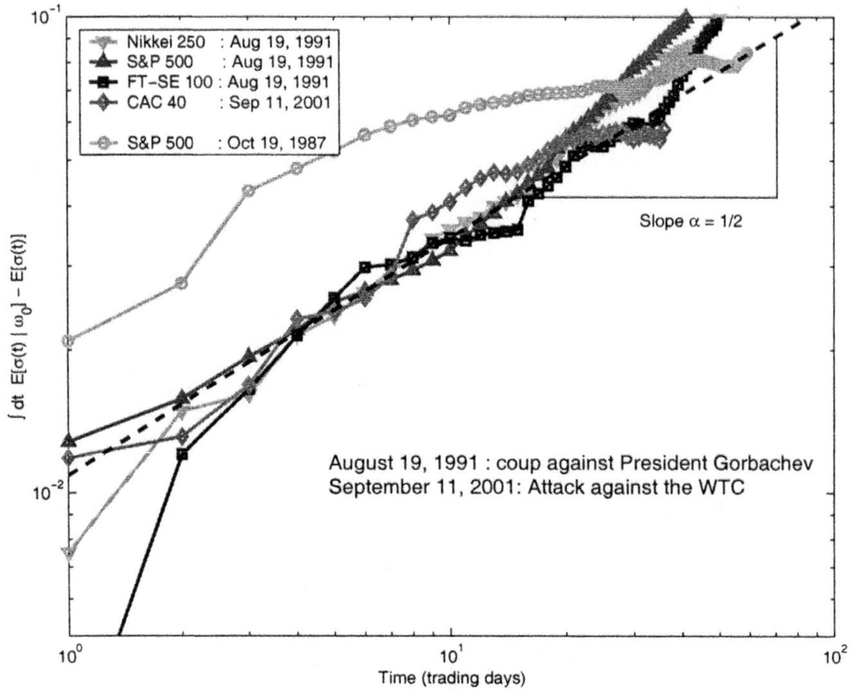

Fig. 1. Cumulative excess volatility at scale $\Delta t = 1$ day, that is, integral over time of $\mathrm{E}_{\mathbf{exo}}\left[\sigma^2(t)\middle|\,\omega_0\right] - \overline{\sigma^2(t)}$, due to the volatility shock induced by the coup against President Gorbachev observed in three British, Japanese and USA indices and the shock induced by the attack of September 11, 2001 against the World Trade Center. The dashed line is the theoretical prediction obtained by integrating (12), which gives a $\propto \sqrt{t}$ time-dependence. The cumulative excess volatility following the crash of October 1987 is also shown with circles. Notice that the slope of the non-constant curve for the October 1987 crash is very different from the value 1/2 expected and observed for exogenous shocks. This crash and the resulting volatility relaxation can be interpreted as an endogenous event.

4 "Conditional linear response" to an endogenous shock

Let us now consider the natural evolution of the system, without any large external shock, which nevertheless exhibits a large volatility burst $\omega(t=0) = \omega_0$ at $t = 0$. From the definition (1) with (2,3,4) and (5), it is clear that a large "endogenous" shock requires a special set of realization of the "small news" $\{\eta(t)\}$. In order to quantify the response in such case we can evaluate $\mathrm{E}\left[\sigma^2(t)\middle|\,\omega_0\right] = \mathrm{E}\left[e^{2\omega(t)}\middle|\,\omega_0\right]$. Since the $\omega(t)$ is a Gaussian process, $\omega(t)$ conditional on ω_0 remains Gaussian, so that

$$\mathrm{E}\left[\sigma^2(t)\middle|\,\omega_0\right] = \mathrm{E}\left[e^{2\omega(t)}\middle|\,\omega_0\right] \tag{13}$$
$$= \exp\left(2\mathrm{E}[w(t) \mid \omega_0] + 2\mathrm{Var}[\omega(t) \mid \omega_0]\right). \tag{14}$$

Due to the still Gaussian nature of the condition log-volatility $\omega(t)$, we easily obtain using (2) and (3),

$$\mathrm{E}[\omega(t) \mid \omega_0] = \mathrm{E}[\omega(t)] + \frac{\mathrm{Cov}[\omega(t), \omega_0]}{\mathrm{Var}[\omega_0]} \cdot (\omega_0 - \mathrm{E}[\omega_0]), \tag{15}$$
$$= (\omega_0 - \mu) \cdot \frac{C(t)}{C(0)} + \mu, \tag{16}$$

and

$$\mathrm{Var}[\omega(t) \mid \omega_0] = \mathrm{Var}[\omega(t)] - \frac{\mathrm{Cov}[\omega(t), \omega_0]^2}{\mathrm{Var}[\omega_0]}, \tag{17}$$
$$= C(0)\left(1 - \frac{C^2(t)}{C^2(0)}\right). \tag{18}$$

Let us set:

$$e^{2\omega_0} = e^{2s}\overline{\sigma^2(t)} \Rightarrow \omega_0 - \mu = s + C(0). \tag{19}$$

By substitution in (14), we obtain thanks to (2) and (3),

$$\mathrm{E}\left[\sigma^2(t)\middle|\,\omega_0\right] = \overline{\sigma^2(t)} \exp\left[2(\omega_0 - \mu) \cdot \frac{C(t)}{C(0)} - 2\frac{C^2(t)}{C(0)}\right], \tag{20}$$
$$= \overline{\sigma^2(t)} \left(\frac{T}{t}\right)^{\alpha(s)+\beta(t)} \tag{21}$$

where

$$\alpha(s) = \frac{2s}{\ln(\frac{Te^{3/2}}{\Delta t})} \tag{22}$$
$$\beta(t) = 2\lambda^2 \frac{\ln(t/\Delta t)}{\ln(Te^{3/2}/\Delta t)} \tag{23}$$

We can see that within the range $\Delta t < t << \Delta t e^{\frac{|s|}{\lambda^2}}$, $\beta(t) << \alpha(s)$ and Eq. (21) leads to a power-law behavior:

$$\mathrm{E}\left[\sigma^2(t)\middle|\,\omega_0\right] \sim t^{-\alpha(s)} \tag{24}$$

Notice that $|s|\lambda^{-2}$ provides directly the logarithmic scaling range over which the power-law can be observed. Since $\lambda^{-2} \sim 20$ this range can quickly extend over the whole domain $[\ln(\Delta t), \ln(T)]$.

Along the same line, we can also compute the conditional variance $\mathrm{Var}\left[\sigma^2(t)\middle|\,\omega_0\right]$. After a little algebra, we get:

$$\mathrm{Var}\left[\sigma^2(t)\middle|\,\omega_0\right] = \overline{\sigma^2(t)}^2 \left(\frac{T}{t}\right)^{2\alpha(s)+2\beta(t)} \left[\left(\frac{Te^{3/2}}{\Delta t}\right)^{4\lambda^2} \left(\frac{T}{t}\right)^{-2\beta'(t)} - 1\right] \tag{25}$$

where

$$\beta'(t) = 2\lambda^2 \frac{\ln(T/t)}{\ln(Te^{3/2}/\Delta t)} \tag{26}$$

It is thus easy to obtain the estimate:

$$\sqrt{\mathrm{Var}\left[\sigma^2(t)\middle|\,\omega_0\right]} \leq \mathrm{E}\left[\sigma^2(t)\middle|\,\omega_0\right] \left[\left(\frac{T}{\Delta t}\right)^{2\beta(t)} - 1\right]^{1/2} \sim \sqrt{6\lambda^2 \ln(te^{3/2}/\Delta t)} \cdot \mathrm{E}\left[\sigma^2(t)\middle|\,\omega_0\right] \tag{27}$$

We thus conclude, that, for s large enough ($\Rightarrow \alpha(s)$ large enough):

$$\frac{\mathrm{E}\left[\sigma^2(t)\middle|\,\omega_0\right] - \overline{\sigma^2(t)}}{\sqrt{\mathrm{Var}\left[\sigma^2(t)\middle|\,\omega_0\right]}} \sim \frac{1}{\sqrt{6\lambda^2 \ln(te^{3/2}/\Delta t)}} \tag{28}$$

We see that over the first decade $\Delta t \leq t \leq 10\Delta t$, the deviation of the conditional mean volatility from the unconditional volatility $\overline{\sigma^2(t)}$ is greater than the conditional variance.

Expressions (24,22) are our two main predictions. These equations predict that the *conditional* linear response function $\mathrm{E}\left[\sigma^2(t)\middle|\,\omega_0\right]$ of the volatility decays as a power law $\sim 1/t^{\alpha}$ of the time since the endogenous shock, with an exponent $\alpha \approx \left(2\omega_0 - \ln(\overline{\sigma^2(t)})\right) \frac{\lambda^2}{C(0)}$ which depends linearly upon the amplitude ω_0 of the shock. Note in particular, that α changes sign: it is positive for $\omega_0 > \frac{1}{2}\ln(\sigma^2(t))$ and negative otherwise. This power-law behavior is very different from the universal response:

$$\mathrm{E}\left[\sigma^2(t)\middle|\,\omega_0\right] \sim e^{2\tilde{K}_0 t^{-1/2}} \tag{29}$$

given by (12) of the volatility to an external shock. exogenous shocks have a stingingly distinct signature since it is the *excess* volatility (Eq. (12)) that approximately relaxes as a power law with an universal exponent $\alpha = 1/2$.

In figure 2, we have reported some empirical estimates of conditional volatility response for S&P 100 intradaily series corresponding to 5 minute close prices during the period ranging from 04/08/1997 to 12/24/2001 (figure 2(a)). The original intraday squared returns have been preprocessed in order to remove the U-shaped variance modulation associated to the intraday variations of market activity. In figure 2 (b) one can see that the MRW logarithmic covariance of the log-volatility $C_{\Delta t}(\tau)$ (Eq. (3)) provides a good fit of the empirical log-squared return covariance in a range of scale that extends from few minutes to one month. In Fig. 2(c) are plotted the estimated conditional volatility responses for $s = 1, 0, -1$ in double logarithmic representation for $\Delta t = 40$ minutes. The linearity of such curves show the s-dependent power

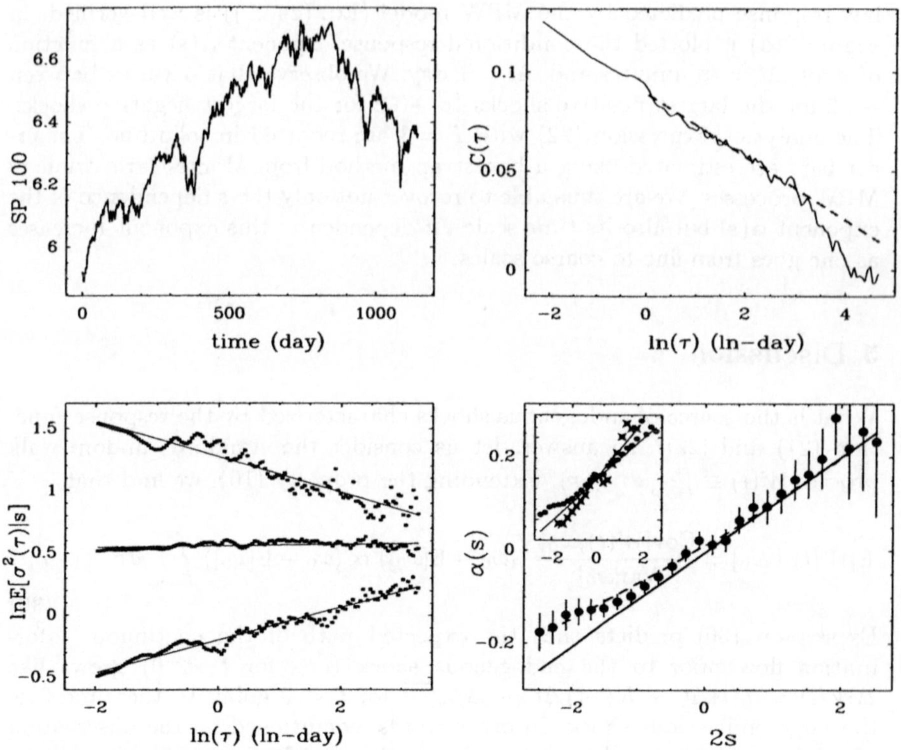

Fig. 2. Measuring the conditional volatility response exponent $\alpha(s)$ for S&P 100 intradaily time series as a function of the endogenous shock amplitude parameterized by s, defined by (19). (a) The original 5 minute intradaily time series from 04/08/1997 to 12/24/2001. The 5 minute de-seasonalized squared returns are aggregated in order to estimate the 40 minutes and daily volatilities. (b) 40 minute log-volatility covariance $C_{40}(\tau)$ as a function of the logarithm of the lag τ. The MRW theoretical curve with $\lambda^2 = 0.018$ and $T = 1$ year (dashed line) provides an excellent fit of the data up to lags of one month. (c) Conditional volatility response $\ln\left(\mathrm{E}_{\mathrm{endo}}\left[\sigma^2(t)\middle|\, s\right]\right)$ as a function of $\ln(t)$ for three shocks with amplitudes given by $s = -1, 0, 1$. (d) Estimated exponent $\alpha(s)$ for $\Delta t = 40$ minutes ($\bullet$) as a function of s. The solid line is the prediction corresponding to Eq. (22). The dashed line corresponds to the empirical MRW estimate obtained by averaging over 500 Monte-Carlo trials. It fits more accurately for negative s (volatility lower than normal) due to the fact that the estimations of the variance by aggregation over smaller scales is very noisy for small variance values. The error bars give the 95 % confidence intervals estimated by Monte-Carlo trials of the MRW process. In the inset, $\alpha(s)$ is compared for $\Delta t = 40$ minutes ($\bullet$) and $\Delta t = 1$ day ($\times$).

law response predicted by the MRW model (Eq. (24,22)) is well verified. In Figure 2(d) is plotted the conditional response exponent $\alpha(s)$ as a function of s for $\Delta t = 40$ minutes and $\Delta t = 1$ day. We observe that α varies between -0.2 for the largest positive shocks to $+0.2$ for the largest negative shocks. The analytical expression (22) with $T = 1$ are reported in solid line. The error bars are estimated using a bootstrap method from Monte-Carlo trials of MRW processes. We are thus able to recover not only the s dependence of the exponent $\alpha(s)$ but also its time scale Δt dependence: this exponent increases as one goes from fine to coarse scales.

5 Discussion

What is the source of endogenous shocks characterized by the response function (21) and (22)? To answer, let us consider the standard random walk process $W(t) \equiv \int_{-\infty}^{t} d\tau\ \eta(\tau)$. Extending the property (16), we find that

$$\mathrm{E}[W(t) \mid \omega_0] = \frac{\mathrm{Cov}[W(t), \omega_0]}{\mathrm{Var}[\omega_0]} \cdot (\omega_0 - \mathrm{E}[\omega_0]) \propto (\omega_0 - \mathrm{E}[\omega_0]) \int_{-\infty}^{t} d\tau\ K(-\tau)\ . \tag{30}$$

Expression (30) predicts that the expected path of the continuous information flow prior to the endogenous shock (i.e., for $t < 0$) grows like $\Delta W(t) = \eta(t)\Delta t \sim K(-t)\Delta t \sim \Delta/\sqrt{-t}$ for $t < 0$ going to the time 0 of the large endogenous shock. In other words, conditioned on the observation of a large endogenous shock, there is specific set of trajectories of the news flow that led to it. This specific flow has an expectation given by (30). This result allows us to understand the distinctive features of an endogenous shock compared to an external shock. The later is a single piece of very bad news that is sufficient by itself to move the market significantly according to (12). In contrast, an "endogenous" shock is the result of the cumulative effect of many small bad news, each one looking relatively benign taken alone, but when taken all together collectively along the full path of news can add up coherently due to the long-range memory of the log-volatility dynamics to create a large "endogenous" shock. This term "endogenous" is thus not exactly adequate since prices and volatilities are always moved by external news. The difference is that an endogenous shock in the present sense is the sum of the contribution of many "small" news adding up according to a specific most probable trajectory. It is this set of small bad news prior to the large shock that not only led to it but also continues to influence the dynamics of the volatility time series and creates the anomalously slow relaxation (21) and (22).

In this respect, this result allows us to rationalize and unify the many explanations proposed to account for the Oct. 1987 crash: according to the present theory, each of the explanations is insufficient to explain the crash;

however, our theory predicts that it is the cumulative effect of many such effect that led to the crash. In a sense, the different commentators and analysts were all right in attributing the origin of the Oct. 1987 crash to many different factors but they missed the main point that the crash was the extreme response of the system to the accumulation of many tiny bad news contributions. To test this idea, we note that the decay of the volatility response after the Oct. 1987 crash is well-described by a power law $1/t^{0.3}$ [26]. This value of the exponent significantly smaller than 0.5 clearly qualifies the Oct. 1987 crash as endogenous in the sense of our theory of "conditional linear response."

It is also interesting to compare the prediction (21) and (22) with those obtained with a linear autoregressive model of the type (4), in which $\omega(t)$ is replaced by $\sigma(t)$. FIGARCH and ARFIMA models fall in this general class. It is easy to show in this case that this linear (in volatility) model predicts the same exponent for the response of the volatility to endogenous shocks, independently of their magnitude. This prediction is in stark contrast with the prediction (21) and (22) of the log-volatility MRW model. The later model is strongly validated by our empirical tests presented in our two figures.

References

1. Mikhailov AS, Calenbuhr V (2002) From cells to societies. Springer, Berlin Heidelberg New York
2. Barro RJ, EF Fama, DR Fischel, AH Meltzer, R Roll and LG Telser (1989) Black monday and the future of financial markets, edited by Kamphuis RW Jr., RC Kormendi and JWH Watson. Mid American Institute for Public Policy Research Inc. and Dow Jones-Irwin Inc..
3. Orléan A. (1989) Mimetic contagion and speculative bubbles, Theory and Decision 27: 63–92.
4. Johansen A and Sornette D (1999) Critical Crashes, Risk 12: 91–94.
5. Shiller RJ (2000) Irrational exuberance. Princeton University Press, Princeton NJ.
6. Sornette D and A Johansen (2001) Significance of log-periodic precursors to financial crashes, Quantitative Finance 1: 452–471.
7. Camerer C (1989) Bubbles and Fads in Asset Prices, Journal of Economic Surveys 3: 3–41.
8. Adam MC and A Szafarz (1992) Speculative Bubbles and Financial Markets, Oxford Economic Papers 44: 626–640.
9. Feigenbaum JA (2001) A statistical analysis of log-periodic precursors to financial crashes, Quantitative Finance 1: 346–360.
10. Sornette D and JV Andersen (2002) A Nonlinear Super-Exponential Rational Model of Speculative Financial Bubbles, Int. J. Mod. Phys. C 13: 171–188.
11. Johansen A and Sornette D (2002) Endogenous versus Exogenous Crashes in Financial Markets, submitted to the Journal of Economic Dynamics and Control (http://arXiv.org/abs/cond-mat/0210509)

12. White EN (1996) Stock market crashes and speculative manias. In The international library of macroeconomic and financial history, 13. An Elgar Reference Collection, Cheltenham, UK; Brookfield, US.
13. Sornette D, Malevergne Y and Muzy JF (2003) What causes crashes? Risk 16: 67–71.
14. Sornette D and Helmstetter A (2003) Endogeneous Versus Exogeneous Shocks in Systems with Memory, Physica A 318: 577–591.
15. Helmstetter A, Sornette D and Grasso JR (2003) Mainshocks are Aftershocks of Conditional Foreshocks: How do foreshock statistical properties emerge from aftershock laws, J. Geophys. Res., 108 (B1), 2046, doi:10.1029/2002JB001991.
16. Ding Z, Granger CWJ and Engle R (1993) A long memory property of stock returns and a new model, Journal of Empirical Finance 1: 83–106.
17. Ding Z, Granger CWJ (1996) Modeling volatility persistence of speculative returns: A new approach, Journal of Econometrics 73: 185–215.
18. Arneodo A, Muzy JF and Sornette D (1998) Direct causal cascade in the stock market, The European physical Journal B 2: 277–282.
19. Granger CWJ and Ding Z (1996) Varieties of long memory models, Journal of Econometrics 73: 61–77.
20. Baillie RT (1996) Long memory processes and fractional integration in econometrics, Journal of Econometrics 73: 5–59.
21. Müller UA, Dacorogna MM, Davé RD, Olsen RB, Pictet OV and von Weizsucker JE (1997) Volatilities of Different Time Resolutions - Analyzing the Dynamics of Market Components, Journal of Empirical Finance 4: 213–240.
22. Muzy JF, Delour J and Bacry E (2000) Modelling fluctuations of financial time series: from cascade process to stochastic volatility model, The European physical Journal B 17: 537–548.
23. Muzy JF, Sornette D, Delour J and Arneodo A (2001) Multifractal returns and Hierarchical Portfolio Theory, Quantitative Finance 1: 131–148.
24. Mandelbrot BB (1997) Fractals and scaling in finance: discontinuity, concentration, risk. selecta volume E, Springer New York.
25. Bacry E, Delour J and Muzy JF (2001) Multifractal random walk, Physical Review E 64: 026103.
26. Lillo F and Mantegna RN (2001) Omori law after a financial market crash, preprint cond-mat/0111257.

Patterns of speculation in real estate and stocks

Bertrand M. Roehner [1]

Institute for Theoretical and High Energy Physics, University Paris 7

[1] Postal address: LPTHE, University Paris 7, 2 place Jussieu, 75005 Paris, France. E-mail: ROEHNER@LPTHE.JUSSIEU.FR

Abstract Today there is a hot debate in most developed countries regarding the future of real estate markets. Some economists claim that demand is strong and will remain so, others contend that there may be a small correction in the two next years but, surprisingly, few researchers have tried to take a close look at former price peaks to better understand what may be in store. This is the task we undertake in this paper. The present study shows that stock and property price peaks are ruled by similar mechanisms; moreover in both markets the price earnings ratio constitutes a reliable indicator for assessing the level of speculative fever. The paper closes with a prediction about housing prices in San Francisco over the next four years.

In this study I focus my attention on real estate and stock prices during speculative episodes with special emphasis on the former and their relation with the latter. What do I mean by the expression "speculative episodes"? I will take a very pragmatic position without trying to take side in the debate about the existence of speculative bubbles. When real estate or stock price increases are out of all proportion to other economic factors such as (real) household income or (real) GDP, then one has good ground to think that a speculative mechanism is at work. For instance, when the price of apartments in Hong Kong jumps from US$ 2,100 per square meter to over US$ 8,600/sqm (a multiplication by a factor of 4) within six years it is fairly obvious that this is out of pace with overall economic development. Later on in this paper we will give another and more precise criterion based on the price earnings ratio. Incidentally, it should be noted that there is no need to assume that speculative episodes are irrational. The decision of investors are certainly perfectly rational provided one includes into the list of fundamentals the collective factors which build up in those times; more details on this viewpoint can be found in [1] (chapter 3) and [2] (chapter 3).

During the past six years real housing prices have increased by 110 percent in London, 85 percent in Boston, 77 percent in San Francisco. For real estate markets such increases are quite substantial; as a matter of comparison it can be recalled that during the previous speculative episode in 1985-1990, the price increases were about 100 percent in London, 120 percent in Paris and 63 percent in San Francisco and in all these cases it was followed by a phase of falling prices which lasted about five years and erased a substantial part of the former price increases; more details are provided in table 1.

Table 1 Price changes during the current and former price peaks

City	Increase 1984-1990 %	Decrease 1990-1994 %	Increase (prov.) 1996-2002 %
Paris	120	−40	43
London	100	−42	111
San Francisco	63	−20	77
Boston	60	−21	85
New York City	60	−30	62
Orange County	53	−30	78
San Diego	43	−18	81

Sources: London: Hamilton website; Paris: Paris-Notaires; U.S.: National Association of Realtors.
Notes: Basically, the former speculative episode took place between 1984 and 1997 but the precise time intervals for the increase and decrease phases were (slightly) market dependent. The decreases have been computed with respect to the maxima which is why they look smaller than the increases; for London, for instance, the price peak is almost symmetrical.

The big issue presently is whether there will be a similar scenario. The prevailing opinion among economists or real estate experts is that there may be a small correction of say 2 percent over two years, but few if any foresee a decline as large as 50 percent. If mentioned at all, previous speculative episodes are merely alluded to, and rarely studied in detail as a forerunner of what might be in store for the coming years. This is the program that we try to carry out in this paper.

1 Comparison of speculative episodes in property and stocks

A basic similarity between property and stocks is the fact that both assets provide (i) capital gains when the price of the asset increases (ii) annual earnings in terms of rent or dividends. Naturally, bonds share the same feature as will be explained in more detail later on. It is true that each one of these markets also has many idiosyncrasies. Whereas in "normal" times these idiosyncrasies may be more important than the common features, during speculative episodes, on the contrary, the price-earnings mechanism takes the upper hand which explains their similar behavior. These similarities may be summarized in the following way (these points are more fully developed in the rest of the paper).

1.1 Similarities

1) The price earnings ratio (PER) is a good indicator of the level of speculative fever in a market. Whenever the PER becomes larger than 30, one has good reason to think that a speculative episode in under way.

2) During such an episode, price peaks are roughly symmetrical with respect to the peak price which means that the more the price has increased in the upward phase, the more it will fall in the downward phase. The increase and decrease rates may not be identical however.

3) Once the turning point has been reached and the market is in the downward phase, the dynamics of this relaxation process in sufficiently strong to overcome standard exogenous stimulations. For instance, in 1929-1931 stock continued to fall in the United States in spite of a huge reduction in discount rates (more details can be found in [1] p.186); similarly, in 2000-2002 the NASDAQ market continued to fall in spite of a huge reduction in the Target Federal Funds Rate which dropped from 6 percent to 1.5 percent; in 1990-1992, real estate prices in California and Boston pursued their decline in spite of a reduction in mortgage rates from 10 percent to about 7 percent.

In short, stock and property prices are basically ruled by the same speculative process; the differences which may exist (and which are listed in the next paragraph) rather concern the rate and circumstances of the process.

1.2 Differences

1) Transaction costs are at least 100 times higher for buying a house than for buying stocks; moreover the transaction time is at least 1,000 times longer. Not surprisingly, differences of that scale affect the shape of the price peak. Whereas the shape of stock price peaks is of the "sharp peak - flat trough" sort (more details can be found in [1] and [3] (chapter 7) the shape of housing price peaks rather tends to be of the "flat peak - flat trough" sort. Moreover the price volatility (i.e. the standard deviation of price fluctuations) is much lower for property than for stock prices. On the other hand, a decrease in property prices can extend over a long period of time; for real estate prices in Tokyo the decrease began in 1991 and is still under way, similarly in Hong Kong the decrease which began in 1997 is still continuing.

2) The price of a stock is somewhat arbitrary in the sense that stocks are subject to splits (especially during bull markets), by which their price is divided by two (or sometimes by 3/2). Thus, it has no meaning to say that the price of a General Electric stock is higher than the price of Cisco stock. On the contrary, if one is told that apartment A costs 5,000 euro per square meter, whereas apartment B costs only 2,000 euro/sqm it will be clear for everybody that A is located in a better area and more luxurious building than B. In short, prices of houses or apartment have an objective meaning which can be used by investors in order to optimize their

investment.
Price data show that the amplitude of speculative peaks, that is to say the ratio (peak price)/(initial price), for different price intervals is larger for more expensive homes. For instance, during the 1985-1995 price peak, the amplitude of the peak was equal to 1.6 in Boston but only to 1.3 in Buffalo (New York State) where prices were lower. This effect (of which other examples are to be found further on) is referred to as the price multiplier effect. It provides an important tool for forecasting relative price changes during a speculative episode.

3) Observation shows that real estate prices peak earlier in expensive areas than in cheaper places. For instance, during the 1985-1995 episode in Britain, prices in London reached their maximum 1.5 years earlier than prices in the North West region and about 2.5 years earlier than prices in Scotland. The observation has an interesting implication: if I want to detect the turning point in speculative episode that is currently under way, I should look at the price changes in the most expensive areas, such as the city centers of London or San Francisco.

The last observation leads us to a question that we have not addressed so far; is it possible to *forecast* the time of the turning point? Needless to say, such a prediction would be of great importance for property developers; what makes it very difficult if not altogether impossible, is the fact that exogenous factors may delay the market's reversal. The example of the NASDAQ is a case in point; it has been claimed that the money which was injected in the banking system as a precaution for the transition from 1999 to 2000 helped to prolong the bull market for another 4 to 6 months (Business Line, 8 May 2000). Similarly, a reduction in mortgage rates may give additional momentum to a speculative peak and delay the reversal for a year or more. However, as already emphasized, once prices have begun to fall in several hot markets the fall is likely to continue in a fairly robust way.

Finally, it can be of interest to notice that historically there seems to be a definite relationship between stock and property price peaks especially between 1830 and 1880 in the United States (see the graph on p.119 in [2]). Needless to say the macroeconomic connection between these two phenomena is far from being straightforward, hence this should be seen merely as an heuristic remark.

We now consider in more detail the question of the price earnings ratio.

2 The price earnings ratio (PER) seen as a "universal" index of speculative fever

2.1 Definitions

For the four assets under consideration the PER in year Y_1 will be defined in the following ways:

$$\text{PER(stocks)} = \frac{\text{Average price in year } Y_1}{\text{Dividends earned during } Y_1} \tag{1}$$

$$\text{PER(residential or commercial property)} = \frac{\text{Average price in year } Y_1}{\text{Rent earned during } Y_1} \quad (2)$$

$$\text{PER(bond)} = \frac{\text{Average price in year } Y_1}{\text{Nominal (coupon) interest rate}} \quad (3)$$

For residential property there are several other indicators which are used by real estate experts in order to assess whether a market is "cheap" or "expensive", such as the affordability index which is the percentage of households which can afford a median-priced home or the ratio of price to income or the median number of days it takes to sell a house. In my opinion, the PER is a better indicator for at least two reasons.

- The affordability index refers to people who buy a house in order to live in it (let us call them the "users"). However, for an investor who buys a house in order to sell it with a profit or to rent it with a profit, the affordability index is meaningless (as is also the ratio of price to income). It may be estimated that in normal times in an active market about 20 to 30 percent of the transactions are made by investors (La Vie Française 18 April 1998), but during speculative episodes it is fairly natural to assume that this proportion increases as speculation unfolds. In other words the above indicators become less and less useful at the very time when they would be most needed. On the contrary, the PER applies to investors as well as to users; for if the price increases without a parallel progression in the rent, one is in a situation that cannot last for very long. A dynamics where prices go up whereas rent remain stable or decline is based only on the prospect of capital gains and is therefore highly fragile and unstable.
- Indicators such as the affordability index do not apply to offices and of course even less to stocks. The great virtue of the PER is that it applies to all assets which provide an annual income in addition to capital gains.

The PER of stocks is not always defined in the way of expression (1). Sometimes the dividends earned in a past year are replaced by the expected dividends for the forthcoming year. Such projections are fairly hazardous in normal times, but during a speculative episode they are even more uncertain because the business climate may change very quickly. Instead of being a useful yardstick such an expected PER can be quite misleading. Another variant is to compute the PER by quarter instead of annually; however, as dividends may undergo substantial fluctuations, the PER will fluctuate in the same way and, as a result, lose much of its usefulness as a yardstick of speculative fever.

The title of this section suggests that the PER is a universal index for speculative fever; in order to prove that point we need to check whether or not the range interval of the PER for various speculative markets is the same. This is the purpose of the next section.

2.2 Range interval of the PER in various markets

• **Stocks** The PER has been used for a long time by stock market analysts and it is well known that from 1871 to 1996 the PER for all U.S. stocks averages about 14 (Sunday Business, January 16, 2000). Observation tends to show that whenever the PER tends toward 25 one is in speculative territory which means that a correction occurs sooner or later. The correction may last a few months as in 1987 or several years as in 1929. The decisive factor as to the duration of the correction is the length of the upward phase (as we explain in more detail later on), however once the speculative peak is at its end whether or not the market will take off again depends upon the economy rather than on the stock market per se. In short, for stocks at global market level, the critical value of the PER is somewhere between 20 and 25.

Remark The reader may object that in early 2000 the PER of the NASDAQ was about 200 which is way above the 20-25 range. As a matter of fact, for stocks there is no "physical" bound for the PER. If no company makes any profit (which was almost the situation of the NASDAQ at the time) the denominator in expression (1) is almost equal to zero and the PER tends toward infinity. However, the case of the NASDAQ was quite exceptional, it was speculation in its wildest form.

• **Real property** In order to compute the PER for residential or commercial property we need data about prices and rents. For residential property price data are fairly easy to find especially since 1990; yet, rent data are less common. For offices, because they are usually rented rather than bought, the opposite situation prevails: rent data are published by all major commercial property developers whereas price data are more difficult to find. Hong Kong is one the few places for which I was able to find price and rent data for both residential and commercial property. Fig.1 shows price and PER data in Hong Kong between 1992 and 2002. A similar graph could be drawn for commercial property. As can be seen on Fig.1 a PER between 15 and 20 seems to characterize a market which enters into speculative territory. For commercial property the top of the peak was characterized by a PER of 20.8 and the data for 2002,Q4 which are the last available show that it has currently fallen to 11.0.

The very fact that the PER increases in the upward phase of the price peak and declines in the downward phase shows that prices move quicker than rents. This observation has a fairly general validity; for instance, American housing prices have jumped by 30 percent between 1996 and mid-2002 while rents have increased just 10 percent (Fortune, October 13, 2002). This confirms the view we expressed earlier that, as price go up, home buyers increasingly put investment ahead of shelter as the primary reason for the purchase; in short, the market is more and more lead by investors.

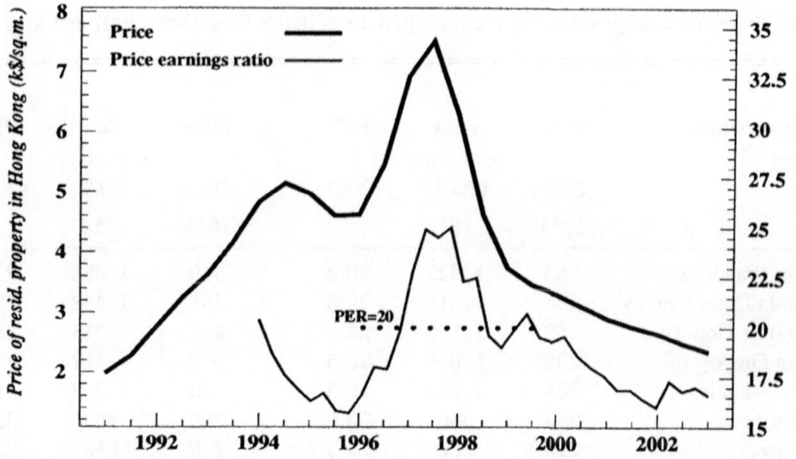

Fig.1: Price and PER in Hong Kong. Thick line: price (left-hand scale); thin line: PER (right-hand scale).The dotted line shows what can be considered as the limit of speculative territory.The prices are expressed in 10^3 US$ per square meter.

Table 2a Price earnings ratio at troughs and peaks of speculative episodes in real estate

City	1989 peak	1995 trough	mid-2002 peak (proxy)
Boston		19	30
Orange County (LA)		17	28
New York		18	25
San Diego	23.4		27.1
San Francisco	25.6	21	27.2

Sources: Fortune October 13, 2002; Foreclosure News September 2002.

Table 2a provides some figures for the magnitude of the PER before and at the top of a speculative peak. As can be seen the PERs at the peak of 1989 match those we observed in Hong Kong in mid-1997. In contrast, the figures for mid-2002 (the last available at a moment when the market is still in its upward phase) are substantially higher than those of 1989. Does this indicate a stronger speculative peak or a change in the way the market operates? Recent history (especially the NASDAQ episode) rather makes me inclined to favor the first hypothesis. The future will tell us which one is true.

Table 2b provides a number of price and rent data for metropolitan areas in the West of the United States. It can be seen that there is a close connection between price levels and PER levels. This confirms the observation made previously in the time

Table 2b Price earnings ratios for metropolitan areas in the West (May 2001 and May 2002)

Metropolitan area	Price (2001) (10^3\$)	Rent per month (2001) (\$)	PER (2001)	Price (2002) (10^3\$)	Rent per month (2002) (\$)	PER (2002)
San Francisco	483	1,922	20.9	540	1,653	27.2
Santa Clara County	448	1,813	20.6	461	1,432	26.8
Orange County	352	1,193	24.6	411	1,215	28.2
San Diego	298	1,056	23.5	362	1,112	27.1
Los Angeles	234	1,199	16.3	276	1,248	18.4
Seattle	244	860	23.6	260	863	25.1
Denver	219	856	21.3	228	858	22.1
Sacramento	176	811	18.1	202	853	19.7
Colorado Springs	165	711	19.3	185	704	21.9
Portland	172	738	19.4	181	745	20.2
Reno	160	732	18.2	181	740	20.4
Riverside/ San B	157	808	16.2	172	856	16.7
Las Vegas	147	726	16.9	157	736	17.8
Phoenix	141	706	16.6	152	707	17.9
Salt Lake City	146	669	18.2	150	673	18.6
Boise (Idaho)	127	720	14.7	140	706	16.5
Albuquerque	132	631	17.4	134	625	17.9
Oklahoma City	95.0	503	15.7	94.4	509	15.5

Sources: Prices: National Association of Realtors; rents: Las Vegas Review Journal (on line edition), July 24, 2002.
Notes: The metropolitan areas have been sorted according to their PER in 2002. The correlation between price and PER is equal to 0.66 in May 2001 and to 0.88 in May 2002.

domain. Let us emphasize that the observation of such a correlation between price and PER provides an understanding of the underlying mechanism that could not be derived from stock price analysis for in this case, as already noted, prices have no intrinsic significance.

• **Bonds** The case of bonds is somewhat peculiar. The reason is simple. In expression (3) the denominator comprises the nominal interest rate of the bond which itself is highly dependent upon the level of long-term interest rates; and the latter are (at least to some extent) under the control of the central bank. If interest rates tend toward zero the PER of government bonds can become very high and remain so as long as interest rates stay low. Such was the case in the United States during the 1930s and 1940s or in Japan during the 1990s. Similarly, since 2001 American interest rates are tending toward zero once again and, as a result, the

Table 3a The price multiplier effect: Amplitude $= a \log(\text{Initial price}) + b$

Region	Years of peak	Number of cases	a	Correlation
California	1989 – 1990	5	0.34 ± 0.4	0.60
North East	1987 – 1991	9	0.45 ± 0.6	0.47
West+Midwest+NE+South (prov.)	2001	32	0.35 ± 0.16	0.61
Paris (prov.)	2002	20	0.10 ± 0.1	0.40
Average			**0.31**	

Sources: Same as for Table 1 plus website of the U.S. Bureau of the Census.
Notes: In the third and fourth cases the abbreviation prov. means provisional (these episodes are still under way).

PER of the Ten-year Treasury bonds has reached levels around 28 that have not been seen for the past forty years. However, between 1932 and 1950 the PER of U.S. government bonds climbed as high as 37 (1936), 48 (1941) or 43 (1950) and remained over 33 all the time between 1935 and 1955 (Historical Statistics of the United States 1974, p.1003).
In short, the PER of government bonds is under the control of the central bank to a much greater extent than the PER of stocks or real property. This explains why for bonds speculative episodes take a somewhat different form than for other assets.

3 The price multiplier effect

The higher the price of an asset, the higher is the amplitude of the price peak. This effect which we called the price multiplier effect has been tested on a broad spectrum of markets belonging to the S-class (namely real property, collector stamps, antiquarian books) in [1] (chapter 6) and [2] (p.170). In all these cases (there is almost no exception) it turns out that there is a relationship of the form:

$$\text{Amplitude} = a \log(\text{Initial price}) + b$$

Note that when price variations are small the logarithm becomes inessential. Table 3a,b provide some additional tests. At time of writing (March 2003) the peak has not yet been reached in most of the markets under consideration which is why a (provisional) amplitude of the peak was computed by using the last price available. In all cases there is a significant relationship between amplitudes and initial prices. It should be noted that during the 1985-1990 episode there were substantial price peaks only in California and the North-East and these two episodes were to some

Table 3b Relationship between initial price and peak time: $t_{\text{peak}} = a\ln(\text{initial price}) + b$

Region	Years of peak	Number of cases	a	Correlation
California	1989 − 1990	6	-1.5 ± 1	-0.78
North East	1987 − 1991	9	-3.0 ± 3	-0.59
Paris	1990 − 1991	20	0.73 ± 0.8	-0.38
Average			**-1.74**	

Sources: Same as for table 3a.

extent independent. In contrast, during the 1996-2003 episode price peaks are more widespread and more synchronized which is why we could consider a global region containing 32 cities. The price multiplier effect clearly implies that speculative trading is an increasing function of the price level, a result which comforts the PER analysis. Investors are so to say attracted toward high price zones in the same way as iron filings are attracted by a magnet.
Table 3b shows that prices in expensive areas peak earlier than prices in cheaper areas.

4 Shape of the price peaks

In this section we describe the shape of price peaks in two different ways.

- At the beginning of the paper we noted that price peaks are roughly symmetrical with respect to their maxima. This, however, is only a first approximation. In fact there are slight differences between the up- and downward phases both in duration and in amplitude. In table 4a these differences are captured by way of two parameters which are the ratio, $T_{\text{up}}/T_{\text{down}}$, of the durations of the two phases and the ratio, p_3/p_1, of the price p_3 at the end of the downward phase to the price p_1 at the beginning of the upward phase.

- Next we want to describe the shape of the up- and downgoing trajectories. For that purpose it is convenient to use the following parametrization (which was introduced in previous works: [1] (p.155) and [3] which is able to represent both peaks or troughs either sharp or flat:

$$p(t) = p_2 \exp\left[-\text{sgn}(\tau)\left|\frac{t-t_2}{\tau}\right|^{\alpha}\right] \tag{4}$$

Peaks are obtained for $\tau > 0$, whereas troughs correspond to $\tau < 0$; moreover, sharp peaks (or sharp troughs) are obtained for $\alpha < 1$, whereas flat peaks (or flat

Table 4a Amplitude and duration of up- and down going phases of real estate price peaks

City or region	Period	Amplitude of peak	$T\text{up}/T_{\text{down}}$	p_3/p_1
Port Said	1898 – 1912	5.1	1.70	1.60
London	1983 – 1993	2.0	1.10	1.13
Paris	1984 – 1997	2.1	1.02	1.28
Tokyo	1982 – 1998	6.50	1.24	1.20
Hong Kong (prov.)	1991 – 2003	5.3	< 1.20	< 1.40
U.S.,West	1972 – 1982	1.59	2.41	1.25
U.S.,West	1984 – 1993	1.47	2.85	1.47
U.S.,North East	1984 – 1993	1.83	1.42	1.44
Boston	1984 – 1992	1.80	0.83	1.36
New York City	1984 – 1992	1.68	0.82	1.27
U.S.,West (prov.)	1997–?	> 1.40	?	?
U.S.,North East (prov.)	1995–?	> 1.38	?	?
San Francisco (prov.)	1996–?	> 1.91	?	?
Average			**1.46**	**1.34**

Sources: Same as for table 3.
Notes: The indication prov. (= provisional) as well as the >, < and the question marks correspond to speculative episodes which are still under way at time of writing.

troughs) correspond to $\alpha > 1$. As already noted housing peaks are usually smoother than stock price peaks, therefore we expect α to be rather larger than one; of course, if instead of using monthly prices one uses annual or quarterly prices, what is in reality a smooth peak may look fairly sharp.

Estimates for the parameters (α_1, τ_1) and (α_2, τ_2) for the up- and downward price trajectories are given in Table 4b.
Several observations are in order. (i) The fit is performed by first linearizing the data (basically by taking the logarithm two times); in most of the cases the fit is fairly good with a coefficient of linear correlation higher than 0.95. However, in two cases (which are signaled by an asterisk) the correlation is below 0.90 and the fit is less accurate: firstly for the downward phase of the case U.S. (North-East) 1984-1993 the correlation is only 0.48 and secondly for the same region after 1994 the correlation is 0.86; in both cases the estimated values of α are abnormally low while those of τ are abnormally high. For these reasons we discarded these two values in computing the average. (ii) The data for the cases labeled U.S. (West) and U.S. (North East) refer to sales of new houses. It turns out that these peak amplitudes are higher than those for existing homes. In other words these estimates may show more speculative fever. (iii) The cases of Port Said, Tokyo and Hong Kong are of particular interest because of the magnitude of the price peaks; they almost resemble stock price peaks.

Table 4b Shape parameters of real estate price peaks

City or region	Period	α_1	τ_1	α_2	τ_2
Port Said	1898 – 1912	0.54	1.3	2.09	3.2
London	1983 – 1993	1.24	6.0	1.00	7.6
Paris	1984 – 1997	0.96	7.5	1.22	10.8
Tokyo	1982 – 1998	1.88	5.9	1.57	4.5
Hong Kong	1991 – 2003	0.62	5.2	0.51	3.0
U.S.,West	1972 – 1982	1.04	16.1	0.93	14.6
U.S.,West	1984 – 1993	1.00	18.4	1.06	9.5
U.S.,North East	1984 – 1993	0.80	14.6	0.22*	11544*
Boston	1984 – 1992	1.30	5.8	0.86	24.5
New York City	1984 – 1992	1.07	6.6	1.01	14.8
U.S.,West	1997–?	0.59	34.8	?	?
U.S.,North East	1995–?	0.31*	318*	?	?
San Francisco	1996–?	0.75	13.0	?	?
Average		**0.98**	**11.2**	**1.14**	**10.3**

Sources: Same as for table 3.
Notes: The asterisks mark cases for which the coefficient of linear correlation of the data is less than 0.90; as a result these estimates are less reliable and were skipped in the computation of the overall average.

Predictions The main usefulness of the former estimates is to permit predictions for the downward phase. Such a prediction has been made in the spring of 2000 for the NASDAQ market and published in January 2001 ([1] p.176); Fig. 2a shows a comparison between the prediction and the actual course of the NASDAQ. Fig. 2b proposes a prediction for housing prices in San Francisco.

5 Conclusion

Why are real estate prices of great importance for the economy? Macroeconomic effects are channeled through several mechanisms; let us briefly summarize three of them. (i) Usually, the upward phase of a speculative episode is characterized by high transaction volumes; we did not study this facet of the phenomenon in the present paper but some evidence can be found in [1] (p.102-107). Obviously high volumes are important for new houses, but even for existing houses they boost consumption through refurbishment and purchase of furniture. When one realizes that in developed countries the building industry (and related sectors such as furniture, financing sector and so on) represents about 15 percent of the GDP (in the case of the United States, the contributions of the construction, furniture and lumber industries plus the

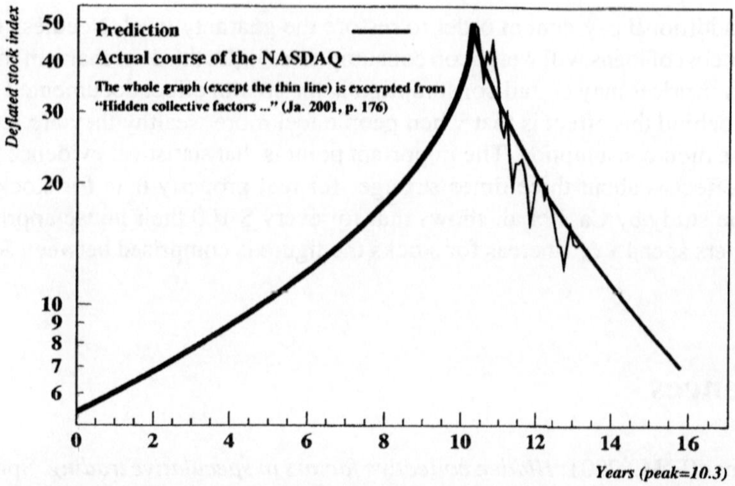

Fig.2a: Prediction for the course of the NASDAQ. The prediction was made in the spring of 2000 during the writing of a book which was published by Springer-Verlag in mid-January 2001.

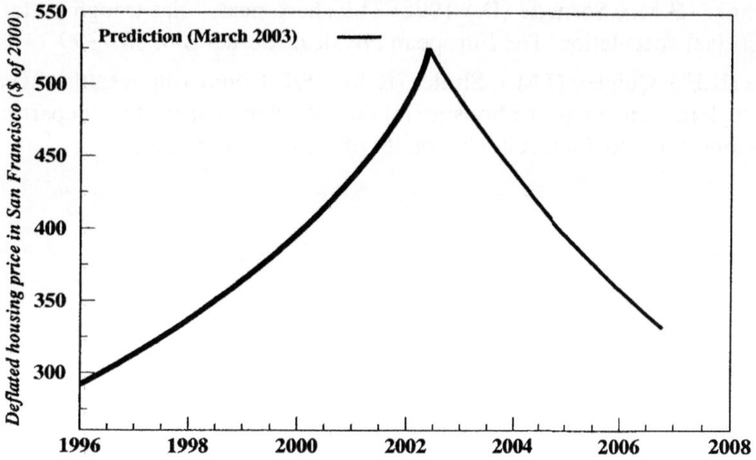

Fig.2b: Prediction for the course of housing prices in San Francisco. The thick line represents an adjustment (using parametrization (4)) of the actual price increase during the upward phase of the speculative episode. The prediction has been made by using the same parametrization with $\alpha_2 = 0.93$ and $\tau_2 = 10.1$ which were derived from the decrease phases during previous speculative episodes which occurred in the West (see table 4b).

financial services related to real estate add up to 16.8 percent of the GDP in 2000) it becomes clear that this effect can have a powerful effect on economic growth. (ii) Usually a mortgage is guaranteed by the value of the house it permitted to buy; if the price of the house falls below the required amount the borrower will be asked to make an additional payment in order to restore the guaranty level. Needless to say, the higher cost of loans will weight on consumption. (iii) A third mechanism through which lower prices may curtail consumption is the wealth effect documented in [4]. The idea behind this effect is that when people feel more wealthy they are tempted to increase their consumption. The important point is that statistical evidence shows that this effect is about three times stronger for real property than for stocks; basically, the study by Case et al. shows that for every $ 100 their house appreciates homeowners spend $ 6, whereas for stocks the figure is comprised between $ 2 and $ 3.

References

[1] Roehner (B.M.) 2001: *Hidden collective factors in speculative trading*. Springer-Verlag. Berlin.

[2] Roehner (B.M.) 2002: *Patterns of speculation*. Cambridge University Press. Cambridge.

[3] Roehner (B.M.), Sornette (D.) 1998: The sharp peak - flat trough pattern and critical speculation. The European Physical Journal B 4, 387-399.

[4] Case (K.E.), Quigley (J.M.), Shiller (R.J.) 2001: Comparing wealth effects: the stock market versus the housing market. Working paper of the department of economics NO. E01-308. University of California, Berkeley.

Generalized Technical Analysis. Effects of transaction volume and risk

Marcel Ausloos[1] and Kristinka Ivanova[2]

[1] GRASP, B5, University of Liège, B-4000 Liège, Euroland
[2] Pennsylvania State University, University Park PA 16802, USA

Summary. We generalize the momentum indicator idea taking into account the volume of transactions as a multiplicative factor. We compare returns obtained following strategies based on the classical or the generalized technical analysis, taking into account a sort of risk investor criterion.

Key words. Econophysics; Moving Average; Technical Analysis; Momentum; Investment Strategies

1 Introduction

First we recall classical technical analysis methods of stock evolution. We recall the notion of moving averages and momentum indicators. We present a generalization of momentum indicators based on classical physics principles, taking into account not only the price of a stock but also the volume of transactions. Next we compare the returns obtained following strategies based on the classical technical analysis and the generalized technical analysis. The cases of four stocks quoted on NASDAQ, four stocks traded on NYSE, three major financial indices and the price of Gold will serve as illustrations. We consider the volume of transactions and the daily closing price of these stocks and indices for the period Jan. 01, 1997 to Dec. 31, 2001. Daily closing price signals $y(t)$ are plotted in Fig.1(a-d) for stocks quoted on NASDAQ, i.e. CSCO, SUNW, AMAT, and MSFT; in Fig.1(e-h) for stocks traded on the NYSE, i.e. GE, AOL, PFE, and GFI; in Fig.1(i-k) are three financial indices: (i) NASDAQ, (j) S&P500), (k) DJIA; the price of Gold is in Fig.1 (l).

2 Classical Technical Analysis

Technical indicators like the *moving average* and the *momentum* are part of the classical technical analysis and much used in efforts to predict market

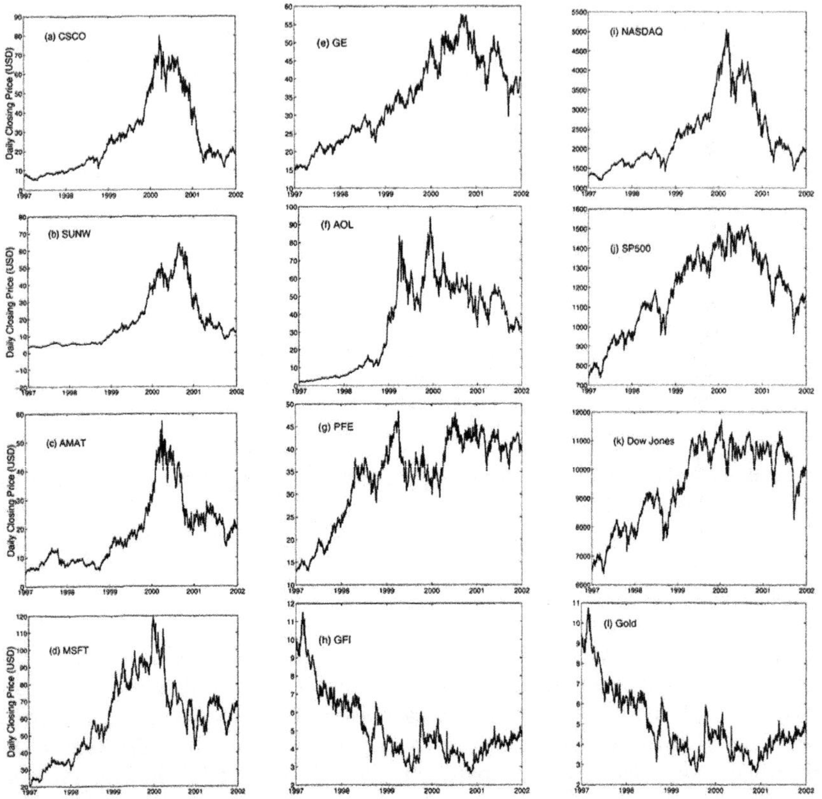

Fig. 1. Daily closing price $y(t)$ for the period Jan. 1, 1997 to Dec. 31, 2001 of stocks traded on NASDAQ – (a) CSCO, (b) SUNW, (c) AMAT, (d) MSFT, or traded on NYSE – (e) GE, (f) AOL, (g) PFE, (h) GFI; three major financial indices – (i) NASDAQ, (j) S&P500, (k) DJIA; the price of Gold (l)

movements [1]. One question is whether these techniques provide adequate ways to read the trends.

Consider a time series $y(t)$ given at N discrete times t. Let us recall that the series (or signal) moving average $M_\tau(t)$ over a time interval τ is defined by

$$M_\tau(t) = \frac{1}{\tau} \sum_{i=t}^{t+\tau-1} y(i-\tau) \qquad t = \tau+1, \ldots, N \tag{1}$$

i.e. the average of y over the last τ data points. One can easily show that if the signal $y(t)$ increases (decreases) with time, $M_\tau(t) < y(t)$ ($M_\tau(t) > y(t)$). Thus, the moving average captures the trend of the signal given the period of time τ. The intersections of the price signal with a moving average can define so-called lines of resistance or support [1]. The intersections between two moving averages, the so-called "death cross" and "gold cross" in empirical finance [1], are usually used to identify points of drastic changes in the trend of the signal. The *cross density* characterizes the signal roughness [2].

The so called momentum indicator [1] is another instrument of the technical analysis and we will refer to it here as the *classical momentum indicator* (CMI). The classical momentum indicator of a stock is a *moving average of the momentum* defined over a time interval τ as

$$R_\tau^\Sigma(t) = \sum_{i=t}^{t+\tau-1} \frac{y(i) - y(i-\tau)}{\tau} \qquad t = \tau+1, \ldots, N \qquad (2)$$

The classical momentum indicator (CMI) R_τ^Σ for time interval, $\tau = 21$, i.e. one month is shown in Fig. 2(a) for GE.

3 Generalized Technical Analysis

Stock markets do have another component beside prices or volatilities. This is the volume of transactions (Fig. 2 (a) for GE) which we have considered as the "physical mass" of stocks, in a generalized technical analysis scheme [3]. Remember that the number of shares is constant over rather long time intervals, usually like the mass of an object.

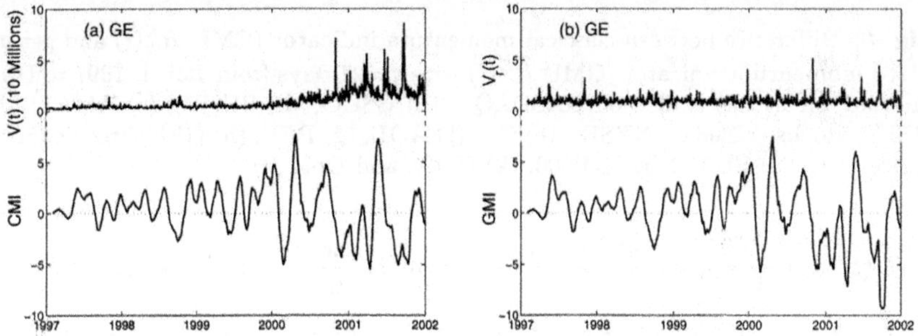

Fig. 2. (a) Volume of transactions $V(t)$ (in 10 Millions) and classical momentum indicator (CMI) $R_\tau^\Sigma(t)$ for τ=21 days, Eq. (2), for General Electric (GE) for the period Jan. 1, 1997 to Dec. 31, 2001; (b) Reduced volume of transactions $V_r(t)$ and generalized momentum indicator (GMI) $\widetilde{R}_\tau^\Sigma(t)$ for τ=21 days, Eq. (3), for General Electric (GE) for the period Jan. 1, 1997 to Dec. 31, 2001

Consider $V(t)$ to be the volume of transactions of a stock with price $y(t)$ at time t. A generalized momentum $\widetilde{R}_\tau$ over a time interval τ can be

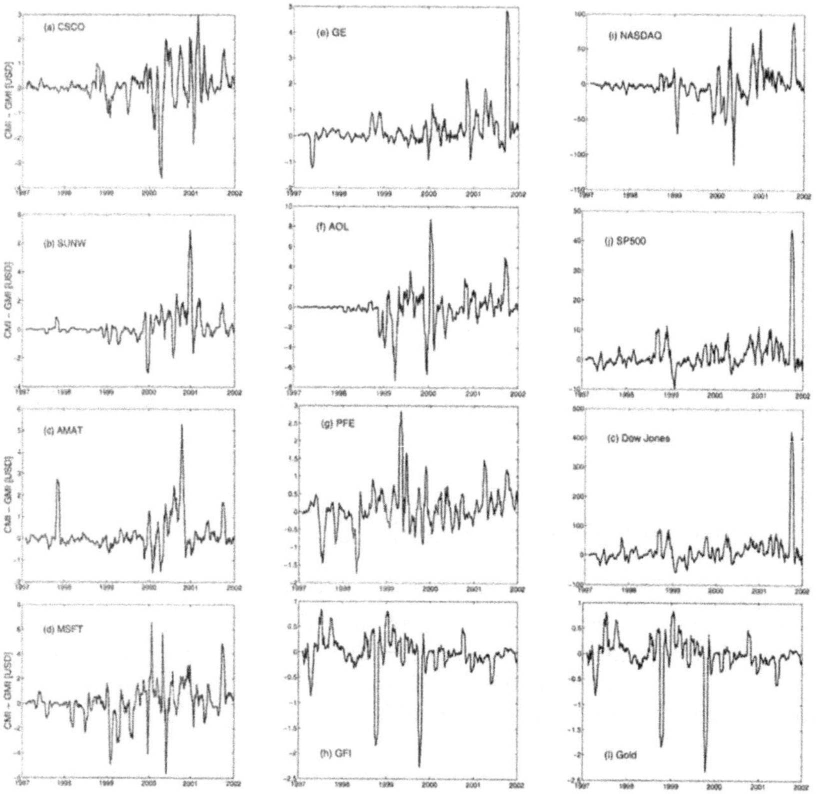

Fig. 3. Difference between classical momentum indicator (CMI) $R_\tau^\Sigma(t)$ and generalized momentum indicator (GMI) $\widetilde{R}_\tau^\Sigma(t)$ with τ=21 days from Jan. 1, 1997 to Dec. 31, 2001 for stocks traded on NASDAQ – (a) CSCO, (b) SUNW, (c) AMAT, (d) MSFT; stocks traded on NYSE – (e) GE, (f) AOL, (g) PFE, (h) GFI; three financial indices – (i) NASDAQ, (j) S&P500, (k) DJIA, and Gold (l)

$$\widetilde{R}_\tau(t) = \frac{V(t)}{< V(t) >_\tau} \cdot \frac{y(t) - y(t-\tau)}{\tau} = m(t)\frac{\Delta x}{\Delta t}, \qquad t = \tau + 1, \ldots, N \quad (3)$$

where the total volume of transactions over the interval τ is $< V >_\tau \tau = \sum_{i=1}^{\tau} V(i)$. In so doing, we introduce some analogy to a *generalized time dependent mass* $m(t)$ that contains some sort of history of the intercorrelations between price and volume of transactions of a stock during the τ interval. The total volume in the denominator is introduced for a normalization purpose. The transaction volume can also be represented as a rescaled volume of transactions $V_\tau(t)$

$$V_\tau(t) = \frac{V(t)}{< V >_\tau} \quad (4)$$

which is plotted in Fig. 2 (b) for GE.

We further consider a *moving average of the generalized momentum* which is called the generalized momentum indicator (GMI)

$$\widetilde{R}_{\tau}^{\Sigma}(t) = \sum_{i=t}^{t+\tau-1} \frac{V(t)}{< V >_{\tau}} \cdot \frac{y(i) - y(i-\tau)}{\tau} \qquad t = \tau + 1, \ldots, N \tag{5}$$

The difference CMI minus GMI for all data of interest is plotted in Fig. 3.

4 Investment strategy

A simple investment strategy can be suggested based on the trends of the market and using both the price per share and the volume of transactions incorporated in the generalized momentum indicator.

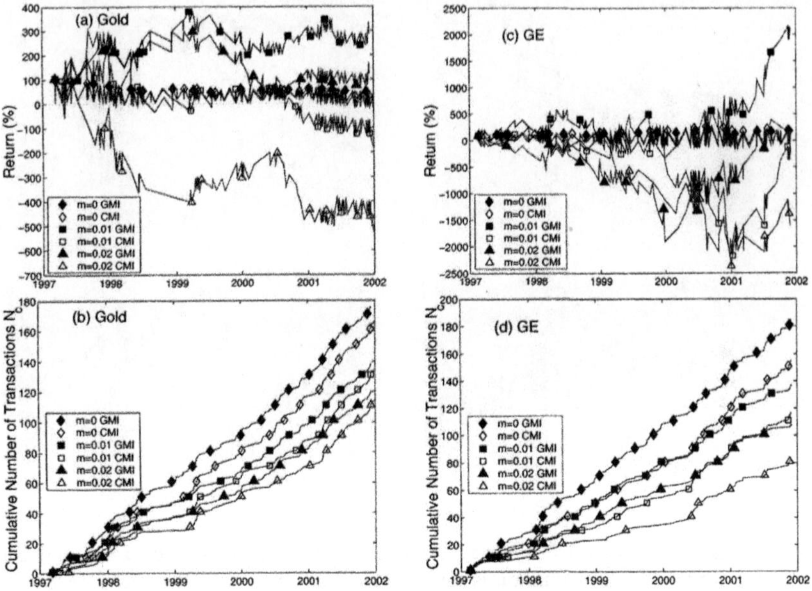

Fig. 4. Comparison of returns (Eq.(6)) in per cent between the generalized momentum indicator GMI-strategy and classical momentum indicator CMI-strategy for different values of m-criterion for (a) Gold and (c) GE. Comparison of the cumulative number of recommended transactions N_c performed when GMI-strategy and CMI-strategy are implemented for different values of m-criterion for (a) Gold and (c) GE

Following basic logics we *buy* at the minima of the GMI and *sell* at the maxima of GMI. We assume that we owe one share of a given stock at time

t_0, which is the first *sell* signal of a stock $y(t_0)$. The accumulated return that one can obtain during a certain period of time can be defined as

$$Z(t) = \sum_{i=1}^{N_c} I(i) y(i) / y(t_0) \tag{6}$$

where N_c is the total number of transactions suggested by the strategy during the period of interest; $I(i) = -1$ at the minima of $R_\tau^\Sigma(t)$ or $\widetilde{R}_\tau^\Sigma(t)$ is producing a *buy* signal to investors, and $I(i) = +1$ at the maxima of $R_\tau^\Sigma(t)$ or $\widetilde{R}_\tau^\Sigma(t)$ is communicating a *sell* signal. Results of investments are plotted in Fig. 4 (a,c) for Gold and GE stocks following GMI- (full diamond) and CMI-strategy (open diamond).

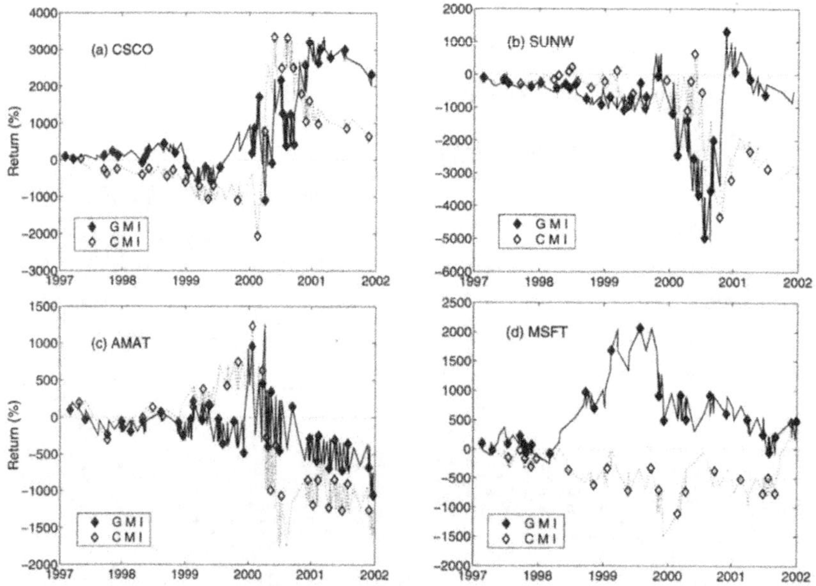

Fig. 5. Retuns obtained following (GMI)-strategy (full diamonds and solid line) and CMI-strategy (open diamonds and dotted line) for τ=21 days and restriction policy parameter m =0.01 for the period Jan. 1, 1997 to Dec. 31, 2001 for stocks traded on NASDAQ – (a) CSCO, (b) SUNW, (c) AMAT, (d) MSFT

In order to take advantage of the knowledge of the dynamics of the market represented by the volume of transactions and included in the GMI we introduce a criterion that will prevent from investment activity (and thus monitor the risk) when the change of the momentum indicator at the local extremum is smaller than a certain percentage of the momentum indicator value at the

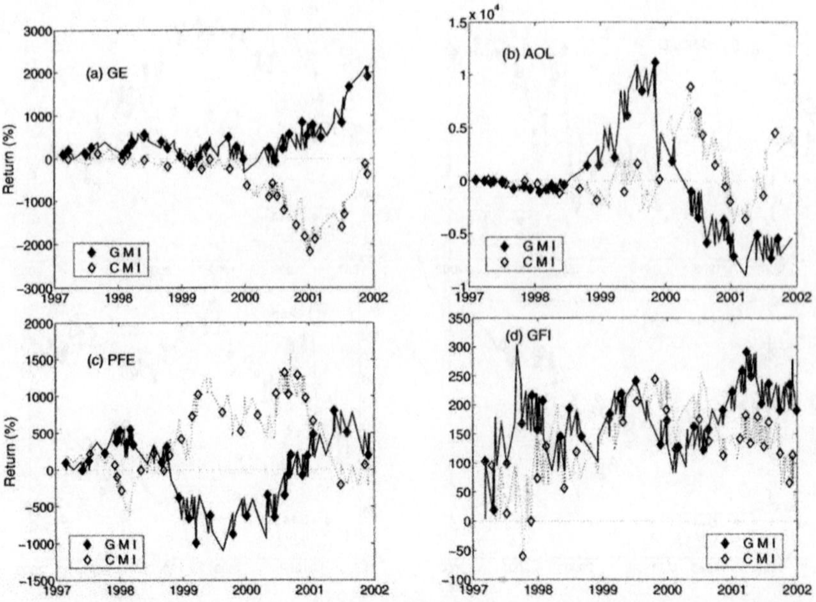

Fig. 6. Retuns obtained following (GMI)-strategy (full diamonds and solid line) and CMI-strategy (open diamonds and dotted line) for τ=21 days and restriction policy parameter m =0.01 for the period Jan. 1, 1997 to Dec. 31, 2001 for stocks traded on NYSE – (a) GE, (b) AOL, (c) PFE, (d) GFI

time just before the extremum. Thus we define a m-criterion for measuring the relative depth of the local extrema as

$$m = \left| \frac{R_\tau^\Sigma(t_{max/min}+1) - R_\tau^\Sigma(t_{max/min})}{R_\tau^\Sigma(t_{max/min})} \right| . \tag{7}$$

Different restrictions on the investment activity expressed in the value of the m-criterion lead to different returns as it is illustrated in Fig. 4 (a,c) for Gold and GE, respectively. This also results in a different number of transactions suggested by the strategies as shown in Fig. 4 (b,d) for for Gold and GE, respectively. It is interesting to notice that for approximately the same number of transactions, and different m-values the GMI-strategy leads to higher returns. Note that an investment strategy should not be excessively *defensive*, e.g. the returns for GE and Gold (marked with full/open triangles (GMI/CMI)) for a relatively high (defensive) m=0.02 are low as compared to the returns for m=0.01, marked as full/open squares (GMI/CMI). To illustrate the performance of the GMI-strategy with respect to the CMI-one for the stocks and indices of interest (data in Fig. 1) we chose the m=0.01 investment strategy and plot all the results in Figs. 5, 6 and 7.

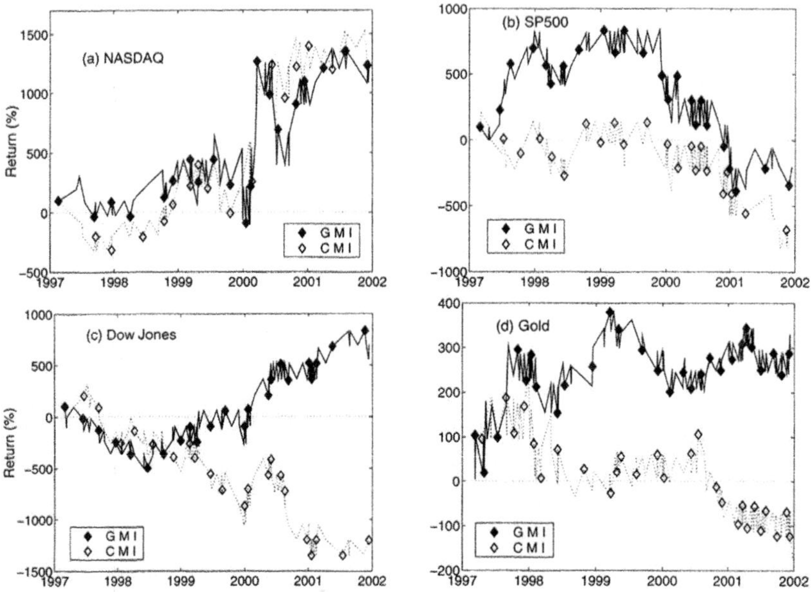

Fig. 7. Retuns obtained following (GMI)-strategy (full diamonds and solid line) and CMI-strategy (open diamonds and dotted line) for τ=21 days and restriction policy parameter m =0.01 for the period Jan. 1, 1997 to Dec. 31, 2001 for three major financial indices – (a) NASDAQ, (b) S&P500, (c) DJIA, the price of Gold (d)

5 Conclusions

We present an investment strategy based on a generalized momentum indicator (GMI) that takes into account complete information from the two faceted market - the share price and the volume of transactions. In addition, our GMI-strategy involves restriction policies defined by the value of a hereabove introduced m-criterion, describing a more or less risky investor. By doing so we obtain profits with the GMI that are higher than those obtained with the (classical momentum indicator) CMI-strategy with the same restriction policies applied.

Acknowledgements We are very grateful to the organizers of the Symposium for their invitation and to the Symposium sponsors for financial support.

References

1. see S.B. Achelis, in http://www.equis.com/free/taaz/
2. Vandewalle N, Ausloos M (1998) Crossing of two mobile averages: A method for measuring the roughness exponent Phys Rev E 58:6832–6834
3. Ausloos M, Ivanova K (2002) Mechanistic approach to generalized technical analysis of share prices and stock market indices Eur Phys J B 27:177–187

Enhancement of the prediction of actual market prices by modifying the regularity structure of a signal

Lyailya M. Karimova , Yerbol B. Kuandykov

Institute of Mathematics, 125, Pushkin str.,480100, Almaty, Kazakhstan.
E-mail: chaos@math.kz

Summary. The work presents approach to enhance and correct the financial time series predictions. The method affects solely singularity structure of a given time series. For this reason it is not sensitive to the concrete model employed for the prediction algorithm and may be incorporated into any of them. In addition to financial time series the method could be employed to enhance any noisy historical database.

Key words: Finance, Prediction, Embedology, Hölder regularity

It is well known that prediction of market prices is a very difficult problem (Hellström and Holmström 1998). Among other factors the difficulties are related to intermittency of different regimes in the market dynamics, e.g., turbulent phase, rising phase, declining phase, etc. (Sornette 2003). The time series of actual market prices are always chaotic, however, the problem is, that the manifestation of chaos for different market regimes is also different. On the other hand the financial time series are noisy, but sources of noise often are unknown, to say nothing about their qualitative characteristics.

For these reasons, a realization of a prediction with the help of topological embedding technique (Cao and Soofi 1999) seems to be difficult. In fact, the standard Takens (1981) algorithm presupposes the existence of the sole universal map of delay coordinates $x_{i+\tau d} = \Phi\left(x_i, x_{i+\tau}, ..., x_{i+\tau(d-1)}\right)$, where d is an embedding dimension and τ is a lag. If embedding parameters are chosen properly, at the most one may expect that Φ is continuous and differentiable. In this situation, supervised artificial neural network can be applied as universal approximator of Φ (Girosi and Poggio 1989).

Nevertheless, the results of forecasting may be far from the actual dynamics. The failure of the predictions can be explained by several reasons. First, it is low determinism of financial time series, so universal model based on embedology algorithm is not applicable (Small 2002). Second, the actual dynamics of financial time series may be described by mapping $x_{i+\tau d} = \Phi\left(x_i, x_{i+\tau}, ..., x_{i+\tau(d-1)} \mid p_1, ..., p_k\right)$, where vectors $p_1, ..., p_k$ are hidden system parameters governing transitions between different dynamical regimes (Bagarinao et al. 1999). The existence of the parameters may explain nonstationary of data. Unfortunately, the problem of actual extracting of the parameters from time series is rather difficult (Güttler et al. 2001). Third, usually the universal mapping Φ is numerically approximated in some class of simple functions or models, which possibly describe highly nonlinear actual dynamics too roughly. In this connection the constant bias arises, which can be used for correction of initial prediction (Judd and Small 2000).

In our contribution the method of prediction correction is based upon recently developed by Lévy Véhel (2002) algorithm of signal enhancement due to improvement of the Hölder regularity function of the initial noisy signal. Hölder regularity is a positive real number $\alpha(x)$, which describes a certain smoothness of the graph of a function at point x. Hölder exponents are known to trace successfully crucial changes of a market conditions (Struzik and Siebes 2002). It is important that Hölder exponent time series are more stationary than actual financial data and keep information about chaotic dynamics. Thus, we can obtain prediction $Y(t+\tau)$ of financial data $Y(t)$ and prediction of their Hölder exponents $\alpha_Y(t+\tau)$ based on embedding technique as it was made in papers of Cao and Soofi (1999) or Tribelsky (2001). Afterwards, the predicted regularity function is shifted upward by a certain constant $\alpha_{\tilde{X}} = \alpha_Y(t+\tau)+\delta, \delta > 0$, where $\alpha_{\tilde{X}}$ is improved Hölder regularity of the forecast in question. Appropriateness of this procedure is confirmed by the fact that for a given noisy signal its local Hölder exponent nearly everywhere is lower than it is in the case of the noise-clean signal. At the next step we face the problem of function reconstruction based on its prescribed regularity (Daoudi et al. 1998), which is accomplished by means of some functional minimization. Such a strategy allows to process data with essential nonregularity and even with singularities, simultaneously preserving information about the initial signal (Lévy Véhel 2002).

Now it is necessary to remind the definition of the Hölder exponent (Mallat and Hwang 1992). In general, let α be a positive real number and $x_0 \in R$ and a function $f(x): R \to R$. If there are a constant C and a polynomial $P(x)$ of order smaller than α so that, for all x in a neighborhood of x_0:

$$|f(x) - P(x - x_0)| \le C|x - x_0|^{\alpha}, \tag{1}$$

then, α is called the regularity index of f at x_0.

The Hölder exponent $\alpha_f(x_0)$ is the supremum of all α such that Eq. (1) holds. If f is n times continuously differentiable at point x_0, then one can use for the $P(x-x_0)$ the n-order Taylor expansion of f at x_0 and thus prove that $\alpha_f(x_0) > n$. Since α_f is defined at each point x, we may associate to $f(x)$ the function $\alpha_f(x)$, which measures the evolution of its regularity. The numerical estimation of the Hölder exponent is rather simple (Mallat and Hwang 1992). One chooses function $\psi(x)$, which is well localized, and which is orthogonal to all the polynomials $P(x)$ up to some order, so that integration of both sides of Eq. (1) with $\psi((x-x_0)/s)$ yields:

$$W(s,x_0)[f]=\frac{1}{s}\int\psi\left(\frac{x-x_0}{s}\right)f(x)dx\sim s^{\alpha_f(x_0)} \qquad (2)$$

at $s\to 0$. The function $W(s,b)[f]$ is called continuous *wavelet transformation* of f, where b and s are translation and scale parameters, respectively (Holschneider 1995).

Often discrete wavelet transformation is applied and b and s need to be discretized. For $s=2^j$ and $b=2^j k$ one can obtain orthonormal basis functions for certain choices of ψ: $\psi_{j,k}(t)=2^{-j/2}\psi(2^{-j}t-k)$, where j indicates the scale, and k indicates a shift along the time axis. Then one gets $X(t)=\sum_{j,k}x_{j,k}\psi_{j,k}(t)$, where $x_{j,k}$ are wavelet decomposition coefficients of function $X(t)$. In this case Eq. (2) transforms into the following

$$\left|x_{j,k}\right|\le C2^{-j(\alpha_x+1/2)}, \qquad (3)$$

where C is constant (Mallat and Hwang 1992). The last expression allows estimating $\alpha_X(t)$. Thus, we can find Hölder exponent at each point of the observational time series Y. Then, our main goal is to obtain signal $\tilde{X}$ with enhanced regularity structure. To do that properly the following conditions must be hold: (i) $\tilde{X}$ and Y must be close in L^2 metric space and (ii) the function of local regularity $\alpha_{\tilde{X}}$ must be prescribed. If $y_{j,k}$ and $\tilde{x}_{j,k}$ are wavelet coefficients of Y and $\tilde{X}$, respectively, then these two conditions can be represented (Lévy Véhel 2002) in formal way like:

$$\sum_{j,k}\left(y_{j,k}-\tilde{x}_{j,k}\right)^2\to\min \qquad (4)$$

subjected to the restriction that

$$\sum_{j=1,n}s_j\log\left|\tilde{x}_{j,(i-1)2^{j+1-n}}\right|=-M_n\left(\alpha_{\tilde{X}}(i)+1/2\right) \qquad (5)$$

Looking for a global minimum of (4)-(5) with standard optimization methods seems to be very difficult, in particular there is no possibility to obtain an explicit formula for coefficients $\{\tilde{x}_{j,k}\}$ estimation. Moreover, the problem is multidimensional and the function to be optimized, as well as bounding conditions are posed in nonlinear form. To solve this problem genetic algorithm is used (Wall 1996).

Results

We apply the above given technique for prediction enhancement of the real-world financial time series. In this case we consider errors of a neural network prediction of time series as a noise of complex structure. To illustrate practical efficiency of the approach we consider data set US dollar - Japanese yen interbank daily exchange rate[1].

[1] http://www.oanda.com/convert/fxhistory

Fig. 1 presents a fragment of the time series dynamics from May 4, 2002 till July 29, 2002 along with its pointwise regularity function evolution, which was estimated by the oscillation based methods by software FracLab[2] with wavelet basis functions scaled at $j=1,2,\ldots,5$.

It is easy to observe that rapid changes of Hölder regularity are in advance to the exchange rate crashes. But unfortunately it cannot be exploited as a precursor for prediction because of edge effect influence when Hölder pointwise regularity is estimated. Nevertheless, it is obvious that the Hölder regularity function adequately captures the overall dynamics observed in such time series (Struzik and Siebes 2002) and it has useful properties to be modeled.

For prediction enhancement according to the signal enhancement technique, we developed the following algorithm. First, for a given initial noisy financial time series $Y(t)$ sampled at equidistant time intervals $t = t_1, t_2, \ldots, t_n$, we obtain the Hölder regularity function, i.e., for every t we estimate pointwise Hölder exponent $\alpha_Y(t)$. With the help of nonlinear analysis standard tools (Hegger and Kantz 1998) we choose optimal embedding parameters for $Y(t)$ and $\alpha_Y(t)$.

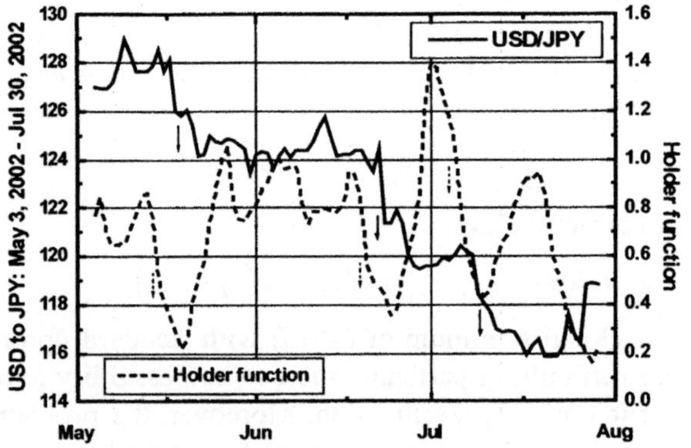

Fig. 1. USD/JPY daily interbank exchange rate and its pointwise estimated Hölder function, May-August, 2002. With solid arrows sharp downfalls of USD are marked and with dashed ones are of Holder function respective variations.

Then, by means of neural network (Tribelsky et al. 2001), we obtain two predictions $Y(t+\tau)$ and $\alpha_Y(t+\tau), \tau > n$. Next, the predicted $\alpha_Y(t+\tau)$ and initial regularity functions $\alpha_Y(t)$ are concatenated and shifted upward by a positive constant δ. The concatenation of those functions is applied for increasing time interval of recovery. Finally, having the improved regularity function $\alpha_{\tilde{X}}$ and predicted $Y(t+\tau)$, we recover the enhanced prediction $\tilde{X}(t+\tau)$ as provided by the signal enhancement algorithm.

[2] http://www-rocq.inria.fr/fractales//

In the end, we illustrate the results of prediction enhancement. Both enhanced and non-enhanced predictions were made simultaneously with the same model. In Fig. 2 the sharp strengthen of JPY against USD over two time intervals are demonstrated. It is seen that the enhanced prediction is more sensitive to dynamical regimes changing. In general, accuracy of the enhanced prediction is also much higher than that for the non-enhanced one. This result was verified by a number of the tests for the different time series fragments with various dynamical behaviors.

The signal enhancement procedure was implemented with the help of the following software. For estimation of Hölder regularity function α_Y we used FracLab Toolbox and Wavelet Toolbox of Matlab. The genetic algorithm was realized using genetic algorithm library Galib (Wall 1996). The parameters of genetics algorithm were as follows. The probability of crossover of individuals was equal to 0.7-0.9, the probability of a genome mutation - 0.0012, the replacement percentage of individuals between generations was amounted to 60%. We used the steady-state scheme of genetic algorithm. As there is no rigorous methods to estimate the shift parameter δ, we empirically found out that for USD – JPY exchange rate time series it equals to 1-1.5. Thus, the developed approach does work and may be employed to improve accuracy of financial predictions.

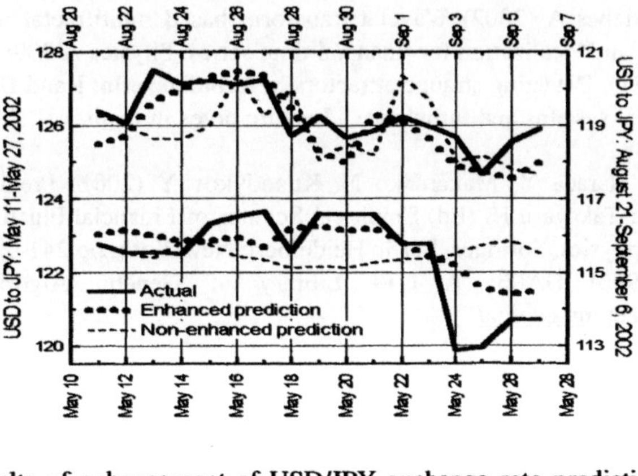

Fig. 2. The results of enhancement of USD/JPY exchange rate prediction with signal enhancement technique. Lower lines correspond to the time interval May 11, 2002 - May 27, 2002, upper lines correspond to the time interval August 21 - September 6, 2002

References

Bagarinao EJ, Pakdamann K, Nomura T, Sato S (1999) Time series-based bifurcation diagram reconstruction. Physica D 130:211-231

Cao L, Soofi AS (1999) Nonlinear deterministic forecasting of daily dollar exchange rates. Intern. J. of Forecasting. 15: 421–430

Daoudi K, Lévy Véhel J, Meyer Y (1998) Construction of continuous functins with prescribed local regularity. Constructive Approximation 014(03) :349-385

Girosi F, Poggio T (1989) Networks and the best approximation property. MIT AI Lab. Thechnical Peport. Memo No 1164. Paper No45. Available http://www.ai.mit.edu/people/poggio

Güttler S, Kantz H, Olbrich E (2001) Reconstruction of the parameter spaces of dynamical systems. Phys Rev E 63:6215-6224

FracLab. Available http://www-rocq.inria.fr/fractales//

Hegger R, Kantz H (1998) Practical implementation of nonlinear time series methods: The TISEAN pakage. arXiv:chao-din/9810005 http://xxx.lanl.gov

Hellström T, Holmström K (1998) Predicting the stock market. Technical report series Ima-TOM-1997-07, pp 1-37

Holschneider M (1995) Wavelets. An analysis tool. Clarendon Press, Oxford

Judd K, Small M (2000) Towards long-term prediction. Physica D 136:31-44

Lévy Véhel J (2002) Signal enhancement based on Hölder regularity analysis. In IMA Volumes in Mathematics and its Applications 132:197-209

Mallat S, Hwang W (1992) Singularity detection and processing with wavelets. IEEE Transaction on Inform Theory 32:617-643

Small M, Tse C (2002) Deterministic dynamics in financial indicators? http://www.eie.polyu.edu.hk/~ensmall/pubs.html

Sornette D (2003) Why stock markets crash.Critical events in complex financial systems.PrincetonUniversity Press.Prenceton and Oxford, pp 3-47

Struzik ZR, Siebes A (2002) Wavelet transform based multifractal formalism in outlier detection and localization for financial time series. Physica A 309: 388-402

Takens F (1981). Detecting strange attractors in turbulence. In: Rand DA, Young LS (Eds) Dynamical systems and turbulence - Lecture notes in math. vol. 898, Springer, Berlin, pp 365–381

Tribelsky M, Harada Y, Makarenko N, Kuandykov Y (2001) Predictability of market prices. In: Takayasu H. (Ed) Empirical Science of Financial Fluctuations. The Advent of Econophysics. Springer, Berlin Heidelberg New York, pp 241–249

Wall M (1996) GAlib: A C++ Library of Genetic Algorithm Components. http://lancet.mit.edu/ga/

New complex approach to market price predictions

Michael I. Tribelsky

Global Business Creation Inc., Hatchery-Shibuya 104,14-1 Sakuragaoka-cho, Shibuya-ku, Tokyo 150-0031, Japan. E-mail: tribel@scroll.apphy.fukui-u.ac.jp

Summary. A new complex approach is applied to predict actual market prices. The main feature of the approach is that it combines methods of nonlinear time series analysis with those of the theory of dynamical chaos and artificial intelligent systems (neural networks, genetic algorithms etc.). As a result, the approach yields very precise predictions.

Key words. Prediction, Market, Chaos, Attractor.

General methodology

The problem of predicting actual market prices is attracting a great deal of attention by researchers all around the world. The importance of the problem is related to obvious benefits, which one could obtain based upon successful predictions. However, the problem is as much difficult as appealing, and for this reason it is still far from completion.

The main difficulties of the issue are as follows:

- Time series of actual price dynamics exhibit highly non-stationary types of chaos.
- As a rule they cannot be described in terms of a single characteristic time, or a discrete set of characteristic times.
- They are noisy. The nature, spectral, and statistical characteristics of the noise are usually unknown. Thus, the noise reduction problem is extremely difficult.
- In addition to purely economic (the so-called *evolutionary*) factors, prices may be substantially affected by *revolutionary* (political crises and natural disasters) and *subjective* factors (rumor, gossips, intentional and/or unintentional disinformation, etc.). While the evolutionary factors, at least in principle, may admit certain algorithmic modeling, both the revolutionary and subjective hardly could be included into any quantitative algorithmic approach.

The above list could be easily extended. On the other hand, the prediction challenge should be solvable. Indirect evidence of its solvability lies in the fact that a number of successful traders made their fortunes in just a few years of trade. A great number of consecutive profitable transactions made by these traders could not be just a result of chance. It means the traders must have a certain statistically correct vision of future price dynamics – i.e., they successfully *predict* the actual price dynamics. Lastly, almost any kind of human intellectual activity, including the successful trades, may be better or worse simulated by a computer.

It has motivated us to try to develop a complex approach, which simulates, to a certain extent, the brain of a successful trader during his decision-making process. Our approach is complex in both narrow and broad meanings of the word. Narrowly, it is complex because in algorithmic treatment of financial data series methods of three different disciplines, namely the dynamical chaos theory, nonlinear time series analysis and the theory of artificial intelligent systems are employed. However, it is complex also in a broad sense, i.e., from the point of view of its practical application. We believe that due to the great difficulty of the problem we cannot and should not restrict the market analysis by utilizing a single method. Therefore we cross-validate results of our predictions, comparing them with those obtained by the traditional methods of fundamental and technical analysis.

Regarding the algorithmic analysis, we deliberately avoid employment of any governing equations, remaining within the framework of pure data processing. In our view, this makes the analysis more robust and free from unjustified assumptions. The main scheme of the developed approach is as follows, the database (in typical cases, daily prices for about consecutive 10 – 15 years) is pre-processed according to the Takens algorithm (1981) [see also, e.g., Cao et al. (1997,1999), Hegger and Kantz (1998), Kennel et al. (1992)]. It allows us to reconstruct in a proper embedding space the so-called topological structure of an attractor, which corresponds to the observed price dynamics in the past. The pre-processed database is inputted into a neural network of a special architecture (a "brain" of our artificial trader). The output of the neural network is the desired prediction. The prediction is made based upon purely geometrical arguments – the neural network finds in the past dynamics those fragments of a phase trajectory, which are closest to the most recent part of it. Then, the network extends the most recent fragment of the trajectory to the future, so that it remains close to the selected fragments of the phase trajectory in past. Naturally, this "skeleton" of the approach is supplemented with a number of "tricks", invented to make predictions more stable and reliable [for more details see our previous publication Tribelsky et al (2001)].

The efficiency of the approach is enhanced with the *trial and error* method. To this end, the entire database is divided into two unequal parts. A large part is used to train the neural network, while a small part, consisting of several of the most recent Takens vectors (*test set*), remains unknown to the computer and should be reconstructed by the trained network. Then we can compare the reconstruction with the test set (we call this comparison *pre-test*). This gives us clues about accuracy of the prediction. Next, we repeat this step several times, each time tuning parameters of the reconstruction (the embedding dimension and lag) and those of the neural network. As a result, we obtain about 10 versions of the predictions that have passed the pre-test with a specified accuracy.

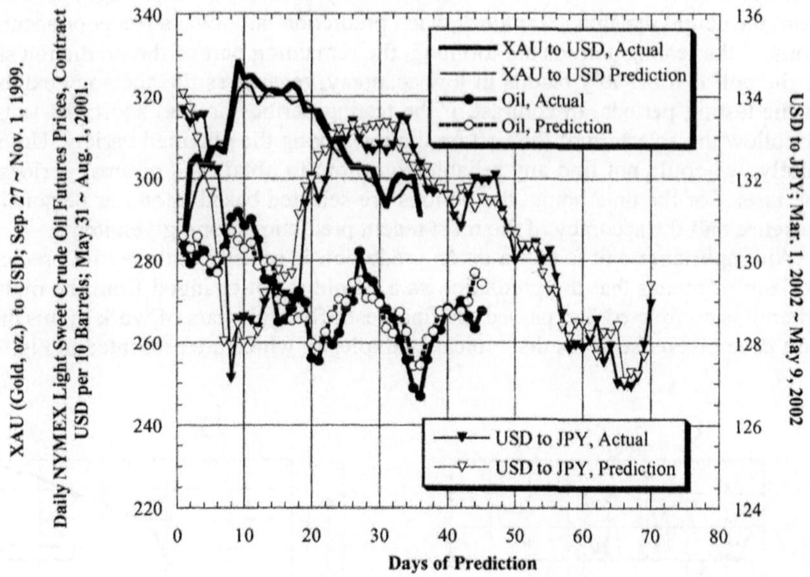

Fig. 1. Typical examples of our predictions[1]. When the predictions are made, the actual dynamics for the predicted periods are not known. Then, day-by-day, the predicted behaviors are compared with the actual performances, when the latter become available in the course of time. Results of the comparison are presented in this figure. The oil price is shown in 10 barrel units to fit the scale of this plot with that for XAU.

In order to select a single actual prediction from these pre-selected versions, the pre-selected predictions undergo record-by-record comparison with the actual price dynamics, when the latter becomes available in the course of time *(final test).* Rejection rates for both the tests are very high – to obtain a single final prediction about 50 – 80 "raw" predictions are required to be inputted into the pre-test. In addition, a certain part of the final prediction is eaten out by the tests – this is the price paid to increase the reliability of the remaining part of the prediction. Both the optimal duration of the pre-test and final test as well as that of the remaining part of the final prediction *(prediction horizon)* depend on a concrete type of price dynamics and may vary significantly not only for different items, but even for the same item in different historical periods of its dynamics.

[1] Daily prices corresponding to data of http://www.eia.doe.gov/emeu/international/Crude2.xls for oil and http://www.oanda.com/convert/fxhistory for both XAU to USD and USD to JPY exchange rates are presented.

It should be stressed that increase of the testing periods does not mean automatic increase of the accuracy of the remaining part of a prediction. The point is that divergence of initially close trajectories is a generic feature of any type of dynamical chaos. Therefore, sooner or later any prediction must begin to deviate from the actual dynamics. It means that a prediction tail always has poor accuracy. Thus, if the testing periods are too long, the remaining part of the prediction shifts to the tail. It inevitably results in low accuracy, regardless the accuracy exhibited in the testing periods. In contrast, if the testing periods are too short, the tests do not allow the selection of the best prediction among the obtained variety. Unfortunately, we could not find any reliable algorithm to obtain the optimal periods for the tests. For the time being, the periods are selected based upon our personal experience and the accuracy of the most recent predictions for a given item.

An important point is that a given prediction is issued for the entire prediction horizon. It means that the prediction as a whole is not changed from the moment when it is completed (i.e. passed the final test). Several years of work in this direction have given rise to results, typical examples of which are presented in Fig. 1.

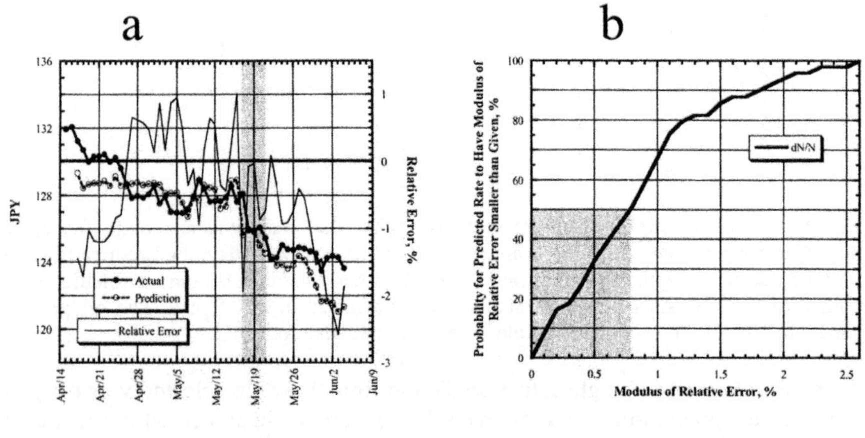

Fig. 2a,b. a Comparison of 49-day-horizon prediction of the exchange rate USD to JPY with the actual rate dynamics and the corresponding relative error; gray box marks a test trade period. **b** Distribution function for the modulus of relative error for the prediction plotted in Fig. 2a; gray box marks the boundary of 50% probability. See the text for details. Historical rates are taken from the same OANDA site as those in Fig. 1.

Another example is shown in Fig. 2a. In the same figure, the relative error of the prediction is plotted. It is seen that for the 49-day-horizon prediction the error margins are -2.6% and 1%, respectively. Statistical treatment of the relative error for this prediction is presented in Fig. 2b. The probability plotted on the vertical axis is calculated as *dN/N*. Here *dN* is the number of predicted records, whose modulus of relative error is smaller than a given value, where the given value is plotted on the horizontal axis. *N* stands for the total number of predicted records (in this case $N = 49$). The upper edge of the gray box indicates $dN/N = 50\%$, while the right edge of it shows the corresponding value of the relative error. It is seen that 50% of the predicted records deviate from the actual values with the modulus of the relative error smaller than 0.8%, and 100% of the predicted records have the modulus of the relative error smaller than 2.6%.

Trades based upon our predictions

It should be stressed, however, that the relative error is not quite an adequate measure of the accuracy for price predictions. Much more important are the so-called directional properties, i.e., those, which indicate the *direction of trends*. For example, for the prediction shown in Fig. 2a in the period May 15 – 17 the modulus of the relative error has a characteristic value greater than that in the period May 23 – 27; however, description of the actual rate dynamics for the former period is *better* than that for the latter. The point is that the rate changes from May 15 to May 17 are rather sharp, while those during the second discussed period are small. Clearly describing the sharp rate changes, the prediction has a little mismatch in time (about one day in average) with the actual dynamics. Due to the sharpness of the changes in this period the mismatch gives rise to a large relative error.

For this reason, the only effective criterion for the accuracy of price predictions is the possibility to employ them for profitable trade. To check this possibility, three test trade sessions have been conducted with the OANDA FXTrade Platform[2]. Results of the trades are summarized in Table 1.

Table 1. Summary of trade sessions*

Sesssion period	Total number of transactions	Number of losing transactions	Gross realized profit (% of initial deposit)	Gross realized loss (% of initial deposit)	Total net profit (% of initial deposit)
May 17 – 21, 2002	90	2	63.66	0.68	62.98
Aug. 7 – Sep. 13, 2002	497	41	204.06	53.39	150.67
Jan. 7 – 25, 2003	126	0	48.56	0	48.56

*The full transection history may be downloaded from the OANDA site and examined by an idependent inspector.

The total net profit gained in 713 transactions conducted during 44 days (the FX market is closed on weekends) constitutes 262.21% of the initial deposit. It should

[2] http://fxtrade.oanda.com

be emphasized that during the last session (Jan. 7 – 25, 2003) the rate of USD was substantially affected by news related to the US – Iraq political crisis. This made the situation extremely unfavorable for analysis with our methods. Nevertheless, the session did not have a single losing transaction. The presented results provide convincing evidence that the developed approach does indeed work and may be successfully employed for actual market trade.

The author has benefited greatly from discussion of this work with J. D. Farmer, Y. Harada, S. Kai, R. Olsen, P. K. Rawlings, D. Sornette and H. Takayasu.

References

Cao L, Mees A, Judd K (1997) Modelling and predicting nonstacionary time series. Int. J. of Bifurcation and Chaos. 7: 1823–1831

Cao L, Soofi AS (1999) Nonlinear deterministic forecasting of daily dollar exchange rates. Intern. J. of Forecasting. 15: 421–430

Hegger R, Kantz H (1998) Practical implementation of nonlinear time series methods: The TISEAN pakage. arXiv:chao-din/9810005 http://xxx.lanl.gov

Kennel MB, Brown R, Abarbanel HDI (1992) Determining embedding dimension for phase-space reconstruction using a geometrical construction. Phys. Rev. A 45:3403–3411

Takens F (1981). Detecting strange attractors in turbulence. In: Rand DA, Young LS (Eds) Dynamical systems and turbulence - Lecture notes in math. vol. 898, Springer, Berlin, pp 365–381.

Tribelsky M, Harada Y, Makarenko N, Kuandykov Y (2001) Predictability of market prices. In: Takayasu H. (Ed) Empirical Science of Financial Fluctuations. The Advent of Econophysics. Springer, Berlin Heidelberg New York, pp 241–249

Inferring of Trade Direction by Imbalance from Intra-Day Data

G.Okita [1], A.Oba [1], H.Takayasu [2]

[1] Nomura Seculities Co. Ltd., Tokyo,2-2-2 Otemachi, Chiyoda-ku, Japan
[2] Sony CSL Inc., 3-14-13 Higashigotanda, Shinagawa-ku, Tokyo, Japan

Intra-day data is very useful in terms of getting information about demand and supply in the market. In this study, I analyzed the effect on trade direction in the future by imbalance of order book information. As a result, it is indicated that imbalance explains the trade direction to the some extent.

1. Model

Price usually goes down, when supply excesses demand, and price goes up, when demand excesses supply. We tried to predict the above price movement by intra-day data. We hereby assume order book information as supply and demand in the market, and analyze how exactly the imbalance of order book information explains the price movement after imbalance. Now, considering transaction price as P(t), depth on Ask and Bid as $V_a(t)$, $V_b(t)$, imbalance as $(V_b(t)-V_a(t))$ / $(V_b(t)+V_a(t))$ and average profit after imbalance as $P(t+\Delta t)-P(t)/P(t)$, we check the relation between imbalance and average profit .(Fig.1)

2. Result

2. 1. Imbalance and average profit

Fig.2 shows the relation between imbalance and average profit until 3 ticks after imbalance. When imbalance becomes more positive, the price goes upward linearly, though its absolute value of profit is small. On the other hand, when the imbalance becomes more negative, the price goes downward. From the above result, it is considered that imbalance of order book information is very useful for predicting the direction of price.

2. 2. Imbalance and bid-hit probability

Fig.3 shows the relation between imbalance and excess bid-hit probability. Excess bid-hit probability is calculated by bid-hit probability after imbalance minus total bid-hit probability. Bid-hit probability is measured by the number of the current price that is on one-tick-before -bid after imbalance divided by the number of the current price that is on the one-tick-before -bid in total transaction. From Fig. 3, imbalance is very useful indicator for trade direction because the more positive the imbalance becomes, the lower the bid-hit probability becomes. In addition, the shape of the graph is also interesting because it means that imbalance side(ask or bid) makes a different effect on bid hit excess probability. If we can confirm the above character on not only Nissan, but also other companies, the empirical model we advocate is considered to be useful for predicting the trade direction.

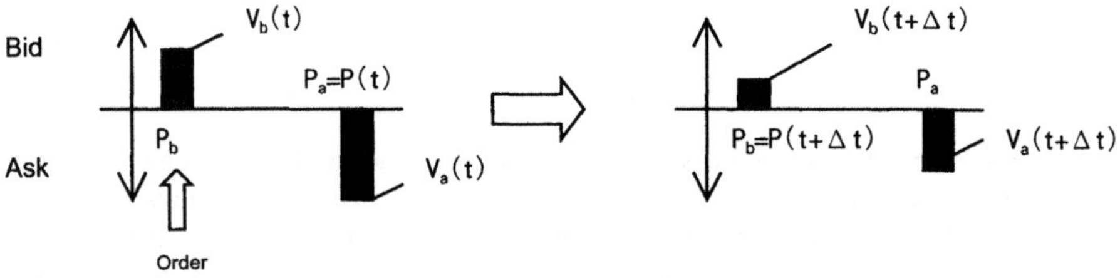

Fig. 1. Model explanation; the above shows price tend to be near bid when $V_b(t) < V_a(t)$.

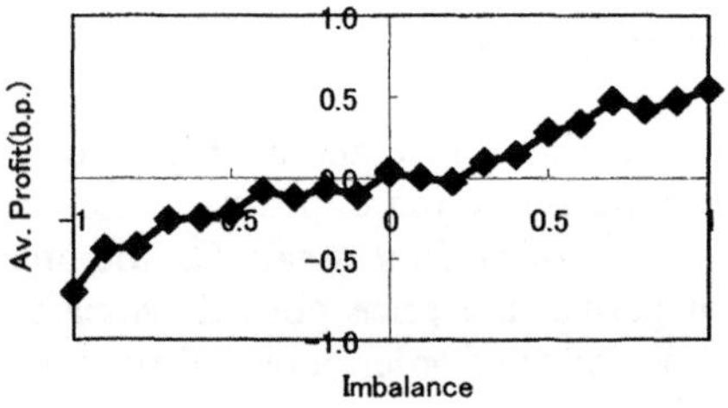

Fig. 2. A relation between imbalance and profit (Nissan). Data is from March 1, 2002 to June 28,2002

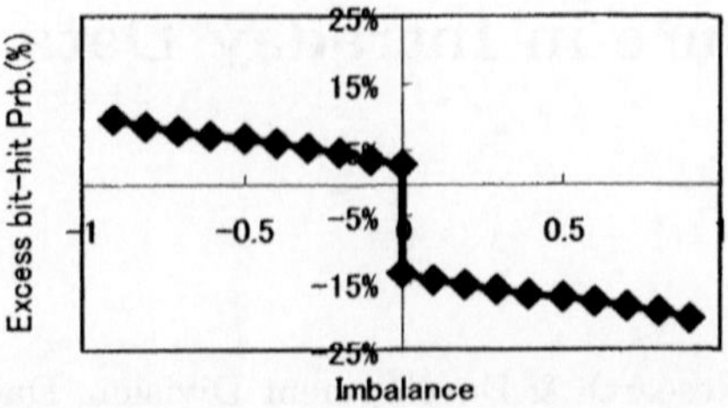

Fig. 3. A relation between imbalance and bit –hit Prb.(%)Data is from March 1, 2002 to June 28,2002

References

Min G. Lee, A.Oba, H.Takayasu ,"Parameter Estimation of a Generalized Langevin Equation of Market Price" , Empirical Science of Financial Fluctuations: The Advent of Econophysics, Springer-Verlag Tokyo 2002

H.Takayasu, M.Takayasu, "Critical fluctuations of demand and supply", Physica A 269(1999) 24-29

H.Takayasu, M.Takayasu, "Econophysics-Toward Scientific Reconstruction of Economy", Nihon Keizai Shinbunsya 2001

Chaotic Structure in Intraday Data of JGB Futures Price

Kohei HAYASHI

Information Technology Research & Development Division, Daiwa Institute of Research Ltd., 15–6 Fuyuki, Koto-ku, Tokyo, 135–8460, Japan.

Summary. The purpose of this study is to examine the deterministic and chaotic structure in intraday data of Japanese government bond (JGB) futures price.

In the first place, the characteristics of the price scales and of the time scales of the data are considered. In JGB futures market, the contracted prices are discrete with the tick size, and there is no transaction while the market is closed (e.g. in the early morning, at lunch time, and at night). Given these factors, two price scales, the price and the deviation from the moving average, and two time scales, the number of transactions and the trading time, are adopted in this paper. And then, the chaos characteristics such as the correlation dimension and the maximal Lyapnov exponent are estimated in the various cases. Finally, the significance of the estimated chaos characteristics are tested statistically with the method of surrogate data.

The result implies the possible existence of chaos when the number of transactions is used as the time scale. This possibility is attributed to the characteristic features of the JGB futures market.

Key words. chaos, correlation dimension, Lyapnov exponent, intraday data

Introduction

In the orthodox financial theory, it is believed that the time series of asset prices are too complicated to predict, and that its price dynamics should be described as stochastic process. This stochastic approach is based on the efficient market hypothesis (EMH), which claims that the market is so efficient that the information is reflected in the price instantaneously. If the market is efficient enough, the asset price behaves stochastically because the events that influence the price occur randomly.

On the other hand, the chaos theory has shown that the deterministic nonlinear system may create complex time series data which looks to be generated by a stochastic process. This fact implies the possibility that the complex time series may be described deterministically and predictable in the short period of time.

Therefore, chaos analysis has been applied to economic phenomena to evaluate whether the process is deterministic or stochastic. In previous works of chaos analysis on financial time series data, daily data were normally used. However, since chaos analysis requires thousands of sample data, several years of daily data are needed to obtain a wholesome result. In that case, it is doubtful that the deterministic structure of the market remains unchanged throughout the period.

In this study, intraday data of Japanese government bond (JGB) futures market, where most participants are institutional investors, are used. Their homogeneity in expertise may characterize the market to be more deterministic than the equity market and others, where the participants are more diverse, including individuals.

Framework of chaos analysis

In this section, we briefly summarize the framework of chaos analysis; to quantify characteristics of the underlying attractor of the phenomenon and to detect signs of a deterministic system.

First, it is necessary to reconstruct the attractor by embedding:

$$\boldsymbol{v}(t) = [x(t), x(t+\tau), \ldots, x(t+(m-1)\tau)] \tag{1}$$

where $x(t)$ is time series data and m and τ denote embedding dimension and time delay, respectively. In this paper, τ is determined with the higher order correlation (Ikeguchi et al. 1997). To be more precise, τ was estimated to be the limit value of the time when the correlation has the first extremum as the order of the correlation increases.

And then, the chaos characteristics, the correlation dimension and the maximal Lyapnov exponent, are estimated.

The correlation dimension D_C is one measure of the fractal dimension (Grassberger and Procaccia 1983). To estimate D_C, the correlation integral $C^m(r)$ is computed:

$$C^m(r) = \lim_{N\to\infty} \frac{1}{N^2} \sum_{\substack{i,j=1 \\ i\neq j}}^{N} I(r - |\boldsymbol{v}(i) - \boldsymbol{v}(j)|) \tag{2}$$

where N is the number of the points of the reconstructed attractor and $I(t)$ is the Heaviside function; $I(t) = 1\ (t \geq 0),\ 0\ (t < 0)$. If $C^m(r)$ behaves as powers of r:

$$C^m(r) \propto r^{d(m)} \Leftrightarrow \log C^m(r) \propto d(m) \log r \tag{3}$$

i.e., if the plot of $\log C^m(r)$ against $\log r$ is liner with a slope of $d(m)$, D_C is the saturated value of $d(m)$ as the embedding dimension m increases.

Meanwhile, the Lyapnov exponents measure the average rate of each axial expansion of the attractor. In this paper, the maximal Lyapnov exponent $\lambda_{\max}$:

$$\lambda_{\max} = \lim_{t\to\infty} \lim_{\epsilon\to 0} \frac{1}{t} \log \left(\frac{|\boldsymbol{v}(t) - \boldsymbol{v}_\epsilon(t)|}{\epsilon} \right) \tag{4}$$

$$|\boldsymbol{v}(0) - \boldsymbol{v}_\epsilon(0)| = \epsilon$$

is estimated with Kantz's algorithm (Kantz 1994).

If D_C is estimated less than m and/or λ is estimated positive, it can be assumed that the phenomenon has deterministic and chaotic structure. But, the possibility that a stochastic process has shown such features accidentally cannot be rejected.

In this connection, the estimated chaos characteristics are tested statistically with the method of surrogate data (Theiler et al. 1992). At first, this method specifies a null hypothesis that the original time series data is created by some linear stochastic process. Then surrogate data sets that are consistent with the null hypothesis are generated, and the chaos characteristics of the original data and of the surrogate data are computed. If the computed value of the original data is significantly different from those of the surrogate data, the null hypothesis is rejected and nonlinearity is detected. In this paper, the surrogate data were created with the iterative AAFT algorithm (Schreiber and Schmitz 1996). Using this algorithm, the surrogate data has the same histogram and auto-correlation as the original data has.

Data and result

The raw sample data is tick data series of JGB futures prices between Sep.25 and Sep.28, 2001 (Fig. 1). Following are the features of the data:

1. The data are sampled with transactions, not at regular timewise intervals.
2. There is no transaction in the early morning, at lunch time and at night.
3. There are outliers in the evening session (3:30p.m. – 6:00p.m.).
4. The prices are discrete with a unit of 0.01 yen (tick size), and the changes in price are ± 0.01 yen or 0 at almost all ticks.

Due to the above features, it becomes necessary to preprocess the data. At first, because of the third feature, the data in the evening session were eliminated. It might be considered that those outliers arised from the different structure of the price formulation specific to the evening session in which only a few investors participate.

And then, in consideration of the first and the second, two time scales were adopted, i.e. the number of transactions (the contracted price at every transaction) and the trading time (the closing price at every one minute). It should be noted that there would be the time instant with no sample price in

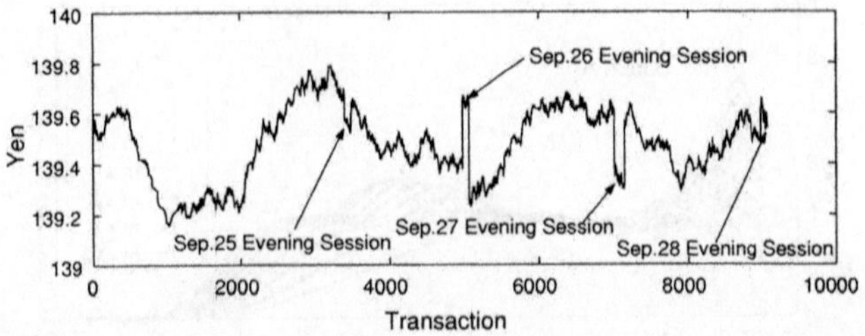

Fig. 1. Tick data of JGB futures prices (Sep.25 – Sep.28).

the latter case, and that the value was interpolated linearly in that event. In addition, the deviation from the moving average was also analyzed in view of the fourth nature, resulting in the four cases shown in Table 1 were adopted as data series.

The result of chaos analysis is displayed in Table 2. D_C converged in Cases A and B (Fig. 2), but the null hypothesis that the process is stochastic was not rejected with the method of surrogate data. In Cases C and D, D_C did not converge, but the null hypothesis was rejected from the aspect of the correlation integral (Fig. 3).

On the other hand, λ_{max}s were estimated to be positive in all cases at any embedding dimension, and the null hypothesis was not rejected only in Case B.

Table 1. Data cases

Case	Time scale	Price scale
A	No. of transaction	Price (the contracted price at every transaction)
B	Trading time	Price (the closing price at every 1 minute)
C	No. of transaction	Deviation from the moving average (nearest 50 steps)
D	No. of transaction	Deviation from the moving average (nearest 150 steps)

Table 2. Result

Case	No. of data	τ	D_C	: N.H.	λ_{max}	: N.H.
A	8532	250	1.9	: Not rejected	Positive	: Rejected
B	1088	20	2.3	: Not rejected	Positive	: Not rejected
C	8532	10	–	: Rejected	Positive	: Rejected
D	8532	10	–	: Rejected	Positive	: Rejected

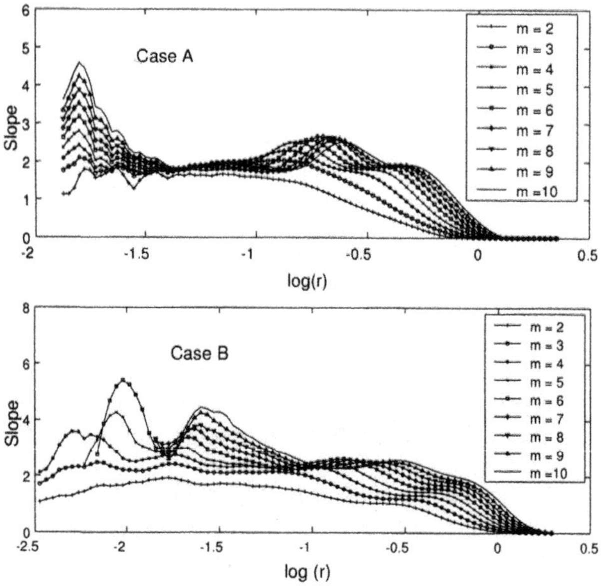

Fig. 2. The slope of the $\log r - \log C^m(r)$ plot in Cases A and B.

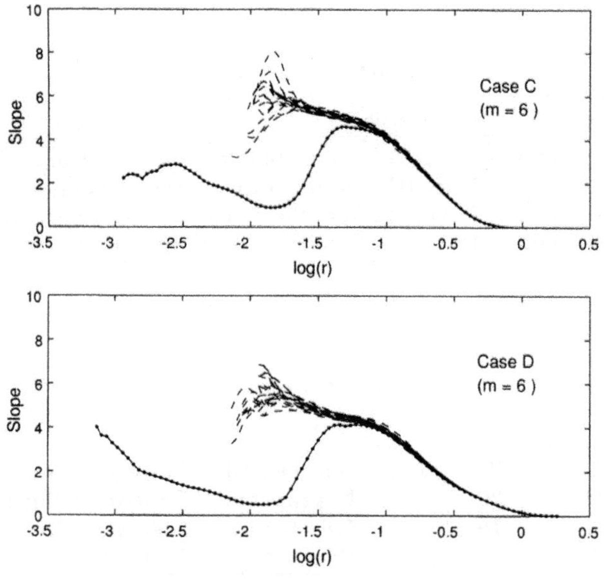

Fig. 3. Comparison of the slope of the $\log r - \log C^m(r)$ $(m = 6)$ plot between the original (solid, circle) and the surrogate data (dashed) in Cases C and D.

Conclusion

In this study, chaos analysis was applied to the intraday data of JGB futures prices in four cases.

In Case B, where the time scale is the trading time, it cannot be asserted that the time series has deterministic structure because the possibility of the stochastic process was not rejected. The result may reflect the loss of the information by resampling tick data into trading-time data. However, regarding Cases A, C, and D, tick data may have chaotic nature.

This result implies that information is not reflected in price immediately but deterministically. In other words, the participants may respond to information deterministically and homogeneously because of their expertise in JGB futures market as mentioned in the introduction.

For future research, the important approach is to identify more precisely the nature of the deterministicity from the perspective of the market microstructure.

References

Grassberger P, Procaccia I (1983) Characterization of Strange Attractors. Physical Review Letters. 50:346-349

Ikeguchi T, Takakuwa M, Aihara K (1997) Reconstructing Attractors with Higher Order Correlation. Proceedings of International Symposium on Nonlinear Theory and Its Applications. 2:1149-1152

Kantz H (1994) A robust method to estimate the maximal Lyapnov exponent of time series. Physics Letters A. 185:77-87

Schreiber T, Schmitz A (1996) Improved Surrogate Data for Nonlinearity Tests. Physical Review Letters. 77:635-638

Theiler J, Eubank S, Longtin A, Galdrikian B, Farmer JD (1992) Testing for nonlinearity in time series: the method of surrogate data. Physica D. 58:77-94

Trend identification and financial trading strategy by using stochastic trend model with Markov switching slope change and ARCH

Morikazu Hakamata[1]

Tokyo-Mitsubishi Asset Management Ltd. morikazuhakamata@hotmail.com

Summary. This paper focuses on two characteristics of the stock price indices — trend slope changes and heteroscedasticity. To capture these characteristics, we propose a stochastic trend model with Markov switching slope changes and ARCH (MS-SC/ARCH), and evaluate the usefulness of this model for a trading strategy. In the empirical analysis using TOPIX, we succeeded in estimating the effective trend slope for trading by using a trading strategy based on the MS-SC/ARCH model.

Key words. ARCH, heteroscedasticity, Markov switching, TOPIX, trading strategy, trend slope.

1 Introduction

In practical financial markets, some practitioners, including dealers and money managers of foreign exchange rates, stocks, commodities, etc., regard "traditional technical analysis" as an important analytical tool. One of the most important principles in technical analysis is that *"asset prices move in trends."* "Trend" in the framework of the traditional technical analysis is obtained by connecting the local bottoms or tops of the asset's price with a straight line. The "Slope" of the trend plays an important role in the practical trading strategy. There are some methods of identifying this unobserved trend in the traditional technical analysis (Neely 1997). However, every method needs not only scientific skill, but also artistic sensitivity based on long experience in financial markets.

The first purpose of this paper is to identify the unobserved trend and its slope of real financial time series by using a nonlinear stochastic approach — the MS-SC/ARCH. The second purpose is to examine whether the trend and its slope obtained by the MS-SC/ARCH model are useful in a practical trading strategy or not, and to analyze the profit profile of this strategy.

2 Model Specification and Trading Strategy

2.1 Model Specification

In this section, we describe the MS-SC/ARCH. Let y_n, $n = 1, 2, \ldots, N$, be the observed financial asset price; e.g. stock prices, exchange rates and commodity prices. The observation model of the MS-SC/ARCH model is given by

$$y_n = t_n + \varepsilon_n, \qquad \varepsilon_n \sim N(0, h_n^2), \tag{1}$$
$$h_n = \gamma_0(1 - s_n) + \gamma_1 s_n + \alpha \varepsilon_{n-1}^2, \tag{2}$$

where $\gamma_1 > \gamma_0 > 0$, and $\alpha \geq 0$, and t_n is a trend component. ε_n is a fluctuation around the trend component t_n defined by the ARCH process with a Markov switching structure. s_n is an unobserved Markov chain, and takes the value of 0 or 1 determined by a first-order two-regime Markov process (Hamilton 1989). When $s_n = 0$, the system is in the low volatility regime. When $s_n = 1$, it is in the high volatility regime. The system models are given by

$$t_n = t_{n-1} + \Delta t_n + v_n, \qquad v_n \sim N(0, \tau^2), \tag{3}$$
$$\Delta t_n = \Delta t_{n-1} + w_n, \qquad w_n \sim N(0, s_n \sigma^2). \tag{4}$$

Here t_n follows the random walk process with a drift (or slope) Δt_n and an innovation term v_n. Δt_n is interpreted as the permanent change in slope, and w_n produces a permanent shift in level with a subsequent undisturbed growth. When s_n=0, Δt_n takes the same value of the previous Δt_{n-1} (i.e. $\Delta t_n = \Delta t_{n-1}$ in eqn (4)), and t_n follows the random walk with the same drift to the previous. When s_n=1, Δt_n follows a random walk process with a normal distribution w_n with zero mean and variance σ^2. Therefore, $s_n = 0$ and $s_n = 1$ indicate the regimes of slope change and no slope change, respectively. We assume that in the MS-SC/ARCH model changes in slope and increases in volatility occur simultaneously depending on s_n.

2.2 Trading Strategy

We consider a trading strategy based on the MS-SC/ARCH model. The investment position is taken depending on the buy and sell signals generated by the MS-SC/ARCH model. When buy (sell) is signaled at time n by the MS-SC/ARCH model, a long (short) position of a financial asset is taken. The same investment position remains until the opposite case of sell (buy) is signaled. When buy (sell) is signaled, the investor closes the previous long (short) position, and takes the short (long) position. The trading signal ψ_n at time n is defined depending on a one-step-ahead prediction of the slope $\Delta t_{n+1|n}$ by the MS-SC/ARCH model. If $\Delta t_{n+1|n} \geq 0$, $\psi_n = 1$, and $\psi_n = -1$ otherwise.

Fig. 1. The original series of the weekly TOPIX

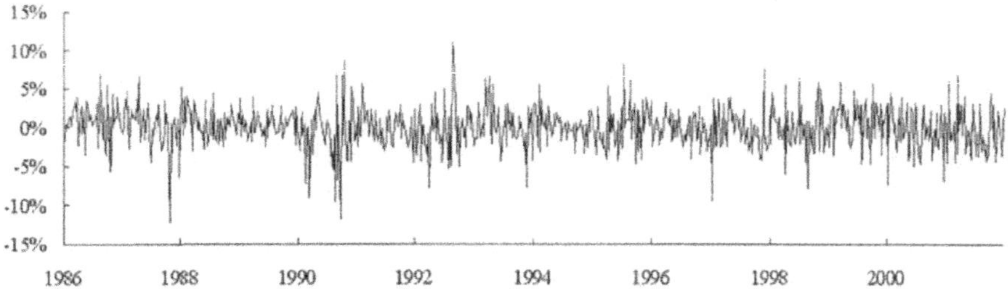

Fig. 2. The logarithmic return series of the weekly TOPIX

3 Application to Real Financial Time Series

The sample data used in this study are the weekly TOPIX over the period from 1986/1 to 2001/12. Fig 1 and 2 show plots of the original and logarithmic return series of the TOPIX, respectively. To evaluate the profitability of the MS-SC/ARCH trading strategy, we estimate the unknown parameters θ using the observations from 1986/1 to 1991/12 as the formation period. Using these parameter estimations, the trading strategy can be constructed based on a one-step-ahead prediction of slope $\Delta t_{n+1|n}$ by the MS-SC/ARCH model in the subsequent simulation period from 1992/1 to 2001/12. According to the trading strategy described in the previous section, we take a long (buy TOPIX) or short (sell TOPIX) investment position.

The mean of the MS-SC/ARCH trading returns is 0.183 percent, and that of sell-and-hold trading returns is 0.093 percent. The t-test for the difference of trading returns is highly significant, rejecting the null hypothesis of equality with zero at the 0.01 critical level. The cumulative sum of the MS-SC/ARCH trading returns in Fig 3 seems to be on the upside. We can say that this MS-SC/ARCH trading strategy succeeds in remaining stable with a positive performance over time.

Fig. 3. Cumulative sums of returns for the MS-SC/ARCH model

4 Conclusions

In this paper, we propose an MS-SC/ARCH model in which the slope of the trend and the ARCH process for the movement around the trend changes simultaneously described by a first-order and two-regime Markov process. In the empirical analysis, we applied the model to the weekly TOPIX. We found sufficient statistical evidence that the trading strategy based on the estimated trend and its slope makes an excess profit over the simple TOPIX sell-and-hold strategy.

References

1. Hamilton J D, (1989) A new approach to the economic analysis of nonstationary time series and the business cycle. Journal of Econometrics 60:357-384
2. Neely C J, (1997) Technical analysis in the foreign exchange market: a layman's guide. Review (Sep), Federal Reserve Bank of St. Louis: 23-38

Pairs-trading strategy: empirical analysis in Japanese stock market

Morikazu Hakamata[1]

Tokyo-Mitsubishi Asset Management Ltd. morikazuhakamata@hotmail.com

Summary. Pairs-trading strategy is one of the market neutral investment strategies in the stock market. We consider a couple of stocks with common trends, and when a spread between these stock prices widens temporally, we take an investment position with one stock long and the other short. Then, we can make a profit on its position if the spread narrows again. In this paper, we propose the pairs-trading strategy by using regression analysis and experimental design. In our empirical analysis in the Japanese stock market, we can succeed in obtaining stable positive returns.

Key words. Pairs-trading strategy, regression analysis, Japanese stock market, empirical analysis.

1 Introduction

Recently, many hedge funds use the pairs-trading strategy in a market-neutral type alternative investment. The basic concept of pairs-trading strategy is very simple: that is, we seek out a couple of stocks, which have a common historical trend, and then when stock prices diverge from the common trend temporarily, we buy one with relative negative divergence and sell the other with relative positive divergence. If the price divergence decreases, we unwind this arbitrage position for realization of its gains. In econometrics and finance, there are a few studies focusing on pairs trading, while Evan et al. (1999) investigate effectiveness of the pairs-trading strategy by using primary descriptive statistics in the U.S. stock market. In this paper, we propose the new pairs-trading strategy by using regression analysis and experimental design (Taguchi and Yokoyama 1993), and implement an empirical analysis of its effectiveness in the Japanese stock market.

2 Pair-Extraction Method

To make stable profit with the pairs-trading strategy, it is most important how to extract pairs from the investment universe. Let $x_{k,n}$ be the k-th stock price or its appropriate transformation at the time of n. We consider the simple linear regression analysis of stock prices, $x_{i,n}$ and $x_{j,n}$, as follows:

$$x_{i,n} = \alpha_{ij} + \beta_{ij} x_{j,n} + \varepsilon_{ij,n}, \qquad n = n_1, n_2, \ldots, n_h, \tag{1}$$

where α_{ij} and β_{ij} are the time-homogeneous constant and intercept coefficients, respectively. The error term $\varepsilon_{ij,n}$ follows a normal distribution with mean zero and variance σ_{ij}^2. Here a cumulative distribution function of $\varepsilon_{ij,n}$ at the time of n_h is defined by

$$\Phi_{ij,h} = \int_{-\infty}^{h} f(\varepsilon_{ij,n_h}) d_{\varepsilon_{ij,n_h}} \tag{2}$$

where

$$f(\varepsilon) = \frac{1}{\sqrt{2\pi}\sigma} \exp\Big(-\frac{\varepsilon^2}{2\sigma^2}\Big), \tag{3}$$

and $\Phi_{ij,h}$ takes a value between 0 and 1. The closer the value of $\Phi_{ij,h}$ is to 0 or 1 means that the spread between stock prices widen at the time of n_h. Three corresponding statistics of such a regression analysis are used as "indicators" to extract more profitable pairs from every pair as follows:

a. Sample size h for estimation range.
b. Coefficient of determination R_{ij}^2 for unobserved co-movement.
c. Cumulative distribution function $\Phi_{ij,h}$ for price deviance from unobserved co-movement [1].

3 Empirical Analysis

We use the members of S&P150 index for a universe for pairs composition. The simulation period is from 1997/1 to 2002/12 (daily data, 1231 days), and there are 59 investment opportunities because a turnover period for pairs portfolios is 20 business days. We construct equally weighted pairs portfolios each time. To evaluate profitability and quality of this pairs-trading strategy, we consider the process of pairs extraction as a series of trading systems, and apply an L8 orthogonal array from experimental design (Taguchi and Yokoyama 1993). We select two levels for each indicator, which are showed in Table 1.

First, as the simulation results of the L8 orthogonal array, we can obtain over five percent return p.a. in every experiment, implying that profitability of

[1] We denote the range of the cumulative distribution function $\Phi_{ij,h}$ with lower and higher bounds, c^1 and c^2, respectively.

Table 1. Indicators for pairs composition and experimental design

Indicators	Level 1	Level 2
a) Sample size h	40	60
b) Coefficient of determination R^2_{ij}	0.70	0.80
c^1) CDF $\Phi^1_{ij,h}$ (lower bound)	0.80	0.90
c^2) CDF $\Phi^2_{ij,h}$ (upper bound)	0.99	1.00

our pairs-trading strategy does not depend entirely upon the selection of the parameter levels. Fig 1 plots the cumulative returns of the best Sharpe-ration experiment. This return series seems to be ever increasing without a downward swing through the simulation period. Second, we analyze the sensitivity of indicators to improve the performance of this pairs-trading strategy. Table 2 summarizes the results of the ANOVA of the L8 orthogonal array for the Sharpe-ration, return and the standard deviation. Significant/non-significant indicators are determined by the p-value. For the Sharpe-ration and return, we should not primarily control a or b separately, but rather both of them at the same time. On the other hand, concerning risk (measured by the standard deviation), b is a most important factor to control.

Table 2. Variance Ratio of ANOVA (p-values are in parentheses)

Indicators	Sharpe-Ratio	Return	Stand. Dev.
a) Sample size h	30.91	1.45	1.18
	(0.031)	(0.352)	(0.391)
b) Coefficient of determination R^2_{ij}	38.59	0.27	10.82
	(0.025)	(0.656)	(0.081)
a × b) Effect of interaction	54.45	10.50	1.39
	(0.018)	(0.083)	(0.360)
c^1) CDF $\Phi^1_{ij,h}$ (lower bound)	4.73	7.85	13.28
	(0.162)	(0.107)	(0.068)
c^2) CDF $\Phi^2_{ij,h}$ (upper bound)	10.97	1.68	0.10
	(0.080)	(0.325)	(0.785)

4 Conclusions

In this paper, we propose the pairs-trading strategy based on a regression analysis and experimental design, and conduct an empirical analysis using data from the Japanese stock market. We find that this strategy is effective and does not depend on setting indicators for pairs extraction. Additionally, the experimental design allows for optimization of the trader's preference by controlling indicators.

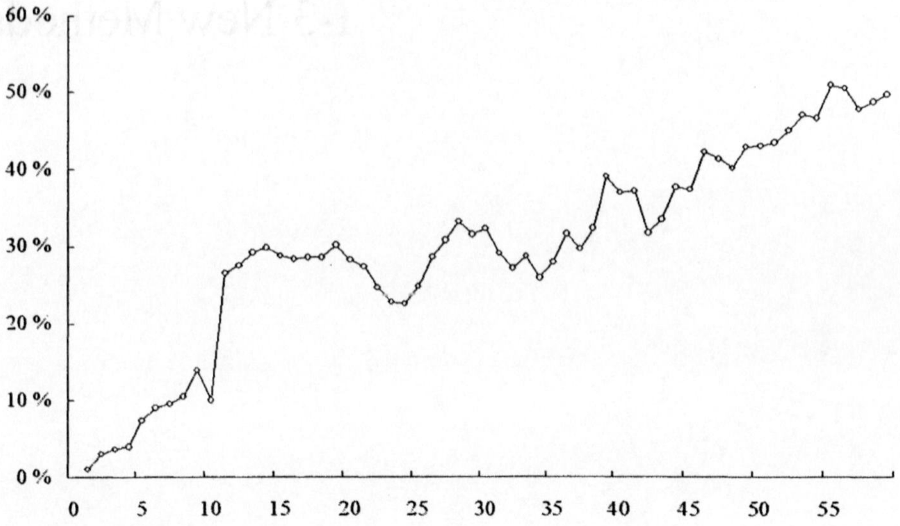

Fig. 1. Cumulative Trading Return

References

1. Evan G, Goetzmann W and Rouwenhorst K (2000) Pairs Trading: Performance of a Relative Value Arbitrage Rule. Yale School of Management Working Paper Series
2. Taguchi G and Yokoyama T (1993) Taguchi methods: design of experiments (Quality Engineering, Vol 4). ASI Press, Tokyo

I-3 New Methods

Self-modulation processes in financial markets

Misako Takayasu

Department of Complex Systems, Future University-Hakodate, 116-2 Kameda-Nakano, Hakodate 041-0803, Japan

Summary. We confirm validity of the new type of stochastic process called the self-modulation process through the analysis of transaction interval data in Yen-Dollar exchange market. In the model the expectation value of the process is continuously modulated by its own moving average. It is proved exactly and generally that point processes with such self-modulation effect generally show the 1/f power spectrum consistent with observations.

Key words. Transaction intervals, moving average, self-modulation process

1. Introduction

Analysis of occurrence of transactions in markets can be regarded as the most basic study of markets' high-resolution statistics. By a naive analysis of tick data of Yen-Dollar exchange market we can easily confirm that the distribution of transaction intervals apparently follows a power law and their autocorrelation has a long tail, however, in the first Nikkei Econophysics Symposium held in 2000, we reported that the occurrence of transaction intervals can be nicely approximated by a non-stationary Poissonian process whose mean value is given by a moving average of transaction intervals for about 150 seconds [1].

In this paper we review this observation facts and introduce a general mathematical model that reproduce the basic properties. The model is called as the self-modulation process (in short SMP) as the mean value that is the most important parameter of the model is modulated by its own traces through the moving average [2].

2. The data analysis

We analyze a set of tick-by-tick data of Yen-Dollar exchange: Bloomberg's data for the 6 months from October '98 to March '99. This data includes records of trading prices with time stamps measured in seconds.

The time intervals of trades are always fluctuating. There is a clear 24-hour pattern as shown in Fig.1. This pattern is obviously due to the working time of dealers, namely, the change of numbers of working dealers is responsible for this movement of the mean trading interval.

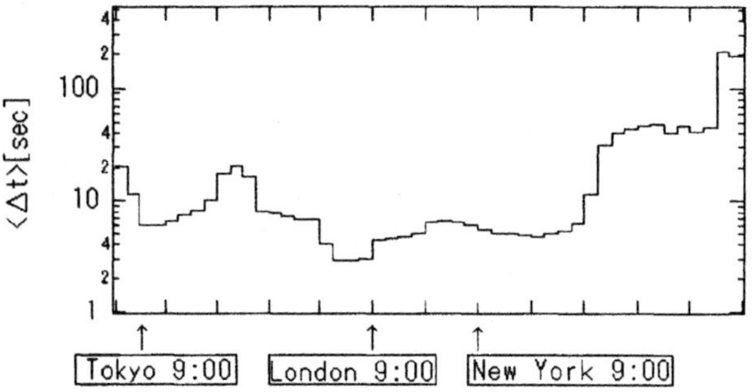

Fig.1 The 24-hour pattern of the mean trading interval.

A general and interesting feature of the trade intensity fluctuations in markets is the long time correlation typically characterized by the 1/f power spectrum [3]. This property can be confirmed also in the Yen-Dollar market as shown in Fig.2. Here, we plot the power spectrum for segmented data from am 7:00 to 12:00 in New York time to eliminate the effect of 24-hour pattern.

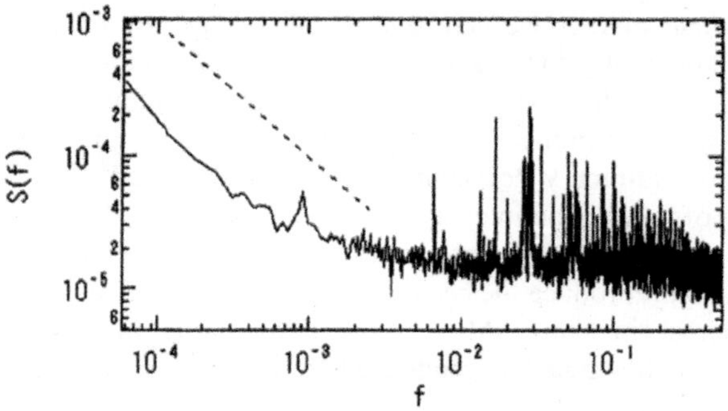

Fig.2 The power spectrum of trade intensity fluctuations for the data of 7:00 to 12:00 by New York time. The dotted line shows the 1/f slope.

An example of short time spontaneous fluctuations of trade intervals is shown in Fig.3 with the change of exchange rate. By comparing these figures carefully we can find a tendency that trade intervals become longer when the market is stable for several minutes and the interval becomes shorter quickly when the exchange rate changes.

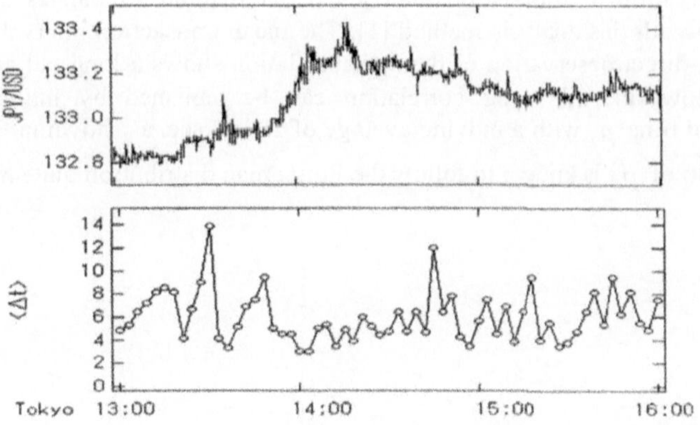

Fig.3 A example of short time change of trade interval represented by a moving average of 3 minutes (below) and the corresponding exchange rates (above).

3. The moving average analysis

Regarding the occurrence of trades as a point process that is characterized by time sequence of events, $\{t_j\}$. Let us introduce a moving average as

$$< \Delta t_j >= \frac{1}{K} \sum_{k=j-K+1}^{j} \Delta t_k \quad , \qquad (1)$$

where the time interval is defined as $\Delta t_{j+1} = t_{j+1} - t_j$ and K denotes the number of events to be averaged. In a basic analysis the value of K is fixed, however, in a variant case we introduce the moving average period T and the value of K is determined as the minimum integer that makes the summation in Eq.(1) becomes larger than T for each time step j. We introduce the normalized time μ that is defined by the following ratio and observe its basic statistics.

$$\mu_j = \Delta t_{j+1} / < \Delta t_j > \quad . \tag{2}$$

By choosing an appropriate moving average we expect that the statistical properties of $\{ \mu_j \}$ becomes simple, independent and identically distributed random variable.

Observation of transaction intervals of Yen-Dollar exchanges has already been done with this analysis method [1]. The mean transaction interval is about 7 sec. and a direct observation of the autocorrelation shows a long tail as shown in Fig.4a. However, this long correlation can be vanished by introducing the normalized time μ_j with a moving average of T=150 sec. as shown in Fig.4b. The distribution of μ_j is known to follow the Poissonian distribution quite nicely.

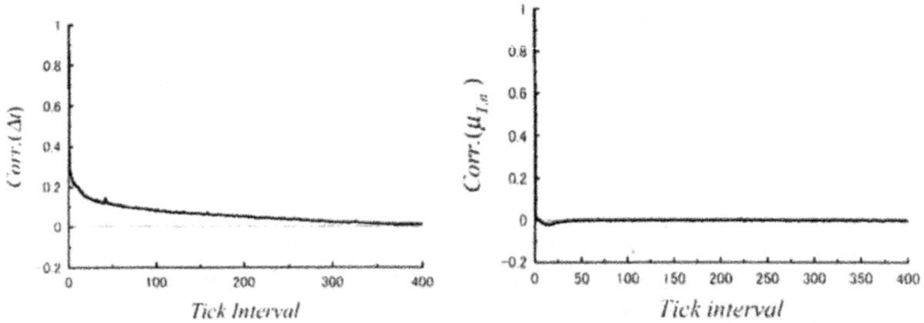

Fig.4 a: The autocorrelation of original trade intervals (Left).
b: Normalized time intervals show almost no correlation.

4. The Self-Modulation Process

Mathematical analysis of the self-modulation process can be done by solving the following type of stochastic equation:

$$\Delta t_{j+1} = \mu_j < \Delta t_j > + f_j \tag{3}$$

where μ_j and f_j are independent random variables. In the special case that the moving average interval is infinitely long the model becomes ordinary Markovian process, while in the case that the moving average is replaced by the previous value, t_j, the process is reduced to the random multiplicative process or random coefficient autoregressive process that has a long history of mathematical study [4, 5]. This type of stochastic process is similar to the point process developed in the study of earthquakes [6], and a general framework similar to this simple model has already been introduced independently in economics [7].

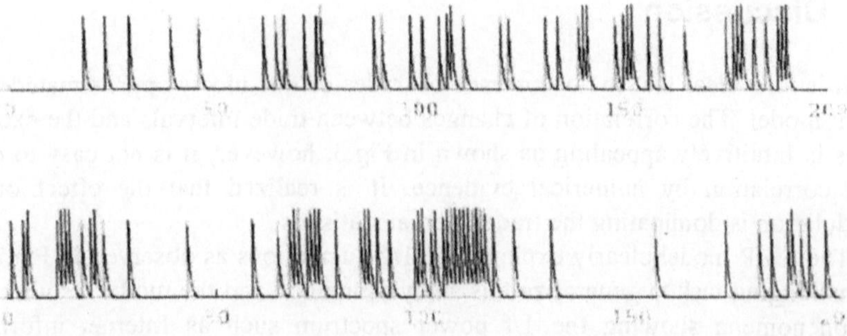

Fig.5. Pulses produced by a normal Poissonian process (above) and by a self-modulation process (below).

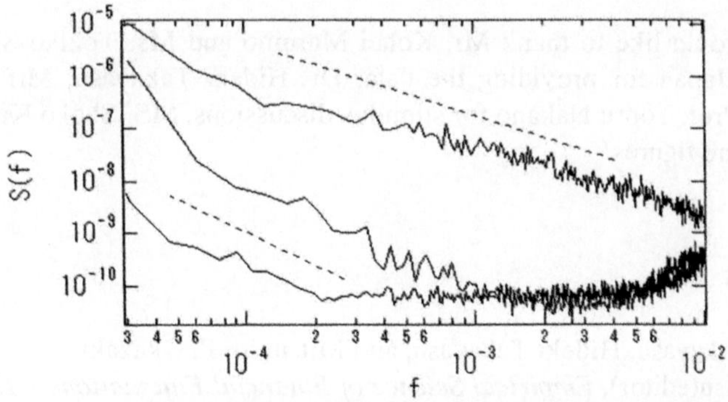

Fig.6. Power spectra of SMP models. *K*=1 (above) shows the 1/f spectrum, *K*=15 (middle) shows an intermediate functional form, and *K*=50 (below) shows a combination of a $1/f^2$ spectrum and the white noise.

As know immediately from examples shown in Fig.5 the effect of self-modulation causes spontaneous clustering of pulses. This effect modifies the power spectrum of the pulse sequences from a white noise to the 1/f fluctuations as demonstrated in Fig.6.

We can prove rigorously that the power spectrum of the self-modulation point process with the shortest moving average follows the 1/f spectrum for the low frequency limit when the mean value of μ_j is unity, a basic relation requiring stationary condition [2]:

$$\overline{\mu_j} = 1 \qquad (4)$$

5. Discussion

It is now clear that the occurrence of trades can be nicely approximated by the SMP model. The correlation of changes between trade intervals and the exchange rates is intuitively appealing as shown in Fig.3, however, it is not easy to clarify this correlation by numerical evidence. It is realized that the effect of self-modulation is dominating the trade interval statistics.

The SMP model clearly explains the 1/f fluctuations as observed in Fig.2. It is a challenging task to generalize this analysis method and the model to other fields of phenomena showing the 1/f power spectrum such as Internet information traffics [8] and automobile traffics [9].

Acknowledgement

The author would like to thank Mr. Kohei Marumo and Ms. Tokiko Shimizu of the Bank of Japan for providing the data, Dr. Hideki Takayasu, Mr. Takayuki Mizuno and Prof. Tohru Nakano for stimulus discussions, MS. Shoko Kurihara for providing some figures.

References

[1] Misako Takayasu, Hideki Takayasu, and Mitsuhiro P. Okazaki, [H.Takayasu(editor), *Empirical Science of Financial Fluctuations – The Advent of Econophysics*, (ISBN 4-431-70316-0) Springer Verlag, Tokyo, 2002], 18-23.

[2] Misako Takayasu and Hideki Takayasu, **Physica** A324(2003), 101-107.

[3] G. Bonnano, F. Lillo, R. Mantegna, **Physica** A280, 2000, 136-142.

[4] D.F. Nicholls and B.G. Quinn, *Random Coefficient Autoregressive Models: An Introduction*, Springer-Verlag(New York)(1982).

[5] H. Takayasu, A.-H. Sato and M. Takayasu, **Phys. Rev. Lett.** 79 (1997), 966-969.

[6] D. Sornette and A.Helmstetter, **Phys. Rev. Lett.** 89 (2002) ,158501-158504.

[7] R.F. Engle and J.R. Russell, **Journal of empirical finance**, 4 (1997) 187--212.

[8] Misako Takayasu, Hideki Takayasu, and Takamitsu Sato, **Physica** A233(1996), 824-834.

[9] M. Kobayashi and T. Musha, **IEEE Trans. Biomed. Eng**. 29(1982), 456-457.

Deterministic and stochastic influences on Japan and US stock and foreign exchange markets. A Fokker-Planck approach

Kristinka Ivanova[1], Marcel Ausloos[2], and Hideki Takayasu[3]

[1] Pennsylvania State University, University Park PA 16802, USA
[2] GRASP, B5, University of Liège, B-4000 Liège, Euroland
[3] Sony Computer Science Laboratories, Tokyo 141-0022, Japan

Summary. The evolution of the probability distributions of Japan and US major market indices, NIKKEI 225 and NASDAQ composite index, and JPY/DEM and DEM/USD currency exchange rates is described by means of the Fokker-Planck equation (FPE). In order to distinguish and quantify the deterministic and random influences on these financial time series we perform a statistical analysis of their increments $\Delta x(\Delta(t))$ distribution functions for different time lags $\Delta(t)$. From the probability distribution functions at various $\Delta(t)$, the Fokker-Planck equation for $p(\Delta x(t), \Delta(t))$ is explicitly derived. It is written in terms of a drift and a diffusion coefficient. The Kramers-Moyal coefficients, are estimated and found to have a simple analytical form, thus leading to a simple physical interpretation for both drift $D^{(1)}$ and diffusion $D^{(2)}$ coefficients. The Markov nature of the indices and exchange rates is shown and an apparent difference in the NASDAQ $D^{(2)}$ is pointed out.

Key words. Econophysics; Probability distribution functions; Fokker-Planck equation; Stock market indices; Currency exchange rates

1 Introduction

Recent studies have shown that the power spectrum of the stock market fluctuations is inversely proportional to the frequency on some power, which points to self-similarity in time for processes underlying the market [1, 2]. Our knowledge of the random and/or deterministic character of those processes is however limited. One rigorous way to sort out the noise from the deterministic components is to examine in details correlations at *different* scales through the so called master equation, i.e. the Fokker-Planck equation (and the subsequent Langevin equation) for the probability distribution function (*pdf*) of signal increments [3]. This theoretical approach, so called solving the inverse problem, based on *rigorous* statistical principles [4, 5], is often the first step in sorting out the *best* model(s). In this paper we derive FPE, directly from the experimental data of two financial indices and two exchange rates series, in terms of a drift $D^{(1)}$ and a diffusion $D^{(2)}$ coefficient. We would like to emphasize that the method is model independent. The technique allows examination of long and short time scales *on the same footing.* The so found analytical form of both drift $D^{(1)}$ and diffusion $D^{(2)}$ coefficients has a simple physical interpretation, reflecting the influence of the deterministic and random forces on the examined market dynamics processes. Placed into a Langevin equation, they could allow for some first step forecasting.

2 Data

We consider the daily closing price $x(t)$ of two major financial indices, NIKKEI 225 for Japan and NASDAQ composite for US, and daily exchange rates involving currencies of Japan, US and Europe, JPY/DEM and DEM/USD from January 1, 1985 to May 31, 2002. Data series of NIKKEI 225 (4282 data points) and NASDAQ composite (4395 data points) and are downloaded from the Yahoo web site (*http : //finance.yahoo.com/*). The exchange rates of JPY/DEM and DEM/USD are downloaded from *http : //pacific.commerce.ubc.ca/xr/* and both consists of 4401 data points each. Data are plotted in Fig. 1(a-d). The DEM/USD case was studied in [3] for the 1992-1993 years. See also [6], [8-10] and [11] for some related work and results on such time series signals, some on high frequency data, and for different time spans.

3 Results and discussion

To examine the fluctuations of the time series at different time delays (or time lags) Δt we study the distribution of the increments $\Delta x = x(t + \Delta t) - x(t)$. Therefore, we can analyze the fluctuations at long and short time scales on the same footing. Results for the probability distribution functions (pdf) $p(\Delta x, \Delta t)$ are plotted in Fig. 2(a-d). Note that while the pdf of one day time delays (circles) for all time series studied have similar shapes, the pdf for longer time delays shows fat tails as in [2] of the same type for NIKKEI 225, JPY/DEM and DEM/USD, but is different from the pdf for NASDAQ.

More information about the correlations present in the time series is given by joint pdf's, that depend on N variables, i.e. $p^N(\Delta x_1, \Delta t_1; ...; \Delta x_N, \Delta t_N)$. We started to address this issue by determining the properties of the joint pdf for $N = 2$, i.e. $p(\Delta x_2, \Delta t_2; \Delta x_1, \Delta t_1)$. The symmetrically tilted character of the joint pdf contour levels (Fig. 3(a-c)) around an inertia axis with slope 1/2 points out to the statistical dependence, i.e. a correlation, between the increments in all examined time series.

The conditional probability function is

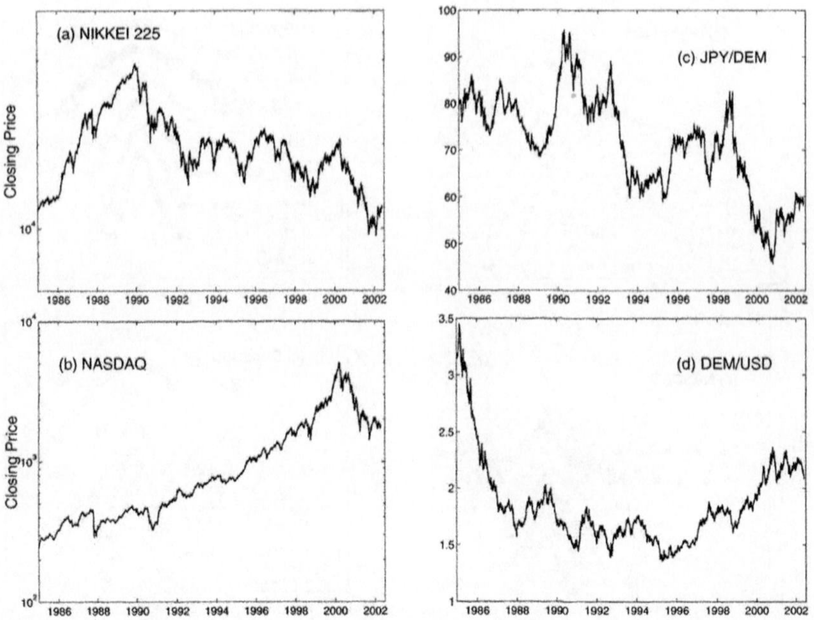

Fig. 1. Daily closing price of (a) NIKKEI 225, (b) NASDAQ, (c) *JPY/DEM* and (d) *DEM/USD* exchange rates for the period from Jan. 01, 1985 till May 31, 2002

$$p(\Delta x_{i+l}, \Delta t_{i+l} | \Delta x_i, \Delta t_i) = \frac{p(\Delta x_{i+l}, \Delta t_{i+l}; \Delta x_i, \Delta t_i)}{p(\Delta x_i, \Delta t_i)} \quad (1)$$

for $i = 1, ..., N-1$. For any $\Delta t_2 < \Delta t_i < \Delta t_1$, the Chapman-Kolmogorov equation is a necessary condition of a Markov process, one without memory but governed by probabilistic conditions

$$p(\Delta x_2, \Delta t_2 | \Delta x_1, \Delta t_1) = \int d(\Delta x_i) p(\Delta x_2, \Delta t_2 | \Delta x_i, \Delta t_i) p(\Delta x_i, \Delta t_i | \Delta x_1, \Delta t_1). \quad (2)$$

The Chapman-Kolmogorov equation when formulated in *differential* form yields a master equation, which can take the form of a Fokker-Planck equation [4]. For $\tau = log_2(32/\Delta t)$,

$$\frac{d}{d\tau} p(\Delta x, \tau) = \left[-\frac{\partial}{\partial \Delta x} D^{(1)}(\Delta x, \tau) + \frac{\partial}{\partial^2 \Delta x^2} D^{(2)}(\Delta x, \tau) \right] p(\Delta x, \tau) \quad (3)$$

in terms of a drift $D^{(1)}(\Delta x,\tau)$ and a diffusion coefficient $D^{(2)}(\Delta x,\tau)$ (thus values of τ represent Δt_i, $i = 1, ...$).

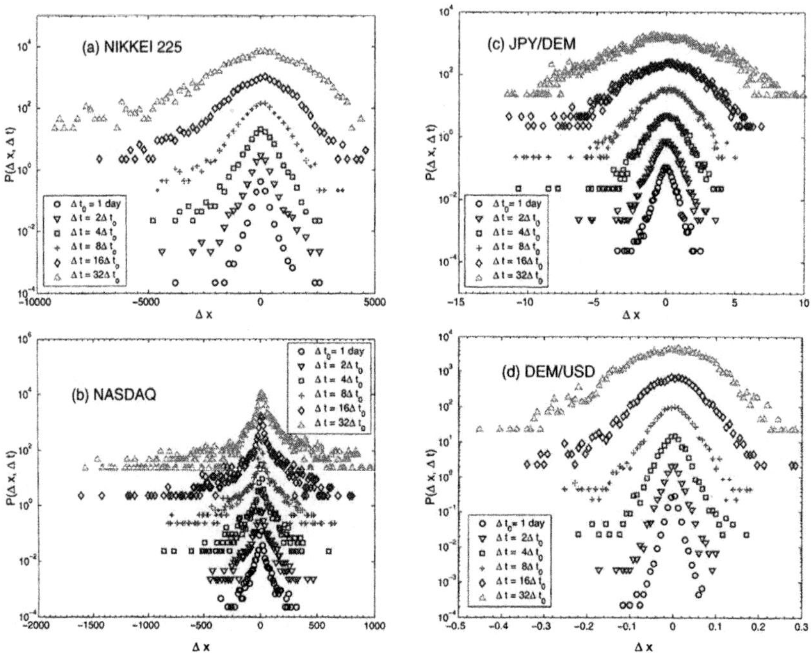

Fig. 2. Probability distribution function $p(\Delta x, \Delta t)$ of (a) NIKKEI 225, (b) NASDAQ, (c) JPY/DEM and (d) DEM/USD from Jan. 01, 1985 till May 31, 2002 for different delay times. Each pdf is displaced vertically to enhance the tail behavior; symbols and the time lags Δt are in the insets. The discretisation step of the histogram is (a) 200, (b) 27, (c) 0.1 and (d) 0.008 respectively

The coefficient functional dependence can be estimated directly from the moments $M^{(k)}$ (known as Kramers-Moyal coefficients) of the conditional probability distributions:

$$M^{(k)} = \frac{1}{\Delta\tau} \int d\Delta x' (\Delta x' - \Delta x)^k p(\Delta x', \tau + \Delta\tau | \Delta x, \tau) \tag{4}$$

$$D^{(k)}(\Delta x, \tau) = \frac{1}{k!} \lim M^{(k)} \tag{5}$$

for $\Delta\tau \to 0$. The functional dependence of the drift and diffusion coefficients $D^{(1)}$ and $D^{(2)}$ for the normalized increments Δx is well represented by a line and a parabola, respectively. The values of the polynomial coefficients are summarized in Table 1 and Fig. 4.

The leading coefficient (a_1) of the linear $D^{(1)}$ dependence has approximately the same values for all studied signals, thus the same deterministic noise (drift coefficient). Note that the leading term (b_2) of the functional dependence of diffusion coefficient of the NASDAQ closing price signal is about

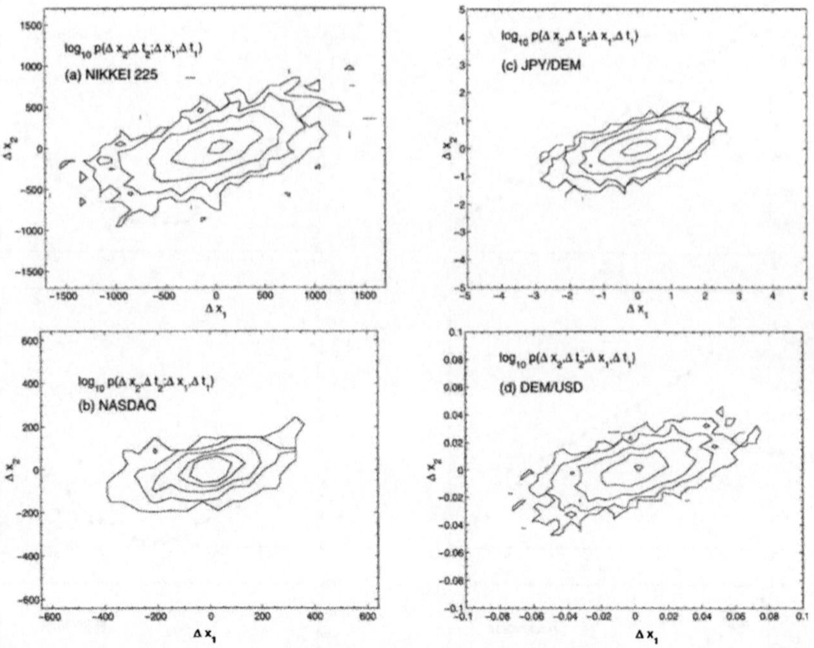

Fig. 3. Typical contour plots of the joint probability density function $p(\Delta x_2, \Delta t_2; \Delta x_1, \Delta t_1)$ of (a) NIKKEI 225, (b) NASDAQ closing price signal and (c) JPY/DEM and (d) DEM/USD exchange rates for $\Delta t_2 = 1$ *day* and $\Delta t_1 = 3$ *days*. Contour levels correspond to $log_{10} p = -1.5, -2.0, -2.5, -3.0, -3.5$ from center to border

Table 1. Values of the polynomial coefficients defining the linear and quadratic dependence of the drift and diffusion coefficients $D^{(1)} = a_1(\Delta x/\sigma) + a_0$ and $D^{(2)} = b_2(\Delta x/\sigma)^2 + b_1(\Delta x/\sigma) + b_0$ for the FPE (3) of the normalized data series; σ represents the normalization constant equal to the standard deviation of the Δt=32 days pdf

	a_1	a_0	b_2	b_1	b_0	σ
NIKKEI 225	-0.54	0.002	0.18	-0.004	0.003	1557.0
NASDAQ	-0.49	-0.0004	0.30	-0.010	0.001	198.11
JPY/DEM	-0.55	0.002	0.17	-0.002	0.004	2.9111
DEM/USD	-0.49	-0.001	0.16	-0.004	0.004	0.0808

twice the leading, i.e. second order coefficient, of the other three series of interest. This can be interpreted as if the stochastic component (diffusion coefficient) of the dynamics of NASDAQ is twice larger than the stochastic components of NIKKEI 225, JPY/DEM and DEM/USD. A possible reason for such a behavior may be related to the transaction procedure on the

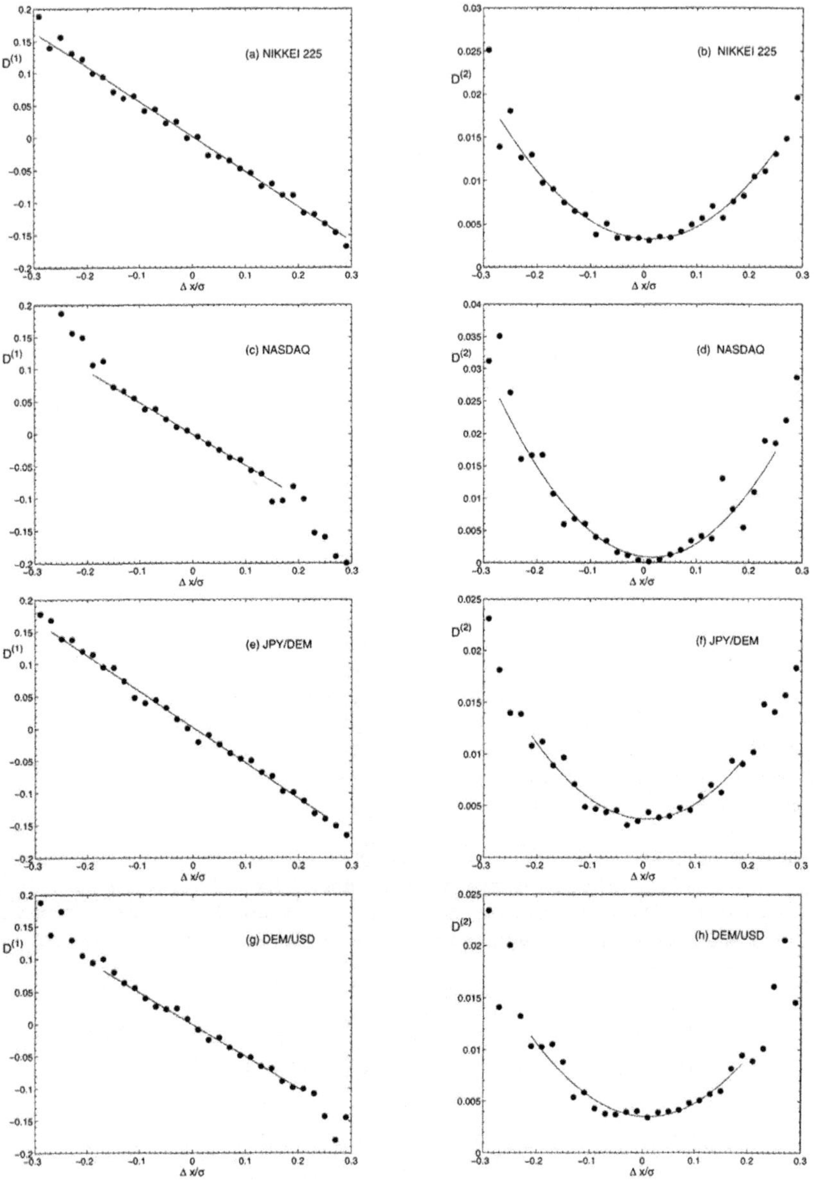

Fig. 4. Functional dependence of the drift and diffusion coefficients $D^{(1)}$ and $D^{(2)}$ for the pdf evolution equation (3); Δx is normalized with respect to the value of the standard deviation σ of the pdf increments at delay time 32 days: (a,b) NIKKEI 225 and (c,d) NASDAQ closing price signal, (e,f) JPY/DEM and (g,h) DEM/USD exchange rates

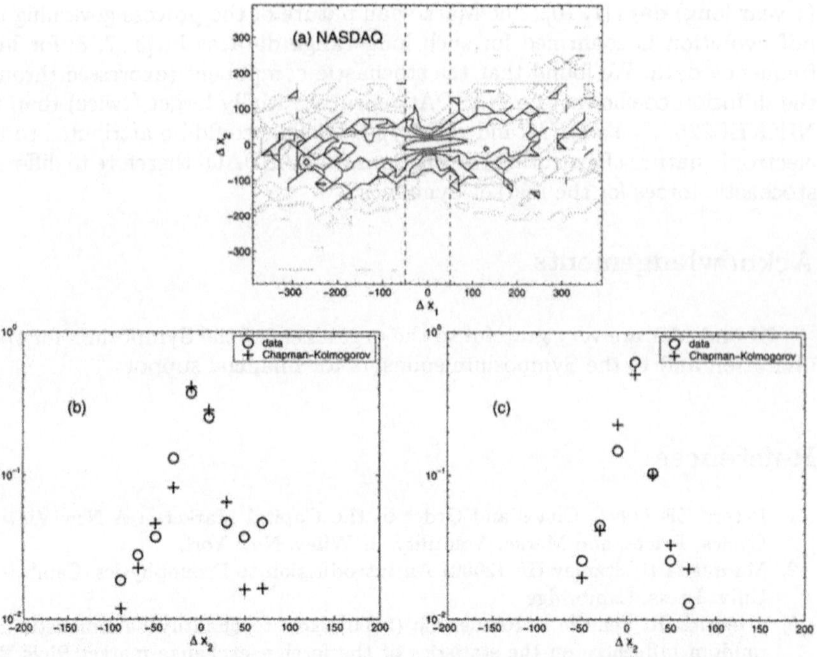

Fig. 5. Equal probability contour plots of the conditional pdf $p(\Delta x_2, \Delta t_2 | \Delta x_1, \Delta t_1)$ for two values of Δt, $\Delta t_1 = 8$ days, $\Delta t_2 = 1$ day for NASDAQ. Contour levels correspond to $log_{10}p$=-0.5,-1.0,-1.5,-2.0,-2.5 from center to border; data (solid line) and solution of the Chapman Kolmogorov equation integration (dotted line); (b) and (c) data (circles) and solution of the Chapman Kolmogorov equation integration (plusses) for the corresponding pdf at Δx_2 = -50 and +50

NASDAQ. Our numerical result agrees with that of ref. [3] if a factor of ten is corrected in the latter ref. for b_2.

The validity of the Chapman-Kolmogorov equation has also been verified. A comparison of the directly evaluated conditional pdf with the numerical integration result (2) indicates that both pdf's are statistically identical. The more pronounced peak for the NASDAQ is recovered (see Fig. 5). An analytical form for the pdf's has been obtained by other authors [6, 10] but with models different from more classical ones [8].

4 Conclusion

The present study of the evolution of Japan and US stock as well as foreign currency exchange markets has allowed us to point out the existence of deterministic and stochastic influences. Our results confirm those for high frequency

(1 year long) data [7, 10]. The Markovian nature of the process governing the pdf evolution is confirmed for such long range data as in [3, 7, 8] for high frequency data. We found that the stochastic component (expressed through the diffusion coefficient) for NASDAQ is substantially larger (twice) than for NIKKEI 225, JPY/DEM and DEM/USD. This could be attributed to the electronic nature of executing transactions on NASDAQ, therefore to different stochastic forces for the market dynamics.

Acknowledgements

KI and MA are very grateful to the organizers of the Symposium for their invitation and to the Symposium sponsors for financial support.

References

1. Peters EE (1994) Chaos and Order in the Capital Markets : A New View of Cycles, Prices, and Market Volatility. J. Wiley, New York
2. Mantegna R, Stanley HE (2000) An Introduction to Econophysics. Cambridge Univ. Press, Cambridge
3. Friedrich R, Peincke J, Renner Ch (2000) How to quantify deterministic and random influences on the statistics of the foreign exchange market Phys Rev Lett 84:5224–5227
4. Ernst MH, Haines LK, Dorfman JR (1969) Theory of transport coefficients for moderately dense gases Rev Mod Phys 41:296–316
5. Risken H (1984) The Fokker-Planck Equation. Springer-Verlag, Berlin
6. Silva AC, Yakovenko VM (2002) Comparison between the probability distribution of returns in the Heston model and empirical data for stock indices cond-mat/ 0211050; Dragulescu AA, Yakovenko VM (2002) Probability distribution of returns in the Heston model with stochastic volatility, cond-mat/0203046
7. Baviera R, Pasquini M, Serva M, Vergni D, Vulpiani A (2001) Eur Phys J B 20:473–479 ; ibid. (2002) Antipersistent Markov behavior in foreign exchange markets Physica A 312:565–576
8. Nekhili R, Altay-Salih A, Gencay R (2002) Exploring exchange rate returns at different time horizons Physica A 313:671–682
9. Matassini L (2002) On the rigid body behavior of foreign exchange markets Physica A 308:402–410
10. Kozuki N, Fuchikami N (2002) Dynamical model of financial markets: fluctuating 'temperature' causes intermittent behavior of price changes, cond-mat/0210090
11. Ohira T, Sazuka N, Marumo K, Shimizu T, Takayasu M, and Takayasu H (2002) Predictability of Currency Market Exchange Physica A 308:368–374

First-Passage Problem in Foreign Exchange Rate

Shoko Kurihara[1],Takayuki Mizuno[1],Hideki Takayasu[2] and Misako Takayasu[3]

[1] Department of Physics, Chuo University, Tokyo 112-8551, Japan
shoko@tkf.att.ne.jp
[2] Sony Computer Science Laboratories, Tokyo 141-0022, Japan
[3] Department of Complex Systems, Future University - Hakodate, Hokkaido 041-8655, Japan

summery We analyze tick-data of yen-dollar exchange rate and study the distribution of first-passage time (in short FPT), which is defined by the time that the rate firstly moves out from a given range. We report that the distribution of FPT is well described by a stretched exponential function with. This fact indicates that the process of yen-dollar exchange fluctuation is much slower than the normal Brownian motion. Applying the same analysis to the data with stable-volatility we find that the stretched exponential properties are partly related to magnitude of volatility.

Key words. Foreign exchange, First-passage time, Stretched exponential distribution

1 Introduction

We consider the first-passage problem in the foreign exchange rates. The first-passage problem is concerned with the event that a diffusing particle or a random walker firstly reaches a specified site. In particular, we are interested in the time that a particle moves a specific distance for the first time.

First-passage problems have been studied in various research fields, such as probability theory, statistical physics, chemical reactions, biology and social sciences [1]. In financial area of study, I. Simonsen et al. and M. H. Jensen et al. studied the distribution of FPT of daily financial prices [2, 3].

We think that the FPT is an important quantity for analysis of markets for the following three reasons: (i) FPT contains multi-time information as it is necessary to check all the data before the FPT event, while the ordinary time correlation function involves only two-time information. (ii) FPT is meaningful for many investors, as it is a popular strategy to set buying and selling thresholds so that FPT gives the waiting time until they can realize the transaction. (iii) Analysis of FPT is directly applicable to option pricing although we do not discuss further.

In this paper, we observe the distribution of FPT in yen-dollar exchange rates analyzing high-precision tick data.

2 Brownian motion

For a given exchange rate width $\Delta p \geq 0$ we calculate the survival probability $S_{\Delta p}(\tau)$, which is defined as the probability that the exchange rate, $p(t)$, starting from $p(t_0)$ keeps staying in the range, $[p(t_0) - \Delta p, p(t_0) + \Delta p]$ up to the time step $t_0 + \tau$. The first-passage probability, $F_{\Delta p}(\tau)$, can be obtained from $S_{\Delta p}(\tau)$ by

$$F_{\Delta p}(\tau) = -\frac{\partial}{\partial \tau} S_{\Delta p}(\tau) \tag{1}$$

Assuming that the probability density of the exchange rate $c(p,t)$ follows the ordinary diffusion equation is given as,

$$\frac{\partial}{\partial t} c(p,t) = D \frac{\partial^2}{\partial p^2} c(p,t) \tag{2}$$

Then $S_{\Delta p}(\tau)$ is estimated as

$$S_{\Delta p}(\tau) \equiv \int_{p(t_0)-\Delta p}^{p(t_0)+\Delta p} c(p,\tau) dp \propto \exp(-\tau/\tau_0), \tag{3}$$

$$\tau_0 \propto (\Delta p)^2 \tag{4}$$

for large τ [1]. So, if exchange rate follows a Brownian motion as often supposed in the financial technology, the probability that the exchange rate stays in the range exponentially decreases.

3 Yen-dollar exchange rates

Here, we analyze the tick-by-tick foreign exchange rate data of yen-dollar consisted of 33,493,606 ticks from 1989 to 2002 provided by CQG. Figure 1 shows $S_{\Delta p}(\tau)$ in semi-logarithmic and double-logarithmic scales. The semi-logarithmic plot shows that its tail is longer than exponential function and the double-logarithmic plot clearly demonstrates that the tail is shorter than a power-law function. We find that $S_{\Delta p}(\tau)$ can always be well approximated by a stretched exponential function of the form,

$$S(\tau) \propto \exp(-(\tau/\tau_1)^{\gamma}) \tag{5}$$

where τ_1 gives a characteristic time and the exponent lies in the range $0 < \gamma < 1$. Comparing with the exponential function of Eq. (3), a stretched exponential decay with $\gamma < 1$ means that real exchange rates diffuse much more slowly than Brownian case. As shown in Figure 2 we observe that the value of γ increases slowly to 1 with the rate range width Δp.

The estimated value of the characteristic time τ_1 introduced in Eq.(5) is plotted in Figure 3 as a function of the range width Δp. The characteristic time tends to be proportional to the square of the width just like Brownian case for large range width, however, we observe non-trivial behaviors for small range width as plotted in Figure 3.

In order to clarify the reason of making the real rate motion different from the Brownian diffusion, we observe the effect of volatility. Here, the volatility is defined by averaging the absolute value of exchange rate difference per day, and Figure 4 shows the volatility fluctuations for the whole data.

Let us focus on the period that the volatility is fairly stable, from the 1688th day (Jan. 2 1996) to the 1882th day (Sep. 30 1996) in Figure 4. The functional form of $S_{\Delta p}(\tau)$ calculated in this period also clearly belongs to a stretched exponential function as shown in Figure 5, however, as known from Figure 6 the exponent γ is apparently larger than that for the whole data in Figure 2. Namely, when volatility is stable, the FPT becomes closer to Brownian case. Although this result shows only a tendency, it is evident that the volatility fluctuations and FPT statistics have a deep relation.

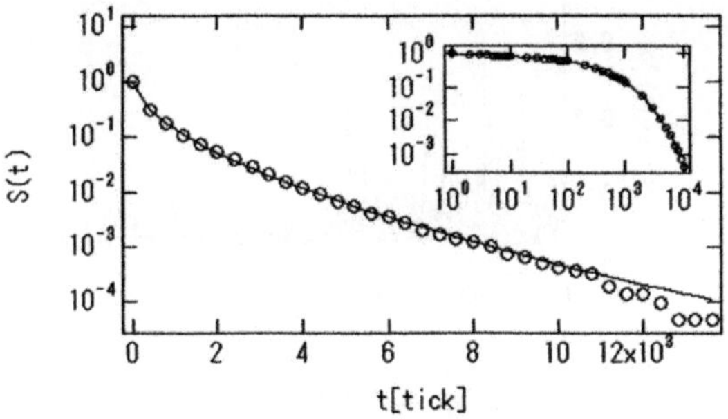

Fig. 1. The survival probability for $\Delta p = 0.3$. The curve shows the stretched exponential fitting.

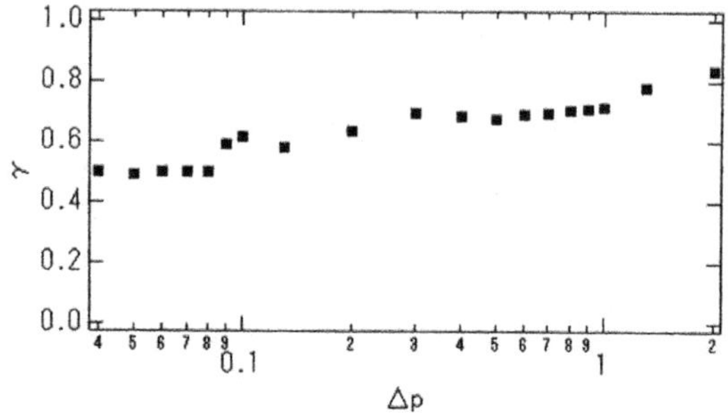

Fig. 2. The range width dependence of the exponent γ.

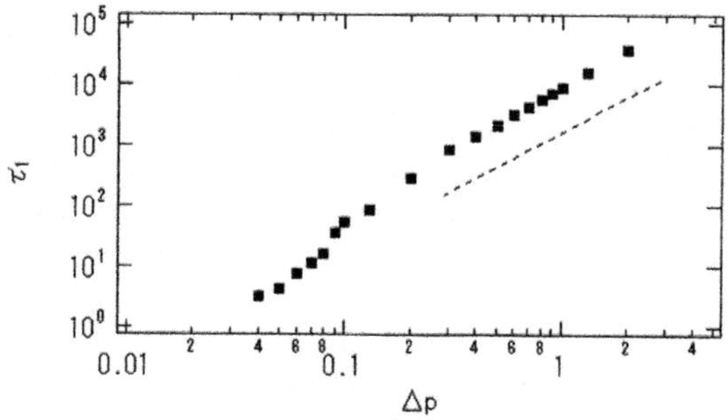

Fig. 3. The range width dependence of the characteristic time. Dashed line shows the slope of Brownian case.

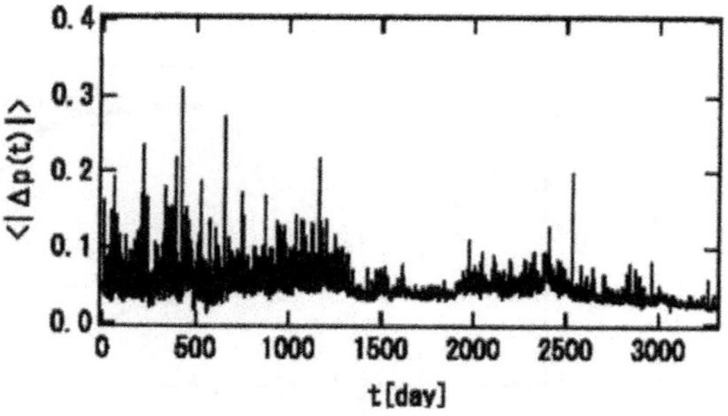

Fig. 4. The daily volatility.
Volatility is defined by daily average of tick-by-tick price change.

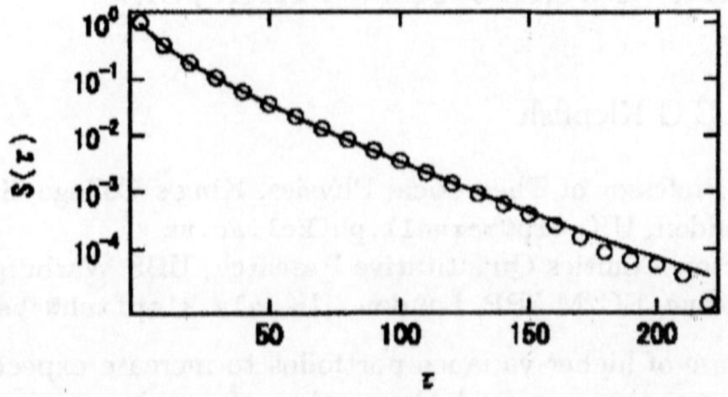

Fig. 5. The survival probability for $\Delta p = 0.05$ where volatility is stable.(1996.1.2~1996.9.30)

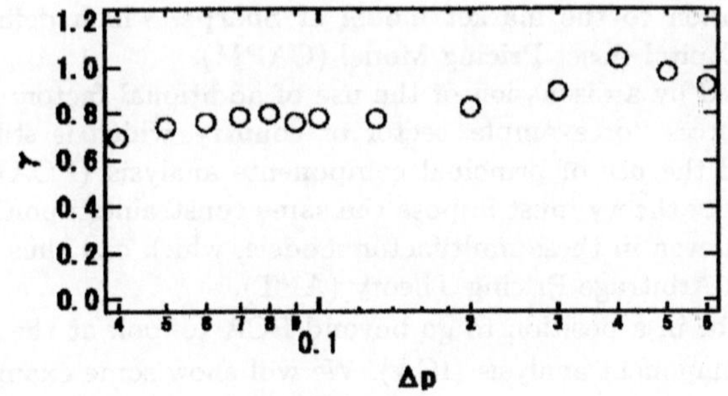

Fig. 6. The exponent γ of a stretched exponential decay(1996.1.2~1996.9.30)

4 Concluding Discussion

We investigated the first-passage time statistics for the high-precision yen-dollar rate. We showed that the survival probability for a given range is well characterized by the stretched exponential function in general, significantly different from Brownian diffusive random walk. It is found that for stable volatility period the FPT properties becomes closer to normal random walk, implying that the volatility is somehow related to the FPT property. It is still an open problem what makes the distribution of FPT to follow the non-exponential function.

References

1. Redner, S., *A Guide to First-Passage Processes*, Cambridge University Press, Cambridge, 2001
2. Simonsen, I., Jensen, M. H., and Johansen, A., Eur. Phys. J. B **27** 583-587 (2002)
3. Jensen, M. H., Johansen, A., and Simonsen, I., cond-mat/0211039

The Analysis of Financial Time Series Data by Independent Component Analysis

E. R. Pike[1] and E G Klepfish[2]

[1] Clerk Maxwell Professor of Theoretical Physics, King's College, Strand, WC2R 2LS, London, UK. erp@maxwell.ph.kcl.ac.uk
[2] Associate Director, Equities Quantitative Research, UBS Warburg, UBS AG, 1, Finsbury Avenue, EC2M 2PP, London, UK. ely.klepfish@ubsw.com

Summary. The use of higher-variance portfolios to increase expected investment return is based upon the empirical observation of a risk premium. This is the expected return $\bar{r}_A$ on an asset A, minus the prevailing risk-free interest rate r_F. We first review concisely the classical methods of efficient portfolio selection based upon the use of variance as a measure of risk. These comprise the Markowitz model and its simplification to the market model of Sharpe, which defines a "market factor" and the Capital Asset Pricing Model (CAPM).

This is followed by a discussion of the use of additional factors to account for specific market forces, for example, sector or country, which is still a subject of investigation, and the use of principal components analysis (PCA) in this problem. The Markowitz theory must impose the same constraints upon the individual portfolio holdings even in these multifactor models, which can thus be formulated within a so-called Arbitrage Pricing Theory (APT).

We will then be in a position to go beyond PCA to look at the newer method of independent component analysis (ICA). We will show some examples of typical features, including simultaneous diagonalisation of second- and fourth-order statistics and suggest ideas for cross fertilisation, for example, with connected moment expansions in physics.

Key words. Principal components, Statistical independence, Cross kurtosis.

1 The Markowitz theory

Let us consider the return of a portfolio P of N assets A_i held in the proportions X_i. The expected total return is $\bar{r}_P = \sum_{i=1}^{N} \bar{r}_i X_i$, where we use overbars to denote non-centred expectation over a defined time period. Markowitz [1] first showed how to diversify a portfolio af assets to minimise total risk, defined as variance of total return, for a given total return. The variance is given by

$$V_P = Var\Big(\sum_{i=1}^{N} \boldsymbol{r}_i X_i\Big) = \sum_{i,j=1}^{N} V_{ij} X_i X_j, \tag{1}$$

where $V_{ij} = \mathrm{Cov}(\boldsymbol{r}_i \boldsymbol{r}_j) = \rho_{ij}\sigma_i\sigma_j$. The ρ_{ij} are the correlation coefficients, and the σs the standard deviations[1]. To minimise V_P we differentiate with

[1] Note that $\mathrm{Cov}(\boldsymbol{r}_i \boldsymbol{r}_i) = \mathrm{Var}(\boldsymbol{r}_i) = \sigma_i^2$ and $\rho_{ii} = 1$

respect to each of the Xs and set the derivatives to zero. We first impose the constraints $\bar{r}_P = \sum_{i=1}^{N} X_i \bar{r}_i$ and $\sum_{i=1}^{N} X_i = 1$. This is done by adding $\mu(\sum_{i=1}^{N} X_i \bar{r}_i - \bar{r}_P)$ and $\lambda(\sum_i X_i - 1)$, both of which have zero value, to V_P

$$V_P = \sum_{i,j=1}^{N} V_{ij} X_i X_j + \mu\Big(\sum_{i=1}^{N} \bar{r}_i X_i - \bar{r}_P\Big) + \lambda\Big(\sum_i X_i - 1\Big), \qquad (2)$$

where μ and λ are Lagrange multipliers to be determined. The derivative of V_P with respect to X_k is

$$\frac{\partial V_P}{\partial X_k} = \sum_{i=1}^{N} (V_{ik} X_i + \mu \bar{r}_k) + \lambda. \qquad (3)$$

Unless we permit negative holdings (short selling) we also need to ensure that $X_i \geq 0$ for all i. The Markowitz portfolio thus requires the solution of

$$\sum_{i=1}^{N} (V_{ik} X_i + \mu \bar{r}_i) + \lambda = 0, \qquad (4)$$

subject to the positivity constraint. This is a problem of quadratic programming and has $N(N+3)/2$ variables. For each desired return the procedure provides the portfolio for which the return $\bar{r}_P$ is achieved with the minimum risk. On a plot of return versus risk ($\bar{r}_P$ versus σ_P) these portfolios constitute the so-called "efficient frontier".

1.1 The market model

The market model of Sharpe [2] introduces an asset sensitivity index, β_A, defined by the hypothesis of a linear relationship between the risk premia of the asset and that of a "market portfolio", M,

$$\bar{r}_A - r_F = \alpha_A + \beta_A[\bar{r}_M - r_F], \qquad (5)$$

where α_A is an "idiosyncratic", non-market, term. Risk relates only to fluctuating components, thus returns are centred and expressed as fractional changes (or logarithmic ratios) which, for example, facilitates comparisons across different currencies. We define the asset returns, per unit of value, as the vector set,

$$r_i(t_j) = \frac{r_i(t_{j+1}) - r_i(t_j)}{r_i(t_j)}; \qquad i = 1, 2, \ldots N; \quad j = 1, 2, \ldots M. \qquad (6)$$

The time variable will be implied when not written explicitly.

Sharpe defines a capital market line (CML) and a market portfolio M, using the concept of the Markowitz efficient frontier and the risk-free interest rate r_F, in the manner shown in fig. 1.

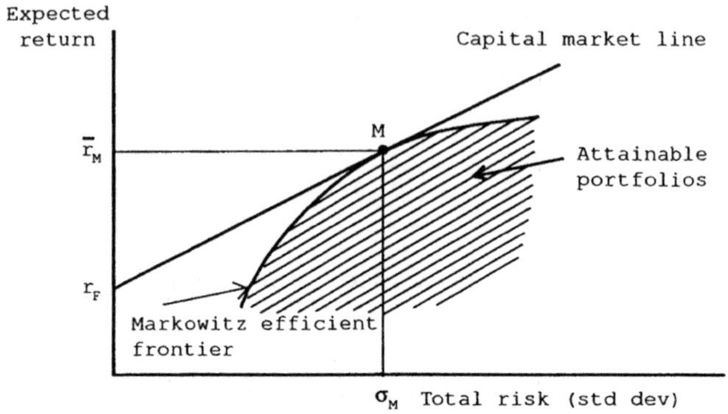

Fig. 1. The capital market line and the market portfolio

All efficient portfolios must lie on the capital market line and are achieved by combining a market portfolio and a risk-free asset in proportions which gives the investor's desired balance between risk and return. This is called the capital asset pricing model. However, in order to make the model more practical Sharpe uses a circular argument in which he assumes that the time-varying returns $\boldsymbol{r}_i$ of each asset in the portfolio follow those of a hypothetical market portfolio $\boldsymbol{r}_M$ in linear ratio β_i in such a way that

$$\boldsymbol{r}_i = \beta_i \boldsymbol{r}_M + \boldsymbol{r}_i^R, \tag{7}$$

where $\boldsymbol{r}_i^R$ is the residual fluctuation of the asset after removing the influence of $\boldsymbol{r}_M$. This residual (self) fluctuation is assumed to be a gaussian distributed, zero-mean fluctuation with variance V_i^R, uncorrelated with the market return.

With these assumptions the Markowitz theory is still valid up to eq.(4) when we may simplify the V_{ik} using eq.(7),

$$V_{ik} = V_{ik}^M + V_{ik}^R = \beta_i \beta_k \sigma_M^2 + \mathrm{Cov}(\boldsymbol{r}_i^R \boldsymbol{r}_k^R), \tag{8}$$

corresponding to market and residual returns, respectively. Sharpe now argues that the market fluctuation is dominant and that we may therefore neglect the cross terms between the self-fluctuations so that $\mathrm{Cov}(\boldsymbol{r}_i^R \boldsymbol{r}_k^R) \approx \sigma_i^2 \delta_{ik}$[2]. The Markowitz system eq.(4) then reduces to $3N + 2$ variables.

1.2 The multi-factor model

In practical applications in which some proxy for the market return such as a share index is used, the CAPM may not fulfill all of its expected aims. Fund managers, for example speak of "tracking errors" which are standard deviations relative to the index. Dissatisfaction with the CAPM has provoked

[2] δ_{ik} is the Dirac delta function =zero if $i \neq k$, and =unity if $i = k$

attempts to generalise it with a pricing model which, applied to a set of assets (including baskets of sets of assets) $A_1, A_2, \ldots A_N$, is given by

$$\boldsymbol{r}_i = \sum_{j=1}^{K} \beta_{ij} \boldsymbol{r}_{f_j} + \boldsymbol{r}_i^R,; \qquad i = 1, 2, \ldots N, \tag{9}$$

where K factors, $f_1, f_2, \ldots f_K$, each with its own sensitivity index, β_{ij}, with respect to asset A_i are introduced [3]. We are now referring all the factor returns and the centred return $\boldsymbol{r}$ to the risk-free rate. A primary classical market factor will be expected to emerge and to be dominant in "absorbing" most of the variance. One attempts to capture the effects of additional individual factors such as market sector, country or style by giving them their own contributions and β values. However, such factors will almost inevitably be coupled so that, for instance, a market sector factor will not be country neutral. The calculation of portfolio risk as a sum of independent contributions from each of the factors is then unsound, since a portfolio calculated from this model would not contain the same assets, in the same proportions, as the exact Markowitz solution.

2 Principal component analysis

A mathematically sound way to proceed is to go back to the raw data to find linear combinations of the assets which are uncorrelated. These may then be used as the factors of a multi-factor model and, in the Markowitz programme, will obey the Sharpe diagonal assumption exactly. There are many such combinations but the set with the minimum number of significant factors with high β values (or the set giving the lowest representational entropy) are called principal components. These are found from the $M \times N$ asset matrix, R, by noting that *any* such matrix may be decomposed into a product UWV^T, where U is an $M \times N$ matrix of orthonormal left singular column vectors, $\boldsymbol{u}$, W is a diagonal $N \times N$ matrix of singular values, w, and V^T is an $N \times N$ matrix of orthonormal right singular row vectors, v. Then, using this decomposition,

$$\boldsymbol{r}_i = \sum_{j=1}^{M} (\boldsymbol{r}_i . \boldsymbol{u}_j) \boldsymbol{u}_j; \qquad i = 1, 2, \ldots N. \tag{10}$$

Thus we have found M uncorrelated factors with returns given by the vectors $\boldsymbol{u}_j$ and β_{ij} values given by the covariances $(\boldsymbol{r}_i . \boldsymbol{u}_j)$. Although this decomposition would allow the Markowitz programme to be carried out fully, it seems that fund managers still would like to be able to truncate the sequence of factors to a small number and to know to what economic or financial reasoning each of them might relate. Perhaps only in this way, where Markowitz himself is superseded, could a fund be said to be "managed" in anything but an essentially mechanical way.

3 Independent component analysis

We noted above that there are many sets of decorrelated components, apart from the principal components of a matrix. We may find them all by performing unitary rotations of the N-dimensional space spanned by the share vectors, for which the principal components form one complete orthonormal basis. Independent component analysis gives up the idea of finding the minimum number of significant components and replaces it by attempting to find that set of decorrelated components which are as statistically independent of each other as possible.

The problem is known in other fields as blind source extraction, the factors being regarded as unknown statistically independent sources which have been mixed by the β matrix to generate the observed time series. Increasingly over the last half decade independent component analysis, has been introduced into the field of finance, [4] [5]. Conferences devoted entirely to blind source separation and ICA are now held annually but applications to finance, as yet however, are still relatively sparse.

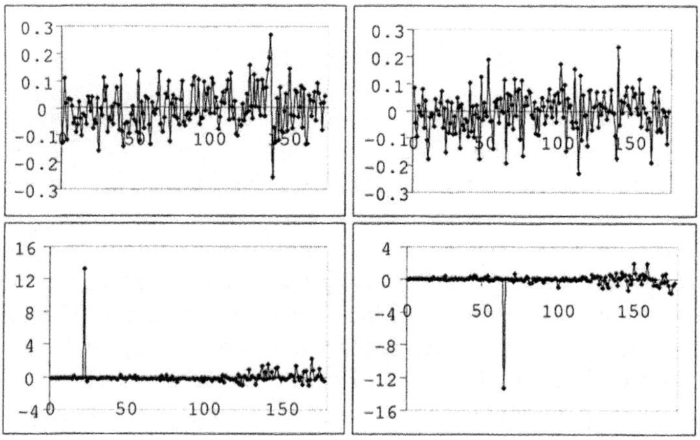

Fig. 2. Top: First two PCs. Bottom: First two Jade ICs

In contrast to PCA, where there are exact results for the decomposition of such a mixing problem into uncorrelated terms, there is as yet no unique algorithm to extract "independent" components, since different cost functions have been used to measure statistical independence. The results obtained to date thus depend on the particular algorithms chosen. For examples we have used three algorithms, Jade [6], FastICA(tanh) and FastICA(g^4) [7] applied to weekly data of 38 Dow Jones industrial sector regional indices (aggregated) between 1st Jan 99 to 30 May 2002.

The low-order ICs display sparse sharp features, see fig. 2, similar to those shown, for example, in [4]. The ICs, say $b[ij]$, simultaneously diagonalise

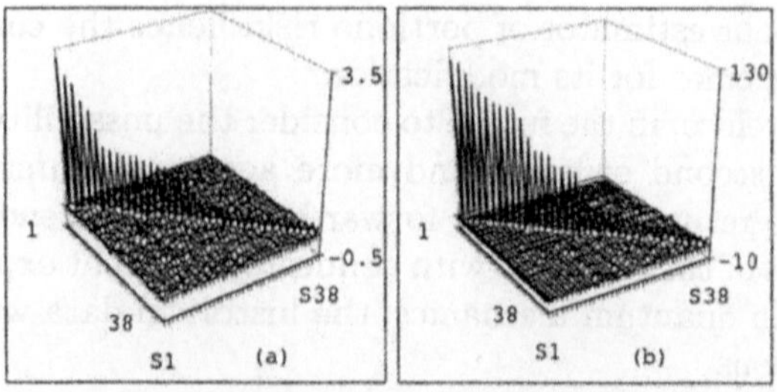

Fig. 3. (a) Covariance of "tanh" ICs (b) Cross-kurtoses of "tanh" ICs

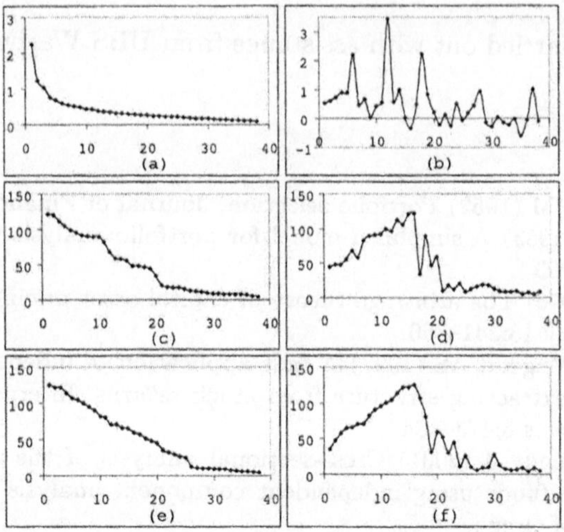

Fig. 4. (a) Variances of PCs (b) Kurtoses of PCs with decreasing variance (c) Kurtoses of tanh FastICs (d) Kurtoses of tanh FastICs with decreasing variance (e) Kurtoses of Jade ICs (f) Kurtoses of Jade ICs with decreasing variance

covariance essentially exactly and, to a good approximation, a cross kurtosis matrix, $\Sigma_{k=1}^{N}\left(b_{ki}^2 b_{kj}^2\right)/M\sigma_i^2\sigma_j^2 - 3$, as shown in fig. 3. They, of course, also form a complete basis for the expansion of any of the 38 constituents. We show some further interesting properties of the PCs and ICs in fig. 4.

We have, as yet, not been able to attempt any serious interpretation of the meaning of these new "factors"; it is clear, however, that the use of fourth and possibly higher-order statistics in portfolio theory merits a considerable amount of further investigation. This is particularly so since historical variance is often a poor estimator of portfolio risk, hence the common introduction of *ad hoc* schemes for its modification.

We wish, therefore, in the future to consider the possibility of using statistics beyond the second order to find more accurate estimators of the dispersion of future returns. One way forward which we intend to pursue is to explore analogies of this problem with connected moment expansions for two-point functions in quantum mechanics, the historical data would then act as trial wave functions.

Acknowledgement

This work was carried out with assistance from UBS Warburg.

References

1. Markowitz HM (1952) Portfolio selection. Journal of Finance 8:77–91
2. Sharp WF (1963) A simplified model for portfolio analysis. Management Science 9:277–293
3. Ross SA (1976) The arbitrage theory of capital asset pricing. Journal of Economic Theory 13:341–360
4. Back AD, Weigend AS (1997) A first application of independent component analysis to extracting structure from stock returns. International Journal of Neural Systems 8:473–484
5. Rojas S, Moody J (2001) Cross-sectional analysis of the returns of iShares MSCI index funds using independent component analysis. Report CSE 610 University of Oregon
6. Cardosa J-F, Souloumiac (1993) Blind beamforming for non-gaussian signals. IEE Proc-F 140:362–370
7. Hyvärinen A (1999) Fast and robust fixed-point algorithms for independent component analysis. IEEE Trans on Neural Networks 10:626–634

I-4 Various Markets

Modeling highly volatile and seasonal markets: evidence from the Nord Pool electricity market

Rafał Weron[1], Ingve Simonsen[2], Piotr Wilman[3]

[1]Hugo Steinhaus Center for Stochastic Methods, Wrocław University of Technology, 50-370 Wrocław, Poland
[2]Department of Physics, The Norwegian University of Science and Technology, N-7491 Trondheim, Norway
[3]Institute of Mathematics, Wrocław University of Tech., 50-370 Wrocław, Poland

Summary. In this paper we address the issue of modeling spot electricity prices. After analyzing factors leading to the unobservable in other financial or commodity markets price dynamics we propose a mean reverting jump diffusion model. We fit the model to data from the Nord Pool power exchange and find that it nearly duplicates the spot price's main characteristics. The model can thus be used for risk management and pricing derivatives written on the spot electricity price.

Key words. Electricity price, Mean reversion, Wavelet transform, Jump diffusion model

Introduction

The power industry's basic function is to convert fuel and primary energy into electricity and transport it to its customers. Because it is very costly to store electricity, the supply has to be matched with demand in real time.

Traditionally, centralized regulation of the electricity supply industry was considered necessary to ensure security of supply and efficient production. Efficiency was achieved through economics of scale. The power sector was characterized by a highly vertically integrated market structure with little competition. However, during the last decade many countries have restructured and deregulated their power sectors to introduce competition. These changes have already taken place in Europe (England and Wales, Norway, Sweden, Finland, Spain, Denmark, the Netherlands, Germany, Poland, Austria), in the Americas (parts of the US, Chile, Argentina, Peru, Bolivia, Colombia) and in the Asia/Pacific basin (Australia, New Zealand, Japan).

Introducing competition was believed to improve cost efficiency, increase diversity of fuel supply and provide additional benefits to the consumer. Microeconomic theory states that the price should decline when competition is introduced to the market (McConnel and Brue 1998). The power markets

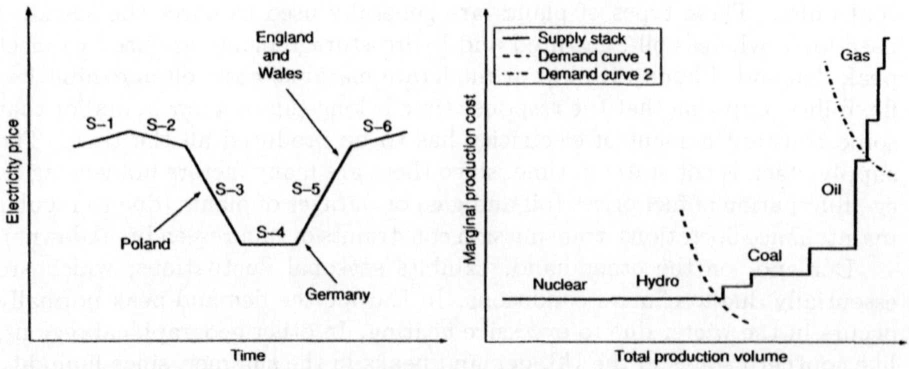

Fig. 1. *Left panel:* Stages of the "road from state to private monopoly" – state monopolies (S-1), declaration of competition (S-2), deregulation and decline of prices (S-3), tranquility period (S-4), mergers and effects of market power (S-5) and formation of private monopolies (S-6). Presumed stages in England and Wales, Germany and Poland are also depicted (see also Szalbierz and Weron 2000). *Right panel:* A schematic supply stack with two potential demand curves superimposed on it. The spot price, given as the intersection between demand and supply, is not very sensitive to demand shifts when the demand is low (curve 1), since the supply stack is typically flat in the low-demand region. However, when demand is high only small increments in demand can have huge effects on the price (curve 2).

in many countries have, however, reacted differently. They have followed (or are following) what could be called a "road from state to private monopoly" (Haas 2000, Szalbierz and Weron 2000) – after an initial fall, the wholesale prices rose back to pre-liberalization levels, see the left panel of Fig. 1.

In a competitive market utilities cannot automatically pass costs to customers. This has the effect of increasing uncertainty and risks born by the investors. Electricity is changing from a primarily technical business, to one in which the product is treated in much the same way as any other commodity, with trading and risk management as key tools to run a successful business. All this calls for adequate models of price dynamics capturing the main characteristics of electricity prices. In what follows we address the issue of modeling spot prices, because spot prices are one of the key factors in strategic planning and decision support systems of a majority of market players and are the underlying instrument of a number of power derivatives.

When supply meets demand

The supply stack is the ranking of all generation units of a given utility or of a set of utilities in a given region. This ranking is based on many factors, such as the marginal cost of production and the response time. The utility will typically first dispatch nuclear and hydro units, if available, followed by coal units. These types of plants are generally used to cover the so-called base load, whereas oil-, gas-fired and hydro-storage plants are used to meet peak-demand. Plants with low or moderate marginal costs often exhibit low flexibility, implying that the response time is long (up to a few hours) or that some constant amount of electricity has to be produced all the time. The supply stack is not static in time, since there are many factors influencing it, eg. fluctuation of fuel prices (oil and gas) or outages of plants (due to regular maintenance operations, transmission constraints or unforeseen breakdowns).

Demand, on the other hand, exhibits seasonal fluctuations, which are essentially due to climate conditions. In Europe the demand-peak normally occurs in the winter due to excessive heating. In other geographical regions, like southern states in the US, demand peaks in the summer, since humidity and heat initiate extensive use of air-conditioning. Electricity demand is also not uniform throughout the week. It peaks during weekdays' working hours and is low during nights and weekends (due to low industrial activity). Moreover, unexpected weather conditions can cause sudden and dramatic shocks with demand typically falling back to its normal level as soon as the underlying weather phenomenon is over.

The spot price is given as the intersection between demand and supply, see the right panel of Fig. 1. It is not very sensitive to demand shifts when the demand is low, since the supply stack typically is flat in the low-demand region. When demand is high, however, only small increments in demand can cause huge price spikes.

Nord Pool

The spot electricity market is actually a day-ahead market. A classical spot market would not be possible, since the system operator needs advanced notice to verify that the schedule is feasible and lies within transmission constraints. The spot is normally an hourly contract with physical delivery and is not traded on a continuous basis, but rather in the form of a conducted once per day auction. It is the underlying of most derivatives.

Several countries have deregulated their power markets and a few power exchanges have been established in the last decade. Yet, typically there are only one or two years of "stationary" data available due to the changes which are constantly taking place in many power markets. In our analysis, therefore, we will use spot prices from Nord Pool which is generally regarded as the most mature and "stable" power market in the world.

The Norwegian deregulation came into force in 1991 when grid owners were compelled to open their grids to competition. In 1993 Nord Pool started its business as a power exchange for the Norwegian market. In 1996 Sweden was integrated into the exchange, followed by Finland (1998) and western Denmark (1999). Nord Pool offers two types of standardized contracts – physically settled spot contracts and financially settled futures, forward, option and other specialized contracts. Every day is divided into 24 hourly spot contracts. Before noon, the previous day, all participants send in their bids for each hour. From the bids Nord Pool obtains the aggregated supply and demand curves. The auction system is used in order to increase liquidity. The system price is calculated as the equilibrium point for each of the 24 hours. It is a theoretical price in the sense that it assumes that no congestions will occur and is the same in the whole Nordic area (Nord Pool 2001).

Spot price dynamics

The goal of this paper is to propose a model for electricity spot price dynamics. The traditional approach to modeling price processes of stochastic (random) nature in finance is to apply diffusion-type stochastic differential equations (SDE's) of the form

$$dX_t = \mu(X,t)dt + \sigma(X,t)dB_t, \tag{1}$$

where $\mu(X,t)$ is the drift, $\sigma(X,t)$ is the volatility (scaling factor) and dB_t are the increments of standard Brownian motion. Some phenomena are modeled by a set of related SDE's, leading to so called multi-factor models. On the other hand, both μ and σ can be defined by SDE's of their own, leading to stochastic drift and/or volatility models. Probably the best known member of this family of processes is geometric Brownian motion (with $\mu(X,t) = \mu X$ and $\sigma(X,t) = \sigma X$), which was used already in the 1960's for modeling stock price movements. In this paper we will construct a model of electricity spot prices in line with the diffusion-type approach.

According to economic theory the price of any good is determined by matching demand with supply (McConnel and Brue 1998). Yet, electricity spot prices exhibit a behavior not observed in other financial or commodity markets. Is the power market a counterexample? Or is electricity so unique that the supply and demand curves intersect in a way causing this odd behavior? We have already given some hints as to the answer to these questions. However, for the sake of completeness, we will shortly review the most important factors leading to such a behavior.

Mean reversion

Energy spot prices are in general regarded to be mean reverting (Schwartz 1997). Among other financial time series spot electricity prices are perhaps the best example of anti-persistent data. In Weron and Przybyłowicz (2000) and Weron (2002) the R/S analysis, detrended fluctuation analysis and periodogram regression methods were used to verify this claim. Here we apply the Average Wavelet Coefficient (AWC) method of Simonsen et al. (1998), that has proven useful in particular when dealing with multi-scale time series (Simonsen 2003). This method utilizes the wavelet transform in order

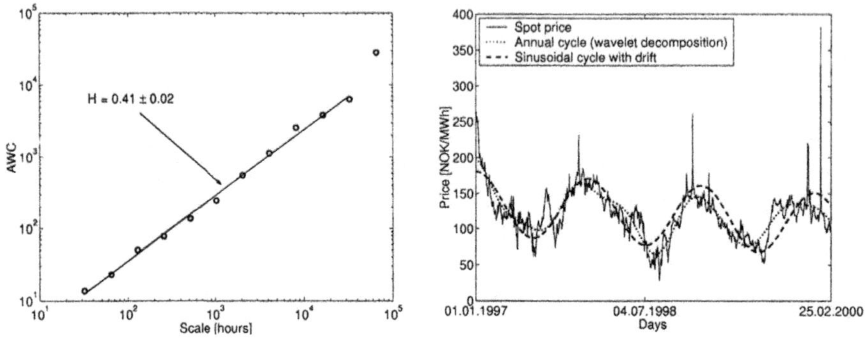

Fig. 2. *Left panel:* The AWC statistics, $W[p](a)$ vs. scale a on a log-log paper, of the hourly Nord Pool electricity spot price since May 4, 1992 until December 31, 2000. The scaling region $a > 24$ hours corresponds to a Hurst exponent $H = 0.41 \pm 0.02$. *Right panel:* Nord Pool market daily average system prices since January 1, 1997 until February 25, 2000. Superimposed on the plot is the annual cycle obtained through a wavelet decomposition technique and its approximation by a sinusoid with a linear trend. Observe that the sinusoid approximates the annual cycle quite well justifying its use in the analysis.

to measure the temporal self-affine correlations, i.e. it measures the Hurst exponent H. This is done by transforming the time series of spot electricity prices p_t into the wavelet-domain, $\mathcal{W}[p](a, b)$, where a denotes the scale parameter, and b is the location (Percival and Walden 2000). The AWC method consists of finding a representative (wavelet) "energy" or amplitude for a given scale a. This is usually done by simply taking the arithmetic average of $|\mathcal{W}[p](a, b)|$ over all location parameters b corresponding to one and the same scale a. We can therefore construct, from the wavelet transform of p_t, the AWC spectrum $W[p](a)$ that only depends on the scale. If p_t is a self-affine process characterized by the exponent H, this spectrum should scale as (Simonsen et al. 1998)

$$W[p](a) = \langle |\mathcal{W}[p](a, b)| \rangle_b \sim a^{H+1/2}. \qquad (2)$$

So, if the signal is self-affine and we plot $W[p](a)$ against a the points should constitute a line with slope $H+1/2$ on a log-log paper. The application of this method to Nord Pool spot prices from the period May 4, 1992 – December 31, 2000 is depicted in the left panel of Fig. 2. For time intervals ranging from a day to almost four years the Hurst exponent $H = 0.41 \pm 0.02$ indicating mean reversion. For time intervals of less than 24 hours, however, H is above 0.5, the level of the Hurst exponent for Brownian motion (Simonsen 2003).

Mean reversion is typically modeled by having a drift term that is negative if the spot price is higher than the mean reversion level and positive if it is lower. To make things simple we will start with probably the simplest mean reverting model originally proposed for describing interest rates dynamics.

The Vasicek (1977) model, also referred to as an arithmetic Ornstein-Uhlenbeck process, is described by the following SDE

$$dX_t = (\alpha - \beta X_t)dt + \sigma dB_t = \beta(L - X_t)dt + \sigma dB_t. \quad (3)$$

This is a one-factor model that reverts to the mean $L = \frac{\alpha}{\beta}$ with β being the magnitude of the speed of adjustment. The second term is responsible for the volatility of the process. The conditional distribution of X at time t is normal with mean $E[X_t] = \frac{\alpha}{\beta} + (X_0 - \frac{\alpha}{\beta})e^{-\beta t}$ and variance $Var[X_t] = \frac{\sigma^2}{2\beta}(1 - e^{-2\beta t})$. These relations imply that $E[X_t] \to L = \frac{\alpha}{\beta}$ as $t \to \infty$. Starting at different points the Vasicek model trajectories tend to reverse to the long run mean and stabilize in the corridor defined by the standard deviation of the process. The equilibrium level L can be also made time dependent to reflect the fact that electricity prices tend to revert to different levels over the year.

Seasonal fluctuations

As we have already mentioned, demand follows seasonal fluctuations, caused mainly by climate conditions. In some countries also the supply side shows seasonal variations in output. Hydro units, for example, are heavily dependent on precipitation and snow melting, which varies from season to season. These seasonal fluctuations in demand and supply are directly translated into the seasonal behavior of spot electricity prices. In the right panel of Fig. 2, we have plotted the Nord Pool market daily average system prices since January 1, 1997 until February 25, 2000. The choice of the period used in the analysis is not incidental – 1996 was a dry year with exceptionally high electricity prices and the rest of the year 2000 was used for out-of-sample testing of the Asian options pricing model (Weron 2003). Superimposed on the plot is the annual cycle obtained through a wavelet decomposition technique, i.e. a technique consisting of removing several layers of noise and leaving out only the large scale wavelets (Percival and Walden 2000, Simonsen et al. 2002). As it turns out, in this period the annual cycle can be quite well approximated by a sinusoid with a linear trend. This is in line with the approach of Pilipovic (1997), who suggests fitting a proper sinusoidal function (eg. a sum of two cosine functions with distinct periods) to spot prices. However, such an approach would not be suitable for the California market where demand and, to some extent, the spot prices are rather flat throughout the year with a hump in the summer (Weron et al. 2001, Nowicka-Zagrajek and Weron 2002). Another method of modeling seasonality consists of fitting a piecewise constant function of a one year period, where for each month one tries to determine an average value out of the whole analyzed time series (Bhanot 2000, Lucia and Schwartz 2002). Although flexible, this method lacks smoothness, which may have a negative impact on statistical inference of the deseasonalized price process.

Jumps

In addition to mean reversion and strong seasonality on the annual, weekly and daily level, spot electricity prices exhibit infrequent, but large jumps. The spot price can increase tenfold during a single hour. This is the effect of non-storability of electricity. Electricity to be delivered at a specific hour cannot be substituted for electricity available shortly after or before, since it has to be consumed at the same time as it is produced. Jumps in the spot prices are an effect of extreme load fluctuations, caused by severe weather conditions often in combination with generation outages or transmission failures. These spikes are normally quite short-lived, and as soon as the weather phenomenon or outage is over, prices fall back to a normal level (Kaminski 1999).

The "jumpy" character of electricity prices, calls for spot price modeling, which is not continuous. One approach, suggested in Deidersen and Trück (2002), reduces to substituting Brownian motion with eg. a positively skewed α-stable Levy motion (Janicki and Weron 1994, Rachev and Mittnik 2000). However, this would lead to purely discontinuous price paths and would limit control of the intensity of the jumps. Another approach is to introduce to eqn. (3) a jump component $J_t dq_t$ (Johnson and Barz 1999, Clewlow and Strickland 2000), where J_t is a random jump size, eg. a lognormal random variable $\log J_t \sim N(\mu, \rho^2)$, and q_t is eg. a Poisson random variable with intensity κ. Eydeland and Geman (2000) propose a similar model, where – to account for the fact that jumps tend to be more severe during high price periods – the jump part is given by $J_t X_t dq_t$.

Electricity prices tend to rapidly revert to their normal level after a jump. In the models mentioned above the price is forced back by the mean reversion after a jump, which may be not fast enough. Geman and Roncoroni (2002) suggest using mean reversion coupled with downward jumps. Alternatively, a positive jump may be always followed by a negative jump to capture the rapid decline of electricity prices after a spike. On the daily level, i.e. when analyzing average daily prices, the latter approach seems to be a better solution since spikes typically do not last more than a day.

The model

As stated previously, the annual cycle can be quite well approximated by a sinusoid (see the right panel of Fig. 2). Instead of making the equilibrium level L time dependent, we incorporate this periodicity into our model in the form of an external, deterministic sinusoidal function $S_t = A \sin\left(\frac{2\pi}{365}(t+B)\right) + Ct$, where A, B and C are constants (obtained through a least squares fit).

Like demand, spot electricity prices are not uniform throughout the week. The intra-week and intra-day variations of demand caused by different level of working activities translate into periodical fluctuations in electricity prices. However, in the present analysis we will not address the issue of intra-day variations and will analyze only daily average prices. We deal with the intra-week variations by preprocessing the data using the moving average technique, which reduces to calculating the weekly profile s_t, i.e. an average week, and subtracting it from the spot prices (see eg. Weron et al. 2001).

Despite their rarity, price spikes are the very motive for designing insurance protection against electricity price movements. This is one of the most serious reasons for including jump components in realistic models of electricity price dynamics. We also do so with our model. Reflecting the fact that on the daily scale spikes typically do not last more than one time point (i.e. one day) we let a positive jump be always followed by a negative jump of about the same magnitude. This is achieved by letting the stochastic part X_t be independent of the jump component $J_t dq_t$. For the sake of simplicity we let J_t be a lognormal random variable $\log J_t \sim N(\mu, \rho^2)$ and q_t be a Poisson random variable with intensity κ. The jump component is estimated from the logarithm of the deseasonalized, with respect to the weekly and annual cycles, prices $d_t = \log(p_t - s_t - S_t)$ through a two-step procedure. First, all jumps – defined as price increments exceeding 3 standard deviations of all price changes – are removed from d_t. Next, the intensity κ and the distribution of the magnitude J_t of the jumps is estimated from these few selected points (7 in the whole analyzed data set).

Putting all the facts together, our model has the following form

$$p_t = s_t + S_t + e^{J_t dq_t + X_t}, \tag{4}$$

where X_t is defined by eqn. (3). The exponent in the last term of eqn. (4) reflects the fact that the marginal distribution of X_t is Gaussian, whereas the deseasonalized, with respect to the weekly and annual cycles, and "spikeless" spot prices can be very well described by a lognormal distribution, see the left panel of Fig. 3. The parameters of the Vasicek SDE (3) are estimated using a Matlab implementation (Cliff 2000) of the Generalized Method of Moments (Hansen 1982).

To verify the adequacy of our model we fitted it to Nord Pool market daily average system prices from the period January 1, 1997 – January 15, 2000, see the right panel of Fig. 2. Next, we simulated price trajectories starting January 15, 2000, see the right panel of Fig. 3. The statistical characteristics of these simulated paths closely resembled the original spot price, allowing us to use the model for pricing Nord Pool's Asian options written on the spot electricity price (Weron 2003).

Conclusions

The liberalization of the power markets has created additional risks and new challenges for players in the market. The uniqueness of electricity – such as highly fluctuating demand, non-flexible supply, transmission congestion

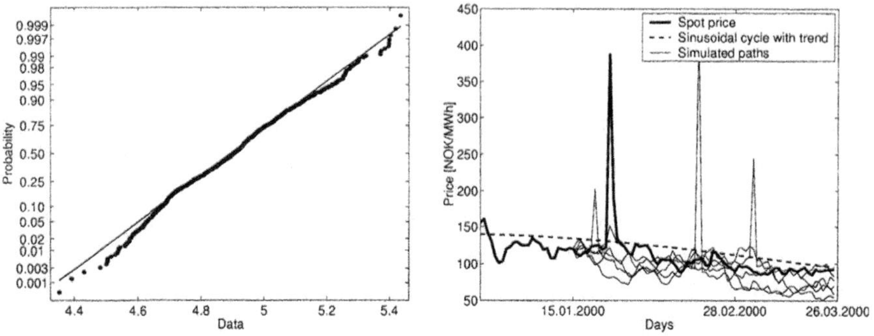

Fig. 3. *Left panel:* The normal probability plot of the logarithm of the deseasonalized, with respect to the weekly and annual cycles, and "spikeless" spot prices (i.e. $\log(p_t - s_t - S_t) - J_t dq_t$). The dots form a straight line indicating a Gaussian distribution. In fact, the Bera-Jarque test (Spanos 1993) indicates that we cannot reject normality at the 2% level, whereas for the logarithm of the deseasonalized (but with spikes) prices the Bera-Jarque test lets us reject normality even at the 0.5% level. *Right panel:* The true spot price trajectory and model simulated paths.

issues and its non-storability – distinguishes the power market from other financial or commodity markets and calls for new models of price dynamics capturing mean reversion, seasonality and price spikes. The number of papers addressing these problems is still scarce and the suggested solutions are usually not universal or unsatisfactory. In this paper we have proposed a model which closely resembles the spot price trajectories from the Nord Pool power exchange. We believe that it can be successfully used to price a number of derivatives in the Nordic power market.

Acknowledgements

Many thanks to SKM Market Predictor AS for providing the data. The first author also would like to thank Nihon Keizai Shimbun for financial support. The last author would like to acknowledge the support from KBN Grant PBZ-KBN 016/P03/99.

References

Bhanot, K. (2000) Behavior of power prices: Implications for the valuation and hedging of financial contracts. The Journal of Risk 2: 43-62.

Clewlow, L., Strickland, C. (2000) Energy Derivatives – Pricing and Risk Management. Lacima Publications, London.

Cliff, M. (2000) GMM libraries for MATLAB. See: www.mgmt.purdue.edu.

Deidersen, J., Trück, S. (2002) Energy price dynamics – quantitative studies and stochastic processes. University of Karlsruhe, Tech. Report TR-ISMW-12/2002.

Eydeland, A., Geman, H. (2000) Fundamentals of Electricity Derivatives in Energy Modeling and the Management of Uncertainty. Risk Books.

Geman, H., Roncoroni, A. (2002) A class of marked point processes for modelling electricity prices. ESSEC Graduate Business School preprint.

Hansen, L.P. (1982) Large sample properties of Generalized Method of Moments estimators. Econometrica 50: 1029-1054.

Haas, R. (2000) When will electricity prices in Europe start to increase again? Vienna University of Technology, IEW preprint.

Janicki, A., Weron, A. (1994) Simulation and Chaotic Behavior of α-Stable Stochastic Processes. Marcel Dekker, New York.

Johnson, B., Barz, G. (1999) Selecting stochastic processes for modelling electricity prices. Energy Modeling and the Management of Uncertainty.

Kaminski, V., ed. (1999) Managing Energy Price Risk. Risk Books, London.

Lucia, J.J., Schwartz, E.S. (2002) Electricity prices and power derivatives: Evidence from the Nordic Power Exchange. Review of Derivatives Research 5: 5-50.

McConnel, C., Brue, S. (1998) Economics. 14th ed., McGraw-Hill.

Nord Pool (2001) Nord Pool Annual Report. Nord Pool ASA.

Nowicka-Zagrajek, J., Weron, R. (2002) Modeling electricity loads in California: ARMA models with hyperbolic noise. Signal Processing 82: 1903-1915.

Percival, D.B., Walden, A.T. (2000) Wavelet Methods for Time Series Analysis. Cambridge University Press, Cambridge.

Pilipovic, D. (1998) Energy Risk: Valuing and Managing Energy Derivatives. McGraw-Hill, New York.

Rachev, S., Mittnik, S. (2000) Stable Paretian Models in Finance. Wiley.

Schwartz, E.S. (1997) The stochastic behavior of commodity prices: Implications for valuation and hedging. The Journal of Finance 52: 923-973.

Simonsen, I. (2003) Measuring anti-correlations in the Nordic electricity spot market by wavelets. Physica A, in press. See also: arXiv:cond-mat/0108033.

Simonsen, I., Hansen, A., Nes, O. (1998) Determination of the Hurst exponent by use of wavelet transforms. Physical Review E 58: 2779-2787.

Simonsen, I., Jensen, M. H., Johansen, A. (2002) Optimal investment horizons. The European Physical Journal B 27: 583-586.

Spanos, A. (1993) Statistical Foundations of Econometric Modelling. Cambridge University Press.

Szalbierz, Z., Weron, A. (2000) From a state monopoly to a private monopoly. When electric energy prices will rise again? Energetyka 12: 634-638.

Vasicek, O. (1977) An equilibrium characterization of the term structure. Journal of Financial Economics 5: 177-188.

Weron, R. (2002) Measuring long-range dependence in electricity prices. In: Takayasu, H. (ed.) Empirical Science of Financial Fluctuations, Springer-Tokyo, pp 110-119.

Weron, R. (2003) Pricing Asian options in the Nord Pool electricity market. HSC Research Report HSC/03/1, see: www.im.pwr.wroc.pl/~hugo/Publications.html.

Weron, R., Przybyłowicz, B. (2000) Hurst analysis of electricity price dynamics. Physica A 283: 462-468.

Weron, R., Kozłowska, B., Nowicka-Zagrajek, J. (2001) Modeling electricity loads in California: a continuous time approach. Physica A 299: 344-350.

Statistical Properties of Commodity Price Fluctuations

Kaushik Matia[1], Yosef Ashkenazy[2], Luis A. Nunes Amaral[1,3], Stephen P. Goodwin[4], and H. Eugene Stanley[1]

[1] Center for Polymer Studies and Department of Physics Boston University, Boston, MA 02215 kaushik@buphy.bu.edu
[2] Department of Earth, Atmospheric, and Planetary Sciences MIT, Cambridge, MA 02139
[3] Department of Chemical Engineering, Northwestern University, Evanston, IL60208
[4] BP Upstream, Chertsey Road, Sunbury-on-Thames, Middlesex TW16 7LN

Summary. Though the statistical properties of price fluctuations for stocks have been studied extensively since the last decade, not many studies have been done on commodity price fluctuations. Here, we perform a comparative study to test whether commodities are statistically similar to stocks with respect to (a) probability distribution and (b) correlation of price fluctuations. We analyze daily returns of spot prices for 29 commodities and daily returns of future prices for 13 commodities over a period exceeding 10 years and compare the results with a database of 2449 stocks over the same period.

1 Introduction

Until now much interest in the study of economic markets has been concentrated on stocks, where a number of empirical findings have been established such as [1, 2] (i) the distribution of price changes $P(x)$ being approximately symmetric and decaying with power law tails $P(x) \sim 1/x^{\alpha+1}$, with $\alpha + 1 \approx 4$; (ii) the price changes being exponentially (short-range) correlated, while the absolute values of price changes ("volatility") are power-law (long-range) correlated [3, 4].

Unlike stock and foreign exchange markets, commodity markets have received little attention. Recently it was found [5] that commodity markets have qualitative features similar to those of the stock market. This similarity is intriguing because of the special features unique to the commodity market such as: (i) most commodities require storage; (ii) most commodities require transportation to bring them to the market from where they are produced; and (iii) it is plausible that commodities may exhibit a slower response to change in demand because their price depends on the supply of the actual object.

Because of the stronger constraints, affecting commodity markets, mentioned above, one might surmise that commodity prices show larger fluctuations than stock prices. In fact, exponents of power law tails of probability distributions of the returns of spot prices [6] of commodities such as cotton and wheat have been reported [7] to be Lévy-stable i.e., $0 < \alpha < 2$, whereas the returns of future prices of commodities such as potatoes have been reported [8] to be outside the Lévy-stable domain i.e., $\alpha > 2$.

The multifractal (MF) spectrum reflects the n-point correlations providing more information about the temporal organization of price fluctuations than two-point correlations. Previous work reports a broad MF spectrum of stock indices and foreign exchange markets [9]. Two recent models [10] explain the observed MF properties by assuming that price changes are the product of two stochastic variables, one being uncorrelated and normally distributed and the other being correlated and log-normally distributed. The price changes predicted by these models do not have the power law probability distribution [2, 5] observed empirically, and thus destroying the temporal organization by shuffling the price changes significantly narrows the MF spectrum.

2 Probability distribution of price fluctuation

First we test whether the probability distribution function (PDF) of commodity price fluctuations is statistically distinguishable from that of stocks. To this end, we study the fluctuations in the spot price for 29 commodities and in the future price for 13 commodities and compare our results with the statistical properties of daily returns in stock markets. We define the normalized price fluctuation ("return") as $g(t) \equiv (\ln S(t+\Delta t) - \ln S(t))/\sigma$, where $\Delta t = 1$ day, $S(t)$ is the price, and σ is the standard deviation of $\ln S(t+\Delta t) - \ln S(t)$ over the duration of the time series (typically 15 years).

The probability distributions $P(g_i > x)$ of the returns follow power law forms $P(g_i > x) \sim \frac{1}{x^{\alpha_i}}$, where α_i is outside the Lévy-stable domain $0 < \alpha_i < 2$.

Figures 1(a), (b), (c), and (d) display Hill estimates [11] of α_i for the spot price of 29 commodities and future price of 13 commodities (see [5] for more detail). For the spot prices the average exponents are

$$\overline{\alpha}_{\text{spot}} \equiv \frac{1}{29}\sum_{i=1}^{29} \alpha_i = \begin{cases} 2.3 \pm 0.2 \text{ positive tail} \\ 2.2 \pm 0.1 \text{ negative tail.} \end{cases} \tag{1}$$

For the future prices the average exponents are

$$\overline{\alpha}_{\text{future}} \equiv \frac{1}{13}\sum_{i=1}^{13} \alpha_i = \begin{cases} 3.1 \pm 0.2 \text{ positive tail} \\ 3.3 \pm 0.2 \text{ negative tail.} \end{cases} \tag{2}$$

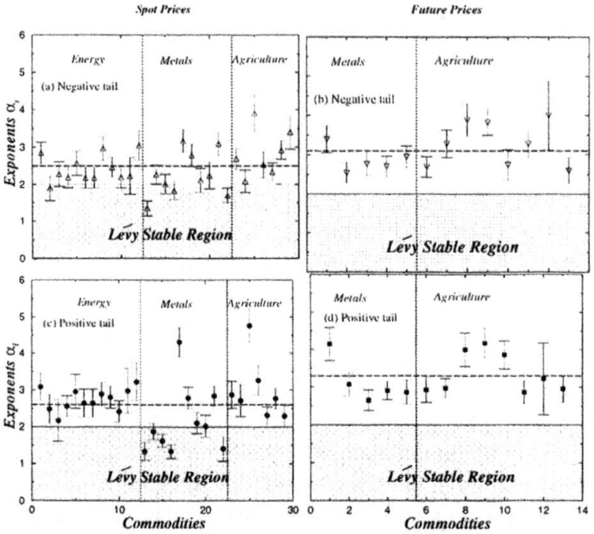

Fig. 1. Exponents α_i of the negative tail of commodity (a) spot prices and (b) future prices, where i indexes the 29 commodity spots and 13 commodity future prices analyzed. Exponents α_i of the positive tail for commodity (c) spots prices and (d) future prices. We employ Hill's method [11] to estimate the exponent α_i of each probability distribution in the range $x \geq x_{\text{cutoff}}$, with $x_{\text{cutoff}} = 2$. The dashed lines show the average values defined in Eq.(1), (2). Shaded regions indicate the range of Lévy-stable exponents, $0 < \alpha < 2$. Note that the mean exponent of spot prices is smaller than the mean exponent of the future prices, and that both are outside the Lévy-stable domain.

3 Correlations of price fluctuation

3.1 Two-point correlations

We next study the temporal correlations of the returns. The average autocorrelation function of commodities $\overline{C}(\tau) \equiv \frac{1}{N}\Sigma_i^N < g_i(t)g_i(t+\tau) >$, where N=29 for spot prices and N=13 for future prices decays exponentially as $e^{-\tau/\tau_c}$. We find that $\tau_c^{\text{spot}} = 2.3$ days and $\tau_c^{\text{future}} < 1$ day. To further quantify time correlations, we use the detrended fluctuation analysis (DFA) method [12]. The DFA exponent $\hat{\alpha}_{\text{DFA}}$ gives information about the correlations present. If $C(\tau) \sim \tau^{-\gamma}$, then $\hat{\alpha}_{\text{DFA}} = (2-\gamma)/2$, while if $C(\tau) \sim e^{-\tau/\tau_c}$, then $\hat{\alpha}_{\text{DFA}} = 1/2$ [12]. We find that $\hat{\alpha}_{\text{DFA}} = 0.51 \pm 0.05$ and $\hat{\alpha}_{\text{DFA}} = 0.50 \pm 0.05$ for spot and future prices respectively, consistent with the exponential decay of $\overline{C}(\tau)$. We also observe that $|g_i|$, the absolute value of returns (one measure of volatility), are power law correlated with

$$\hat{\alpha}_{\text{DFA}} = \begin{cases} 0.63 \pm 0.05 \text{ spot prices} \\ 0.60 \pm 0.05 \text{ future prices,} \end{cases} \tag{3}$$

which implies a power law decay of the autocorrelation of the absolute value of returns with

$$\gamma = \begin{cases} 0.74 \pm 0.1 \text{ spot prices} \\ 0.80 \pm 0.1 \text{ future prices.} \end{cases} \tag{4}$$

Note that γ for commodities is larger than γ for stocks [3].

3.2 Higher order correlations

We use the multifractal detrended fluctuation analysis (MF-DFA) technique [13] to study the MF properties and thus the different orders of temporal correlations, of the returns for stocks and commodities. The scaling function of moment q, $F_q(s)$ [13] follows the scaling law $F_q(s) \sim s^{\tau(q)}$.

First we perform a shuffling procedure on the time series of price fluctuations for stocks and commodities by randomly exchanging pairs. This shuffling procedure preserves the distribution of the returns but destroys any temporal correlations (see [14] for more details).

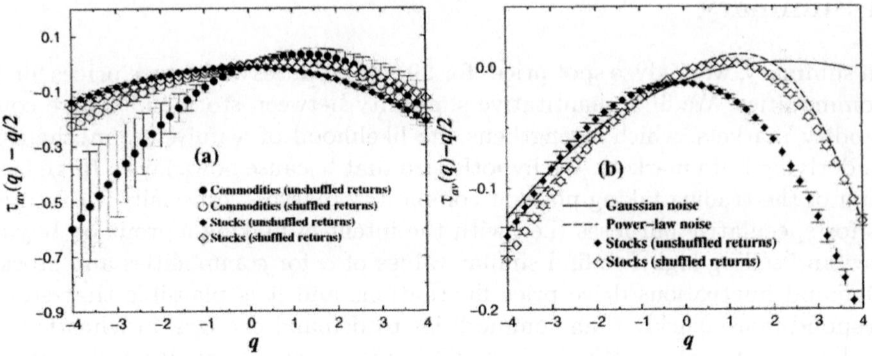

Fig. 2. (a) $\tau_{\text{av}}(q)$ for returns and shuffled returns for 29 commodities and 2449 stocks. To better visualize the results we plot $\tau_{\text{av}}(q) - q/2$ instead of $\tau_{\text{av}}(q)$. The exponents $\tau_{\text{av}}(q)$ are calculated for window scales of 10 – 100 days. After shuffling, $\tau_{\text{av}}(q)$ are comparable for both stocks and commodities. (b) $\tau_{\text{av}}(q)$ spectrum of the returns and shuffled returns for stocks, compared with uncorrelated surrogate data with Gaussian probability distributions and power law probability distributions (with power law exponent $\alpha \approx 3$). After shuffling, $\tau_{\text{av}}(q)$ for stocks becomes comparable with $\tau_{\text{av}}(q)$ of the surrogate data obtained for the power law probability distribution.

Next, we analyze the MF properties of the returns of stocks and commodities before and after shuffling of the returns. Figure 2(a) displays the averages separately $\tau_{\text{av}}(q) \equiv \frac{1}{N}\sum_{i=1}^{N}\tau_i(q)$, for $N = 29$ commodities and $N = 2449$ stocks. Note that (i) the scaling exponents, $\tau_{\text{av}}(q)|_{q<0}$ significantly differ for commodities and stocks, whereas $\tau_{\text{av}}(q)|_{q>0}$ are similar, suggesting that commodities are similar to stocks for the large fluctuations and differ for the small fluctuations; (ii) we observe that after shuffling the returns, $\tau_{\mathrm{av}}(q)$ for stocks hardly changes for $q < 0$, but $\tau_{\mathrm{av}}(q)$ for commodities changes and becomes comparable to stocks for the entire range of q.

In order to study the contribution of the power law tails of the returns on the MF spectrum, we generate surrogate data sets (i) with a normal distribution and (ii) with power law tails with $\alpha \approx 3$ (as observed empirically [2, 5]). Figure 2(b) displays $\tau_{\mathrm{av}}(q)$ averaged over 2449 realizations of surrogate data, each with 3000 data points. The $\tau_{\mathrm{av}}(q)$ of the surrogate power law distributed data is very close to the $\tau_{\mathrm{av}}(q)$ of stocks after shuffling. This indicates that a significant part of the $\tau_{\mathrm{av}}(q)$ spectrum of stocks and commodities comes from the power law distribution of the returns. Note that there is a small difference in $\tau_{\mathrm{av}}(q)$ of stocks before and after shuffling, indicating that the power-law distribution of the returns is not the only source of multifractality, but that there is also a relatively small contribution due to the temporal organization of returns. For commodities, this temporal organization is more dominant.

4 Summary

In summary, we analyze spot prices for 29 commodities and future prices for 13 commodities. We find quantitative similarity between stock and future commodity markets, which strengthens the likelihood of a universal mechanism underlying both markets. We hypothesize that because nowadays a large fraction of the trading taking place at commodity markets, especially for futures, is for speculative purposes (i.e., with the intent of making a profit by buying low and selling high) we find similar values of α for commodities and stocks. Demand fluctuations drive price fluctuations and it is plausible that stocks respond more quickly than commodities to demand changes. Stochastic perturbations, together with the immediate price response to demand changes, may weaken the existing higher order temporal organization. Commodities, on the other hand, have a slower response. Thus, small or short-time perturbations are felt less by commodities than by stocks. We conjecture that the more homogeneous returns of stocks explain the difference between the MF properties of stocks and commodities.

We also find that commodities have a broader MF spectrum than stocks. A major contribution to multifractality is the power law tail of the probability distribution of the returns. Moreover, the MF spectra of stocks and commodities are partly related to the power law probability distribution of returns and partly to the different orders of temporal correlations present.

We thank S. V. Buldyrev, X. Gabaix, P. Gopikrishnan, S. Havlin, J. W. Kantelhardt, V. Plerou, A. Schweiger, and especially P. R. King for helpful discussions and suggestions, and BP/Amoco for financial support. Y. A. thanks the Bikura fellowship for financial support.

References

1. J. P. Bouchaud and M. Potters, *Theory of Financial Risk* (Cambridge University Press, Cambridge 2000); R. N. Mantegna and H. E. Stanley, *An Introduction to Econophysics: Correlations and Complexity in Finance* (Cambridge University Press, Cambridge 2000).
2. M. M. Dacorogna *et al.*, J. Int'l Money and Finance **12**, 413 (1993); G. Weisbuch *et al.*, Econ J. **463**, 411 (2000); J. P. Nadal *et al.*, in *Advances in Self-Organization and Evolutionary Economics*, edited by J. Lesourne and A. Orlian (Economica, London, 1998), p. 149; P. Gopikrishnan, V. Plerou, L. A. N. Amaral, M. Meyer, and H. E. Stanley, Phys. Rev. E **60**, 5305 (1999); Y. Liu, P. Gopikrishnan, P. Cizeau, M. Meyer, C.-K. Peng, and H. E. Stanley, Phys. Rev. E **60**, 1390 (1999); V. Plerou, P. Gopikrishnan, L. A. N. Amaral, M. Meyer, and H. E. Stanley, Phys. Rev. E **60**, 6519 (1999);
3. R. N. Mantegna and H. E. Stanley, Nature **376**, 46 (1995); P. Gopikrishnan *et al.* Phys. Rev. E **60**, 5305 (1999); P. Gopikrishnan *et al.* Eur. Phys. J. B **3**, 139 (1998); Y. Liu *et al.* Phys. Rev. E **60**, 1390 (1999); V. Plerou *et al.* Phys. Rev. E **60**, 6519 (1999);
4. For stock markets the autocorrelation function decays exponentially to the noise level with time scales of minutes [3].
5. K. Matia, L. A. N. Amaral, S. Goodwin, and H. E. Stanley, Phys. Rev. E **66**, 045103 (2002).
6. A commodity can have two types of prices attached to it, namely, the spot and future prices. See `http://www.investorwords.com/`.
7. B. B. Mandelbrot, J. Business **36**, 394 (1963); *ibid* **40**, 393 (1967); W. Working, Wheat Studies **9**, 187 (1933); *ibid* Wheat Studies of the Stanford Food Institute **2**, 75 (1934); R. Weron, Physica A **285**, 127 (2000); B. M. Roehner, Eur. Phys. J. B **8**, 151 (1999); *ibid* **13**, 175 (2000).
8. D. Gribbin et al. J. Futures Markets **12**, 475 (1992); P. Kofman in *Advances in Futures and Options Research Vol. 6*, edited by D. and R. Trippi (JAI Press, London); P. Kofman and C. de Vries, Review of Futures Markets **8**, 244 (1989); S. Yang and B. Brorsen, J. Futures Markets **13**, 175 (1993).
9. B. B. Mandelbrot, Sci. Am. **280**, 70 (1999); A. Bershadskii, J. Phys. A: Math. Gen. **34**, L127 (2001); Z. R. Struzik, Physica A **296**, 307 (2001); L. Calvet and A. Fisher, Rev. Econ. Stat. **84**, 381 (2002); *ibid.* J. Econometrics **105**, 27 (2001); X. Sun, H. Chen, Z. Wu, and Y. Yuan, Physica A **291**, 553 (2001); E. Canessa, J. Phys. A: Math. Gen. **33**, 3637 (2000); M. Ausloos and K. Ivanova, Comput. Phys. Commun. **147**, 582 (2002); K. Ivanova and M. Ausloos, Eur. Phys. J. B **8**, 665 (1999); M. Bernaschi, L. Grilli, D. Vergni, Physica A **308**, 381 (2002); J. P. Bouchaud, M. Potters, M. Meyer, Euro. Phy. J. B **13**, 595 (2000).
10. E. Bacry, J. Delour, and J. F. Muzy, Phys. Rev. E **64**, 026103 (2001); B. Pochart and J.-P. Bouchaud, preprint cond-mat/0204047.
11. B. M. Hill, Ann. Stat. **3**, 1163 (1975).
12. C. K. Peng, *et.al.* Phys. Rev. E **49**, 1685 (1994).
13. J. W. Kantelhardt, S. Zschiegner, E. Koscielny-Bunde, A. Bunde, S. Havlin, and H. E. Stanley, Physica A **316**, 87–114 (2002).
14. K. Matia, Y. Ashkenazy and H. E. Stanley, Europhys. Lett. **61**, 422 (2003).

International trade: Do two poles attract each other? *

Liming Liang[1], Jingjing Wang[2]

[1]Institute for Science Technology and Society, Henan Normal University, Xinxiang, 453002, China
[2]Star Echo Media & Marketing Co. Ltd., 98-100 Taojin Road, Dongshan District, Guangzhou, 510095, China

Summary. Based on the import-export data of international merchandise trade within OECD and among the seven country groups, the cooperation preference of countries of different trade capacities is analyzed. A new method of comparing the observed value to the expected value is used.

Key words. International trade, Cooperation, Pole rule

Introduction

It is a universal rule in magnetism that two poles of the opposite nature attract each other and two poles of the same nature repulse. This is also true in electricity: the electrostatic attraction and repulsion. Sometimes social phenomena are extremely similar to physical phenomena. Over the past two years we have seen the similarity in two studies in China.

In a case study on Chinese computer science and control theory, we found that the combination of old scientists with young scientists was the commonest form of scientific collaboration, while inside the circle of young scientists, as well as that of old scientists, the preference of collaboration was very weak. This conclusion is based on the comparison of the observed value to the expected value [1].

In another study on China's inter-regional collaboration pattern of co-authored papers, we found that scientists in either middle or low productive regions were reluctant to seek partners from their own region, but instead from the high productive regions, and vice versa. The study was based on the measured frequency distribution of China's inter-regional co-authored papers and the comparison of the observed value to the expected value [2].

We regard international trade as a form of cooperation -- an economic cooperation. We posit that the most active countries in international trade and the least

* The research was funded by the National Natural Science Foundation of China (project 70073007)

active countries could act as the two poles. Under the universal rule, the two poles should attract each other, or in other words, the trade preference between most active and least active countries should be stronger than that between other combinations, such as the middle active and the least active countries, the most active and the middle active countries, or inside the circle of the countries at the same active-level. This is a very interesting topic, but we aware that the research method needs special attention as the absolute import or export data could not fully reflect the cooperation preference. A new method is applied to reveal the preference behind the phenomenon.

Data and methods

We used the import-export data of 30 OECD countries over the period of 1995-2000[3], and that of 7 country groups from 1997 to 1999. The seven country groups are, according to WTO's classification, North America, Latin America, Western Europe, C. / E. Europe/ Baltic States/CIS, Africa, Middle East and Asia. The import-export data of the seven country groups are listed in Table 1[4]. The table containing the data of the 30 OECD countries is not given here, since it would be too big to be shown in one page.

Table 1. Network of world merchandise trade by seven country groups, 1997-1999 (observed value):

	Destination								
Origin	North America	Latin America	Western Europe	C./E. Europe /Baltic States/ CIS	Africa	Middle East	Asia	World	Share
North America	1034.6	430.4	525.3	22.6	36.6	69.6	614.4	2733.5	0.171
Latin America	481.7	163.3	120.2	8.6	9.8	9.6	59.7	852.9	0.053
Western Europe	630.9	174.1	4798.4	381.7	182.2	183.0	561.8	6911.9	0.433
C./E. Europe/ Baltic States/CIS	27.3	10.5	340.5	200.6	7.6	13.7	46.3	646.4	0.041
Africa	52.8	9.4	177.6	4.4	31.1	5.2	47.6	328.1	0.021
Middle East	65.3	7.7	101.5	3.9	20.7	39.0	218.5	456.5	0.029
Asia	1033.9	108.4	731.4	43.5	63.2	112.7	1924.7	4017.7	0.252
World	3326.4	903.7	6794.9	665.2	351.0	432.7	3472.8	15946.7	1.000
Share	0.209	0.057	0.426	0.042	0.022	0.027	0.218	1.000	

The simple illustration of the new method we used in the analysis of OECD data is as below:

First, we build an import-export matrix "A" of observed value with the left column as export countries (origins) and the upper row as import countries (destinations). The elements of the matrix represent the import-export amounts between any two countries, denoted as x_{ij} ($i \neq j$). In general, $x_{ij} \neq x_{ji}$, thus A is an asymmetric matrix with no diagonal element. Let x_i be the sum of observed values in i-th row, i.e. the total export amount of the country "i" to other countries, and y_j the sum of observed values in j-th column, i.e. the total import amount of country "j" from other countries. Therefore, x_i and y_j could represent, in some degree, the export and import capability. T is the total trade amount of OECD. Obviously,

$$\sum_i x_i = \sum_j y_j = T$$

Second, we calculate the expected value corresponding to the observed value x_{ij}, denoted as X_{ij} ($i \neq j$), and build the matrix "B" to represent the expected value. Since B is also an asymmetric matrix with no diagonal elements, the calculation of X_{ij} is somewhat complicated. X_{ij} can be split into two symmetrical parts as X_{ij}' and X_{ij}'' , and $X_{ij} = (X_{ij}' + X_{ij}'')/2$. X_{ij}' could be obtained mainly in the basis of the reassignment of export amounts among the import countries according to their holistic import abilities, and X_{ij}'' is the contrary reassignment. When $i \neq j$

$$X_{ij}' = x_i[y_j/(T - y_i)] \qquad X_{ij}'' = y_j[x_i/(T - x_j)]$$

It is easy to prove, that $\sum_{i,j} X_{ij} = T$.

Third, we divide the 30 OECD countries averagely into three groups: most active import-export countries, medium import-export countries and least active import-export countries. Thus, Matrix A, as well as B, would be cut into nine cells: most-most, most-medium, most-least, medium-most, medium-medium, medium-least, least-most, least-medium and least-least cells. Then we calculate the ratio of observed value to expected value of each cell. The larger the ratio is, the stronger the cooperation preference is in this cell.

The matrix of observed value of merchandise trade among seven country groups is the matrix with diagonal elements, so it is easier to get the corresponding matrix of expected value.

Analyses and results

Two cases have been examined. Table 2 and Table 3 show the ratios of the observed to the expected values of the international merchandise trade within 30 OECD countries and among 7 country groups, respectively.

The following points could be seen in Table 2.

1. The three ratios in the diagonal are all larger than 1. The ratio of least-least cell is as high as 1.96. It indicates that the cooperation preference between two

countries of the same level is stronger.

2. The ratios of most-least and least-most cells are all much smaller than 1. It tells that the two polls are not attracted each other, but repulse.

3. The trade relationship between medium and least active countries is closer. The observed value of the export merchandise of the medium countries to the least active countries is 1.341 times of its expected value.

Table 2. Ratio of observed value to expected value of international merchandise trade among 30 OECD countries (1995-2000):

	Destination		
Origin	Most	Medium	Least
Most	1.045	0.932	0.833
Medium	0.909	1.120	1.341
Least	0.880	1.041	1.960

Table 3 contains more information than Table 2 and reveals some characteristics of trade cooperation among seven country groups as followed.

1. Generally, the cooperation relation within the same group is the closest since all the ratios at the diagonal are larger than one. The ratio in the cell of C./E.Europe/Baltic States/CIS- C./E.Europe/Baltic States/CIS is as high as 7.44. The second highest ratio is 4.302 in the cell of Africa-Africa. Comparatively, the ratios in the cells of Western Europe-Western-Europe and North America-North America are smaller.

2. Geographical proximity is the most important factor affecting the international trade. North America-Latin America, Western Europe-C./E.Europe/Baltic States/CIS, Western Europe-Africa, Asia-Middle East are all adjoining regions and they also are all closely collaborative couples.

3. The most aloof collaborators are North America and C./E. Europe/Baltic States/CIS. Their two ratios are 0.198 and 0.202. The next pair is Latin America-C./E. Europe/Baltic States/CIS, with the ratios 0.243 and 0.285. Apart from the geographical factor, the phenomenon could be caused by other reasons, subject to further studies.

4. We noticed that the ratios in the cells crossed by North America and Western Europe are all less than a half. It may suggest that the two trade regions of high trade ability repulse to each other.

Table 3. Ratio of observed value to expected value of Intra- and inter-regional merchandise trade of seven country (region) groups (1997-1999):

Origin	Destination						
	North America	Latin America	Western Europe	C./E.Europe/ Baltic States/CIS	Africa	Middle East	Asia
North America	1.815	2.779	0.451	0.198	0.608	0.938	1.032
Latin America	2.708	3.379	0.331	0.243	0.523	0.413	0.321
Western Europe	0.438	0.444	1.629	1.324	1.197	0.976	0.373
C./E.Europe/Baltic States/CIS	0.202	0.285	1.236	7.440	0.532	0.781	0.329
Africa	0.771	0.506	1.270	0.323	4.302	0.584	0.666
Middle East	0.686	0.297	0.522	0.203	2.055	3.148	2.198
Asia	1.234	0.476	0.427	0.260	0.714	1.034	2.200

Discussion

Our study does not support the hypothesis that two poles in international trade attract each other. But there is also not enough evidence to reject the hypothesis. The analysis of international trade preference is difficult because various factors and variables are knitted together. We believe that, before any conclusion could be made, at least two tasks should be done: to analyze the trade relationship of each of all the countries in the world, one by one, not by country group; and to create a new method to eliminate the influence of geographical proximity on trade cooperation.

In addition, other attempts on the "pole rule" could classify countries by GDP, or other economic indicators.

The method we used in this paper could also be used in the analyses of other economic relationships, such as the flow of international investment and cross-border project cooperation. We also wondered whether two poles attract each other in these economic activities.

Reference

[1] L. Liang, Y. Guo, M. Davis (2002). Collaborative patterns and age structures in Chinese publications, Scientometrics, Vol.54, No.3, pp473-489, 2002

[2] L. Liang, L. Zhu (2002). Major factors affecting China's inter-regional research collaboration: Regional scientific productivity and geographical proximity, Scientometrics, Vol.55, No.2, pp287-316, 2000

[3] www.oecd.org/EN/document .

[4] World Trade Organization. International Trade Statistics 2000, pp 162-163

Emergence of power-law behaviors in online auctions

B. Kahng[1], I. Yang[1], H. Jeong[2] and A.-L. Barabási[3]

[1] School of Physics and Center for Theoretical Physics, Seoul National University, Seoul 151-747, Korea, kahng@phya.snu.ac.kr
[2] Department of Physics, Korea Advanced Institute of Science and Technology, Taejon
[3] Department of Physics, University of Notre Dame, Notre Dame, IN 46556, USA

Summary. We downloaded all online-auction bidding histories closing on a single day on eBay, including 264,073 auctioned items, involved by 384,058 distinct agents, finding emerging behaviors in electronic biddings. The data analysis indicates that the total number of bids placed in a single category and the number of distinct auctions frequented by a given agent follow power-law distributions, implying that a few agents are responsible for a significant fraction of the total bidding activity on the online market. To confirm the validity of our findings in different markets and time spans, we collected the data over a one year period from eBay's Korean partner, auction.co.kr, and confirmed the findings obtained from the eBay data. This domination of online auctions by a few unusually active minorities may be a generic feature of all online mercantile processes.

Electronic commerce (E-commerce) is any type of business or commercial transaction that involves information sharing via the Internet. Online auctions represent a rapidly expanding segment of such E-commerical transactions. Internet technology provides bidders with increased convenience, both geographic and temporal, relative to traditional auctions. It has also lowered the costs of organizing an auction and of participating as a bidder, which appears to be causing auctions to be used for more and more transactions over time [1, 2]. For example eBay, the largest consumer-to-consumer auction site, boosts over 40 million registered consumers, and has grown in revenue over 100,000 percent in the past five years. With the rapidly increasing number of agents, the role of individuals diminishes and self-organizing processes among individual bidders increasingly dominate the market's behavior [3, 4].

Recently self-organizing features of complex systems have attracted many attentions in the statistical physics community because they contain diverse cooperations among numerous components of a system, resulting in patterns and behaviors which are more than the sum of the individual action of each component. While many systematic studies have been carried out to understand such emerging patterns in various systems, little attention has been paid to online auctions. Here we collect online auction data and show that the bidding pattern made by hundreds of thousands of agents leads to unexpected emerging behavior. We found that the total number of bids placed in a single category by a given agent follows a power-law distribution, rooted in the finding that an agent that makes frequent bids up to a certain moment is more likely to bid in the next time interval. Moreover, we find that the number of distinct items frequented by a given agent also follows a power-law distribution. The power-law behaviors imply that a few powerful agents bid more frequently and on more distinct items than others.

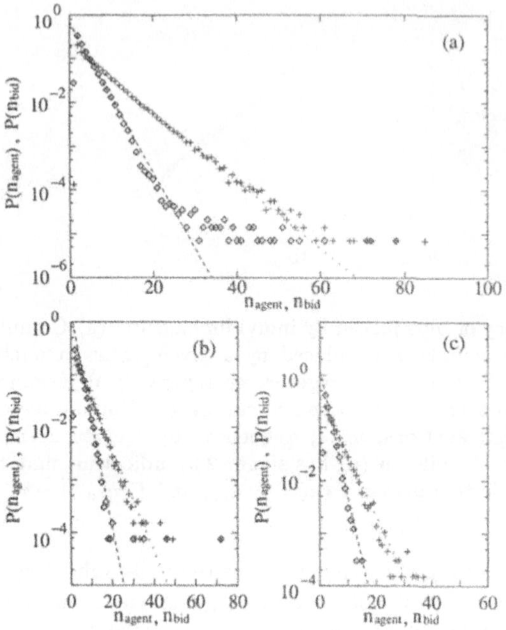

Fig. 1. Bid and agent distribution on eBay. (a) Distribution of number of agents ($n_{\mathbf{agent}}$,($\diamond$)) simultaneously bidding on a certain item and number of bids ($n_{\mathbf{bids}}$,(+)) received for an item, obtained by considering all items contained in the 194 categories on individual bids that were collected from auctions ending on July 5, 2001 on eBay. (b) and (c) Agent and bid distribution in the largest (b) and the second largest (c) category on eBay. The largest category by the number of auctioned items contains 21,461 items related to sport trading cards while the second largest category includes 13,610 items related to printed and recorded music. The straight lines correspond to exponential fits and the symbols are the same as in (a).

To be specific, we collected auction data from two different sources. First, we downloaded all auctions closing on a single day on eBay, including 264,073

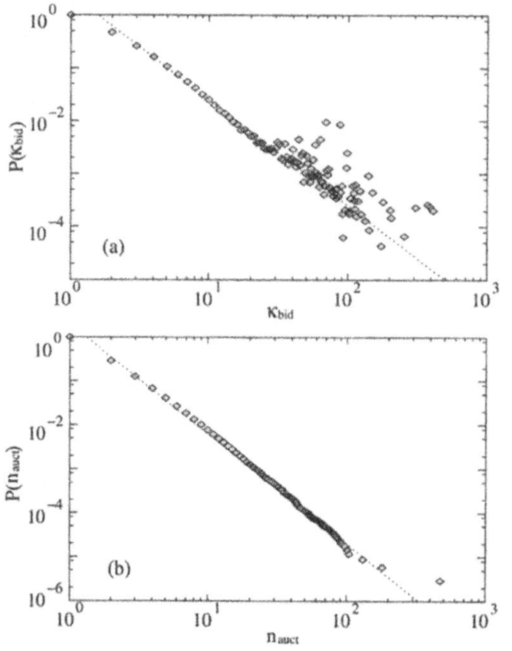

Fig. 2. Frequency of bids placed by individual agents. (a) Cumulative distribution of total number of bids, κ_{bid}, placed by a given agent in auctions in the same category. For each of the 194 categories we separately determined the cumulative distributions and averaged the obtained curves. (b) Cumulative distribution of the number of distinct auctions, n_{auct}, frequented by a given agent. The dotted line on (a) has slope -2 while on (b) has slope -2.5, indicating that the corresponding probability distribution follows $P(\kappa_{\text{bid}}) \sim \kappa_{\text{bid}}^{-3}$ and $P(n_{\text{auct}}) \sim n_{\text{auct}}^{-3.5}$, respectively.

auctioned items, grouped by the auction site in 194 subcategories. The dataset allowed us to identify 384,058 distinct agents via their unique user ID. To verify the validity of our findings in different markets and time spans, we collected data over a one year period from eBay's Korean partner, auction.co.kr, involving 215,852 agents that bid on 287,018 items in 62 subcategories.

Most online auction sites keep a detailed, publicly available record of all bids and identify the bidding agents via an unique login name. It is this transparency of the bidding history that allows us to characterize in quantitative terms the auction process. Each completed auction can be characterized by two quantities: the number of distinct agents bidding on the same item (n_{agent}) and the total number of recorded bids for the item (n_{bids}), where $n_{\text{bids}} \geq n_{\text{agent}}$, as each agent can place multiple bids. In Fig. 1 we show the distribution of n_{agent} and n_{bids} over all auctions recorded on eBay, finding that they both follow $P(n) \sim \exp(-n/n_0)$, where $n_0 = 5.6$ for n_{bids} and $n_0 = 2.5$

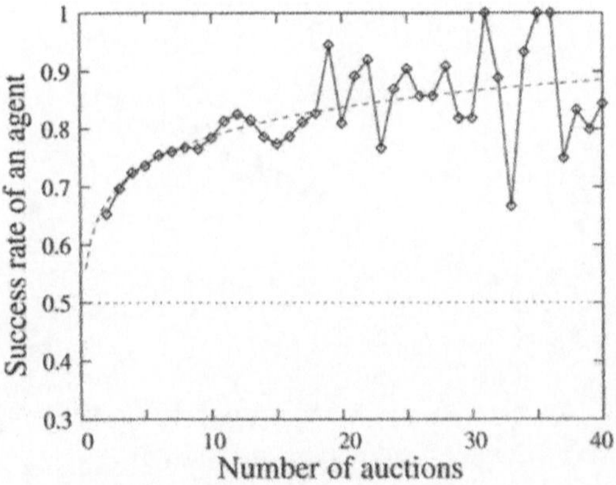

Fig. 3. **The dependence of an agents success rate on the number of auctions the agent participates. For each product subcategory (containing highly similar items) we calculated P_i^{win}, the average of the winning prices for items won by agent i. For the same agent we also calculated P_i^{lost}, the average over the winning price over items in which agent i participated but lost. A successful agent can get a lower price for the items it won than other agents bidding on similar items on parallel auctions, i.e. for a successful agent $P_i^{\text{win}} < P_i^{\text{lost}}$. We find that the success rate of an agent, measured as the function of auctions won at a lower than average price (i.e. the fraction of agents for which $P_i^{\text{win}} < P_i^{\text{lost}}$) increases with the number of auctions these agents participate in. A horizontal dotted line corresponds to the case when there is no correlation between the frequency of bidding and the chances of getting a better price. A numerical fitting indicates that the success rate increases logarithmically (dashed line).**

for n_{agent}. We obtained similar results for the Korean market, with $n_0 = 10.8$ for n_{bids} and $n_0 = 7.4$ for n_{agent}. In Fig. 1b and 1c we show the distribution in two subcategories (sports trading cards and printed, recorded music), finding that they follow the same functional form as the aggregated data. Therefore, the exponential form for the activity distribution appears to be a general feature of all auctions, indicating that the majority of auctions have only a few bidders and auctions with a large number of bids or participating agents are exponentially rare.

To characterize the activity of individual agents we determined the number of bids placed by each agent on each auction. As agents place simultaneous bids on items that closely resemble each other, we denote by κ_{bid} the total number of bids placed by the same agent in auctions in the same category. For example, if several similar computers are sold on separate auctions, agents looking for a computer often bid simultaneously for several or all of them. We find that the distribution of κ_{bid} follows a power law

$$P(\kappa_{\text{bid}}) \sim \kappa_{\text{bid}}^{-\gamma}, \qquad (1)$$

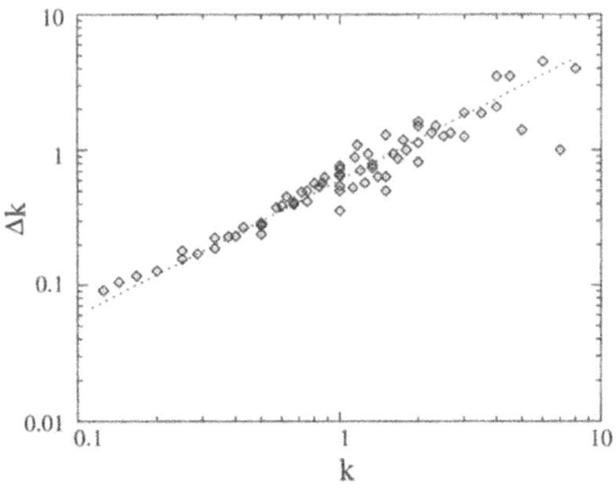

Fig. 4. The origin of the observed power law is in preferential attachment. The figure shows the change in the number of bids placed by an agent i that previously placed k bids, averaged over all items and all times. The linear behavior in the log-log plot indicates that the more bids an agent places up to a given moment, the more likely it is that it will place another bid in the next time interval. The dotted line with slope 1 is shown for visual reference.

where $\gamma = 3$ (Fig. 2a) on both eBay and the Korean auction. A similar power law characterizes the distribution of the number of different auctions, n_{auct}, frequented by individual agents, finding that

$$P(n_{\text{auct}}) \sim n_{\text{auct}}^{-\beta}, \qquad (2)$$

where $\beta = 3.5$ (Fig. 2b). The power-law distribution shown in Fig. 2a implies that while most agents place only a small number of bids, a few agents bid very frequently, placing several hundred bids on the same day. Similarly, Fig. 2b indicates that while most agents participate in a few auctions only, a few agents bid very widely, some placing simultaneous bids on over a hundred distinct items on the same day. Therefore, unknown to most participants, the auction process is dominated by a small number of highly active agents, or *power-agents,* that pursue a very aggressive bidding pattern, placing simultaneously a large number of bids on a wide range of items. These power-agents are responsible for the power-law tail of the distribution shown in Fig. 2. Our measurements indicate that there is a strong correlation between the number of bids placed by an agent on an item and the number of items the same agent bids for, indicating that power agents simultaneously bid frequently and widely.

The observed power law distribution is rooted in the dynamics of the bidding process. Indeed, two processes contribute to the final number of bids placed on a given item: new agents entering the bidding process and agents that already placed a bid increase their bid. We find that this pattern is governed by a process often referred to as preferential attachment, similar to those responsible for the emergence of scaling in complex networks [5]: more bids an agent places on a given item up to a certain moment, more likely is that he/she will place an another bid in the future (Fig. 4). The linearity of the observed relationship is known to lead to power law distributions, as demonstrated by studies in both economics [6, 7] and complex networks [8, 9, 10].

In conclusion, we have collected online auction data and analyzed the statistical properties of emerging patterns created by a large number of agents. We found that the total number of bids placed in a single category and the number of distinct auctions frequented by a given agent follow power-law distributions. Such power-law behaviors imply that the online auction system is driven by self-organized processes, involving all agents participated in a given auction activity. We also uncovered the empirical fact that the more bids an agent places up to a given moment, more likely it is that it will place another bid in the next time interval, which plays an important role in generating the power-law behavior in the bidding frequency distribution by a given agent.

This work is supported by the KOSEF Grant No. R14-2002-059-01000-0 in the ABRL program.

References

1. E. van Heck and P. Vervest, Communication of the ACM **41**, 99 (1998).
2. R. D'Hulst and G.J. Rodgers, Physica A **294**, 447 (2001).
3. R.N. Mantegna and H.E. Stanley, An introduction to econophysics: Correlations and complexity in finance. (Cambridge University Press, Cambridge, 2000).
4. J.P. Bouchard and M. Potters, Theory of financial risks: From statistical physics to risk management. (Cambridge University Press, Cambrdige, 2000).
5. A.-L. Barabasi and R. Albert, Science **286**, 509 (1999).
6. H.A. Simon, Biometrika **42**, 425 (1955).
7. M. Marsili and Y.-C. Zhang, Phys. Rev. Lett. **80**, 2741 (1998).
8. R. Albert and A.-L. Barabasi, Rev. Mod. Phys. **74**, 47 (2002).
9. R. Pastor-Satorras, A. Vazquez and A. Vespignani, Phys. Rev. Lett. **87**, 258701 (2001).
10. M. Newman, Phys. Rev. E **64**, 025102 (2001).

Econophysics vs Cardiophysics: the Dual Face of Multifractality

Zbigniew R. Struzik

Centrum voor Wiskunde en Informatica (CWI)
Postbus 94079, NL-1090 GB, Amsterdam, The Netherlands
email: Zbigniew.Struzik@cwi.nl

Summary. Multifractality in physiological time series and notably in human adult heart rate has been primarily attributed to the Fourier phase ordering of the signal [1]. In contrast, the primary cause for the width of the multifractal spectrum in financial time series has recently been connected with fat-tailed distribution rather than phase ordering [2]. Reducing the source of multifractality of the human heart to nonlinear phase ordering may in fact not be entirely exhaustive or even correct. In the ultimate stress of the heart of the fetus during labour, there is an essential contribution of fat tails of the event distribution. Generally the multifractal spectra of the fetal heart rate during labour are considerably wider than the spectra for adults. We attribute the difference to the large shocks which have previously been shown to exhibit similar behaviour to financial index crashes [3].

Key words: Econophysics, Cardiophysics, Multifractal spectra, Non-linearity

1 Introduction - econophysics versus cardiophysics?

At first glance, there seems to be very little linking human heart rate and the stock index level, unless of course one is a trader or has invested a considerable amount of money... There may, however, be a much closer and physically and fundamentally more meaningful link than meets the eye between heartbeat neuro-regulation and market self-regulation.

There are a number of reasons why one may want to have a simultaneous look at the cardiac and economic systems. Both exhibit complex behaviour classified as scale invariant and multifractal [1, 4, 5, 6, 7]. Both can be seen as systems of a number of coupled oscillators, subject to stochastic regulation. Both consist of antagonistic actors; in the neuro-regulatory cardiac system, in addition to the pacemaker, the periodic pulse generating node, heart rhythm is driven by sympathetic and parasympathetic nodes, in constant imbalance.

The sympathetic system slows down the heart rate, while the parasympathetic one accelerates it. One might think that during activity, only the parasympathetic system would be active, while during rest or sleep only the sympathetic one would be in action. Contrary to this, both systems are at work at any moment in time. In fact, it is a constant, simultaneous 'struggle' of both systems which gives rise to the actual heartbeat ratio, characterised by continuously changing dynamics [1, 8, 9, 10].

This scenario resembles a view of the market with *pessimists* who sell and *optimists* who buy stock options at any given instant. The number of actors changes as the market changes. Some pessimists join the optimist group and vice versa. Additionally, there is leakage of information, or simply similar action, clustering the behaviour of agents in both the pessimistic and the optimistic groups. This leads to a continuously varying number of *active* agents in the system, which may be related to the number of effectively active degrees of freedom of the abstract market system under consideration.

In particular, in the market scenario, polarisation of opinions leads to clusters of agents, effectively reducing the number of *independent* degrees of freedom in the system [11]. The resultant behaviour of the stock becomes (anti-)correlated and bends towards anti-persistent fractional Brownian motion characterised by $H < 0.5$.

The behaviour of the healthy heartbeat is dominated by the coherent, concerted action of the inputs within the nodes of the two antagonistic systems (sympathetic and parasympathetic), leading to an extremely high level of anti-persistence at the range of $H = 0.0$ to $H = 0.1$ (for the normal healthy heart rate at rest). In the case of pathology, exercise or stress, this behaviour becomes less anti-persistent and less (anti-)correlated, potentially indicating less coherence in the functioning of the two antagonistic systems and thus possibly the increase of the effective independent degrees of freedom in the system.

The similarities drawn between both systems of market self-regulation and heart neuro-regulation can be expanded still further. In the condition of extreme stress during labour, the heart rate of the child being born can be shown to undergo a dynamical evolution which can resemble, and be compared to, severe stock market crashes [3].

In this article, we further elaborate on the reasons for comparing the neuroanatomic system of heartbeat regulation with the market. In particular we confront the view suggesting different origins of multifractality in financial time series [2] and physiological time series [1], by demonstrating that both mechanisms coexist in fetal heartbeat during labour.

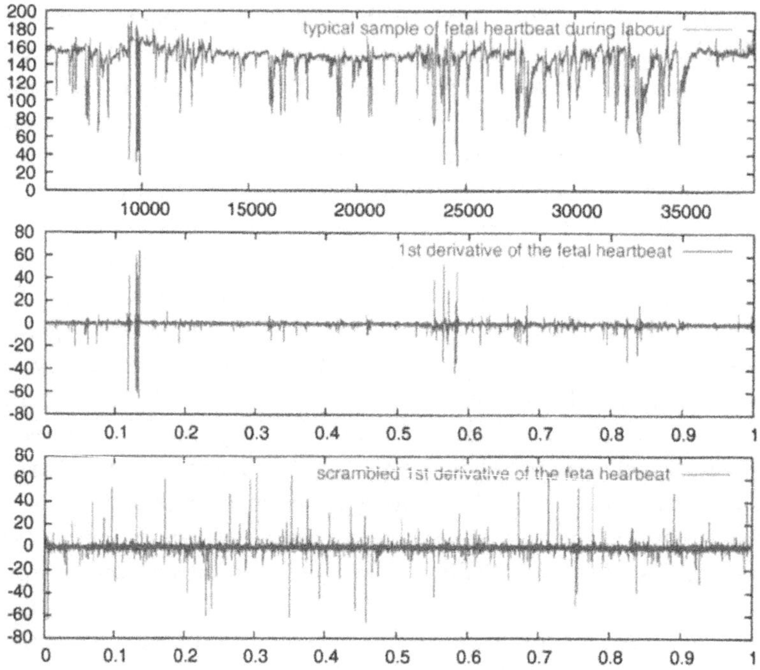

Fig. 1. Top: the input time series: fetal heart rate record. Middle: first derivative of the input time series. Bottom: scrambled first derivative of the input time series.

2 The dual face of multifractality in fetal heartbeat

Multifractality in physiological signals and notably in human adult heart rate has been primarily attributed to the Fourier phase ordering of the signal [1]. In contrast, the primary cause of the width of the multifractal spectrum in financial time series has recently been connected with the fat-tailed distribution rather than phase ordering [2]. In the same ref. [2], the distinction has been made between commodities and stock indexes, indicating that the commodities possess increased level of *higher order* temporal correlations which results in a wider multifractal spectrum than the stock indexes.

Reducing the source of multifractality of the human heart to nonlinear phase ordering only may, in fact, not be entirely exhaustive or even correct. In the ultimate stress of the heart of the fetus during labour, there is an essential contribution of extreme events resulting in fat tails of the event distribution. Generally the multifractal spectra of fetal heart rate during labour are considerably wider than the spectra for adults. We attribute the difference to the large shocks which have previously been shown to exhibit similar behaviour to the financial index crashes [3].

In figure 1 we show typical fetal heart rate during labour, followed by the first derivative of the signal which represents the 'returns' in the middle plot. Next, in the bottom plot of the same figure we show a scrambled version of the 'returns'. For scrambling we used random pairwise replacement of the returns carried out 1000 times the length of the time series, i.e. $10^3\ 2^{15}$ pair replacements. The multifractal spectra have been evaluated using the Wavelet Transform Modulus Maxima (WTMM) method. In figure 2 we show spectra for both the raw time series, for the first derivative - the 'returns' and for the scrambled 'returns'. For comparison, the same procedure has been applied to a random walk sample (the returns of the random walk are white noise).

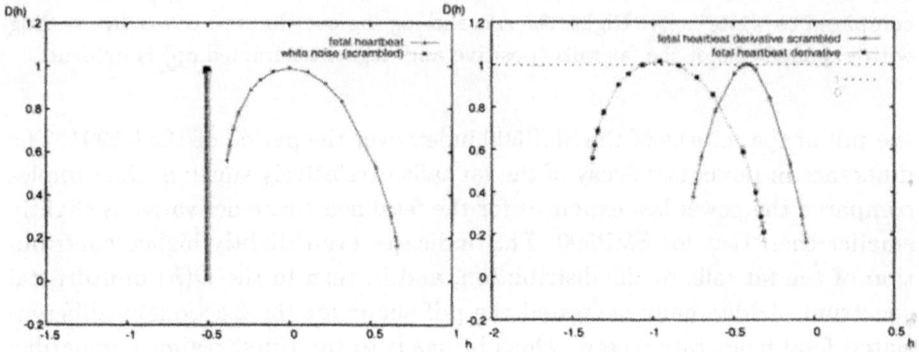

Fig. 2. Left: the MF spectrum of the raw fetal heart rate record and raw white noise. Right: the MF spectrum of the first derivative of the fetal heart rate record and its scrambled version.

As expected, there is no difference in the monofractal behaviour of the white noise. The difference in the behaviour of the multifractal spectrum of the fetal heart rate 'returns' suggests that the phase ordering substantially contributes to the width of the spectrum, as is the case with the healthy adult heart rate. Shown in figure 2 right, there is, however, a substantial non-vanishing residue, which we associate with the long tails in the distribution of the 'returns'. This behaviour resembles the duality of the multifractal properties in heart rate, in analogy to what has recently been discovered in financial time series (in particular of commodities) [2].

The presence of fat tails and their power law-like scaling is evident in the pdf shown in figure 3. Both the pdf of the fetal heart rate derivative and of a white noise record have been shown in linear-log and log-log coordinates. For the log-log plot, the modulus of the time series have been used and for the fetal heart rate it shows exponent 0.3 characterising the fat tails of the pdf. For comparison with finacial time series we plotted the pdf together with

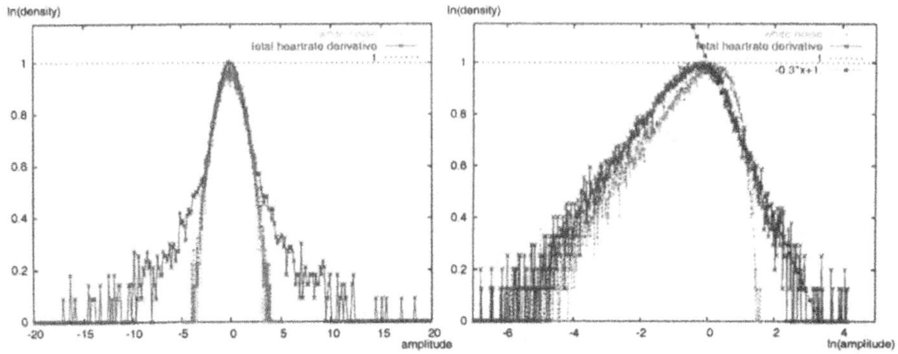

Fig. 3. Left: the fetal heart rate derivative shows fat tails in linear-log coordinates as compared to white noise. Right: the same in log-log coordinates; power law scaling with exponent 0.3 of the fat tails (positive and negative summed up) is evident.

the pdf of the returns of the S&P500 index over the period of 1983-1991. The difference in power law decay of the fat tails is relatively small: in the samples compared the power law exponent for the fetal heart rate derivative is slightly smaller than that for S&P500. This indicates even slightly higher contribution of the fat tails to the distribution, and in turn to the $D(h)$ multifractal spectrum. Additionally we tested the pdf shape for the fractionally differentiated fetal heart rate record, which brings it to the Hurst regime compatible with the stock index returns $H \approx -0.5 + 1$. Shown in figure 4, we observe the formation of two regimes of scaling in the corresponding fetal heart rate pdf. This double regime is analogous to the double scaling regime of the pdf of S&P500 returns [12].

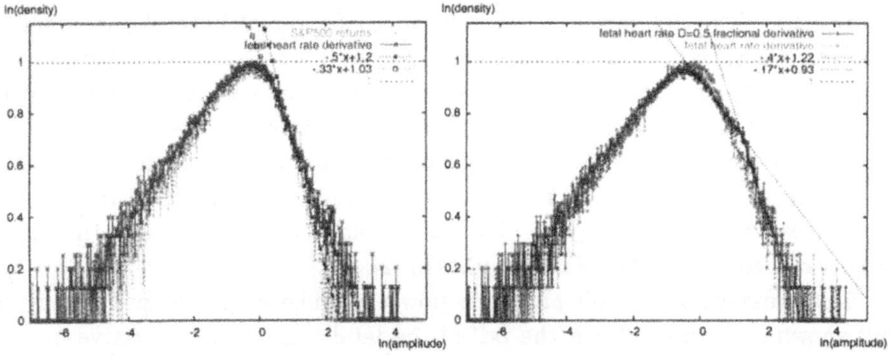

Fig. 4. Left: the fetal heart rate derivative fat tails in log-log coordinates scale with slightly lower exponent than S&P500 record, indicating stronger tail presence. Right: fractionally differentiated fetal heart rate centred at $H \approx -0.5+1$ compatible with financial time series shows clear double scaling regimes.

3 Conclusions

We have elaborated on similarities between heart rate and financial time series. The contribution of extreme events to the event distribution and multifractal spectra has been evaluated. We have confronted the view suggesting different origins of multifractality in financial time series and physiological time series by demonstrating that both mechanisms can coexist in heartbeat regulation in particular in fetal heartbeat during labour.

Acknowledgments

Discussions with Yosef Ahskhenazy and Takeshi Matsumoto are gratefully acknowledged.

References

1. P.Ch. Ivanov, M.G. Rosenblum, L.A. Nunes Amaral, Z.R. Struzik, S. Havlin, A.L. Goldberger and H.E. Stanley, Multifractality in Human Heartbeat Dynamics, *Nature* **399**, 461-465, (1999).
2. K. Matia, Y. Ashkenazy, H.E. Stanley, Multifractal Properties of Price Fluctuations of Stocks and Commodities, *Europhys. Lett.*, **61**, (3), pp. 422-428 (2003).
3. Z.R. Struzik, Econonatology: The Physics of the Economy in Labour, *Physica A*, in press, (2003).
4. A. Fisher, L. Calvet, B.B. Mandelbrot, Multifractality of the Deutschmark/US Dollar Exchange Rate, *Cowles Foundation Discussion Paper*, (1997).
5. M.E. Brachet, E. Taflin, J.M. Tchéou, Scaling Transformations and Probablity Distributions for Financial Time Series, *preprint* cond-mat/9905169, (1999).
6. F. Schmitt, D. Schwertzer, S. Levejoy, Multifractal Analysis of Foreign Exchange Data, *Appl. Stochasic Models Data Anal.* **15**, pp 29-53, (1999)
7. Y. Fujiwara, H. Fuijsaka, Coarse-graining and Self-similarity of Price Fluctuations, *Physica A*, **294** (3-4) pp.439-446, (2001).
8. M. Kobayashi, T. Musha, 1/f Fluctuation of Heartbeat Period, *IEEE Trans Biomed. Eng.*, **29**, 456-457 (1981).
9. C.-K. Peng, J. Mietus, J.M. Hausdorff, S. Havlin, H.E. Stanley, A.L. Goldberger Long-Range Anticorrelations and Non-Gaussian Bahavior of the Heartbeat *Phys. Rev. Lett.*, **70**, 1343-1346 (1993).
10. J.B. Bassingthwaighte, L.S. Liebovitch, B.J. West. *Fractal Physiology*, (Oxford University Press, 1994).
11. M. Zaluska-Kotur, S. Krukowski, A. Orlowski, Niestandardowa Analiza Fluktuacji Kursu Dolara Amerykańskiego, *Proceedings of the 1st Conference on Modelling of the Economical Processes*, 2002, Kielce, Poland (in Polish).
12. P. Gopikrishnan, V. Plerou,L.A. Nunes Amaral, M. Meyer, H.E. Stanley, Scaling of the Distributions of Fluctuations of Finacial Market Indices. Physical Review E, **60** (5), pp. 5305-5316, (1999).

I-5 Models

If the World were a Village of 100 Traders

Y. Ohtaki[a] and H.H. Hasegawa[ab]

Department of Mathematical Science, Ibaraki University[a]
Mito 310-8512, Japan
Center for Statistical Mechanics, University of Texas[b]
Austin, TX78712

Summary. We consider an ideal closed world, a village, in which only 100 traders have economic activities. The assets of the traders change through buying and selling stocks. We simulate the assets under the conservation of both total currency and total number of stocks. The assets are distributed as a stationary Gaussian for almost identical traders. When differences of traders make winners and losers in the stock market, the asset distribution displays power law scaling such as the Pareto law [1].

Of the wealth in this village, 6 people own 59%, -all of them from the United States- 74 people own 39%, and 20 people share the remaining 2%. This statement comes from the recent popular book titled "If the World were a Village of 100 People[1] ." Why do only 6 people own so much wealth? We may find several answers for this question from different reasons such as historical, cultural, political and so on. In this paper, we will consider what kind of wealth distribution is realized in financial markets.

We will consider an extremely simple situation: the world as a closed village, in which only 100 traders have economic activities. The assets of the traders change through buying and selling stocks. In these trades, the total amount of currency and the total number of the stocks are kept constant. Under these constraints the asset distribution may approach to a steady state. If the steady state exists, we may be able to construct a statistical mechanics of financial markets [2].

Statistical Mechanics of Financial Market

In physics, a typical statistical mechanics is based on a gas system. One of the important features of the gas system is that the elements have the same characteristics. For example, hydrogen molecules are completely identical. They have the same mass, size, electric charges and so on. Any particle has a wide possibility to be high/low energy states.

[1] The book "If the World were a Village of 100 People" was originated from the essay "The Global Village" written by Donella H. Meadow.

The mechanism of stability is another important feature in a gas system. A high/low energy particle collides with another particle, which has relatively lower/higher energy. Since two colliding particles exchange relative velocities, the high/low energy particle loses/gains energy. After many collisions, the Maxwell-Boltzmann distribution is realized.

We can find several common features between a stock market and a gas system. In a stock market, a trader corresponds to a molecule in a gas system. Assets play the role of energy. The conservation of total assets means energy conservation. Change of assets of a trader in a trade corresponds to change of energy in a collision.

Furthermore, the mechanism of stability also exists in a stock market, if we rule that a buy/sell order volume is proportional to assets of the trader. The price of a stock becomes higher/lower for a buy demand/sell offer. A trader, who demands/offers greater volume, has a tendency to lose the asset. For identical traders, we can expect the assets to distribute as a stationary Gaussian.

A different behavior is expected when the characters of the traders are widely diverged. Some may prefer frequent trading, others may prefer long-term trading. Some are strong trader and others are weak. When the differences make superiors and inferiors in a stock market, only traders who are fit become winners. This is an essential difference between a finance market and a gas system.

In physics, elements are completely identical. Any element has a wide possibility to be in high/low energy states. This means there is no strong correlation among elements. The number of statistically independent elements is huge, so a Boltzmann type distribution is realized.

In financial markets, only traders who are fit become winners. This means that there are strong correlations among winners. The number of statistically independent elements is effectively small, so a power distribution is realized.

Artificial Stock Market

We consider an ideal closed stock market, in which N traders buy and sell stocks. For simplicity, we choose only one brand of stock. The ith trader has currency $E_i(t)$ and stocks $V_i(t)$ at time t.

The stock market is completely isolated. There is no flow in/out of traders, currency or stocks. Therefore, the number of traders N, total currency $E_{\text{tot}} = \sum_{i=1}^{N} E_i(t)$ and total number of stocks $V_{\text{tot}} = \sum_{i=1}^{N} V_i(t)$ are kept constant.

From a statistical mechanical point of view, the assets of traders play the role of energy in the stock market. The ordinary asset of the ith trader is defined as $X_i \sim E_i(t) + V_i(t)p(t)$ where $p_(t)$ is price of the stock at time t. In that definition, the total assets depend on time. In this paper we choose the following definition:

$$X_i \equiv E_i(t) + V_i(t)p(0).$$

Although this definition is unusual, we give priority to strict conservation of total assets[2].

The assets of the traders change through buying and selling stocks. In a trade at time t,

$$\begin{array}{l} E_i(t) = E_i(t-1) \pm \mu_i(t)p(t) \\ V_i(t) = V_i(t-t) \mp \mu_i(t) \end{array} \quad (1)$$

where $\mu_i(t)$ is the volume of trade for the ith trader and $p(t)$ is the price of the stock at time t. Buyer/Seller pays/receives currency $\mu_i(t)p(t)$ to buy/sell $\mu_i(t)$ stocks. Although assets of the traders change, the total assets are kept constant.

Takayasu-Miura-Hirabayashi-Hamada Model

The traders follow the rules introduced by Takayasu-Miura-Hirabayashi-Hamada [3, 4]. The Takayasu-Miura-Hirabayashi-Hamada (TMHH) model is a deterministic model of the stock market. Although the TMHH model is very simple, its stock price has similar statistical properties as real stock prices.

The stock price is determined as follows,

$$p(t) = \begin{cases} (\max\{B_i(t)\} + \min\{S_i(t)\})/2 & : \max\{B_i(t)\} \geq \min\{S_i(t)\} \\ p(t-1) & : \text{otherwise} \end{cases} \quad (2)$$

where $B_i(t)/S_i(t)$ is a bid/ask price of the ith trader. $\max\{\cdots\}/\min\{\cdots\}$ means the maximum/minimum value of the set.

The bid/ask price of the ith trader is determined by the following rules,

$$\begin{array}{l} B_i(t+1) = B_i(t) + a_i(t) + c_i[p(t) - p(t_{\text{prev}})] \\ S_i(t+1) = B_i(t+1) + L_i \end{array} \quad (3)$$

where L_i, $|a_i|$ and c_i are characterized parameters of the ith trader. The bid/ask price depends on the up/down movement of the stock price $c_i[p(t) - p(t_{\text{prev}})]$ where $p(t_{\text{prev}})$ is the stock price of the previous trade.

The sign of $a_i(t)$ changes as follows:

$$a_i(t) = \begin{cases} -a_i(t-1) & : \begin{cases} \text{when } a_i(t-1) > 0 \text{ for the buyer} \\ \text{or} \\ \text{when } a_i(t-1) < 0 \text{ for the seller} \end{cases} \\ a_i(t-1) & : \text{otherwise} \end{cases} \quad (4)$$

The bid/ask price changes a_i in each step. After the ith trader becomes a buyer/seller, a_i reverses its sign. As a result, the ith trader alternately becomes a buyer and seller. The period is approximately given as $T_i \sim 2L_i/|a_i(t)|$.

Although the stock price in the TMHH model has similar statistical properties as the real stock prices [3, 4], this model misses the volume of trade,

[2] Another way is to consider the following asset distribution rates instead of the asset, $R_i(t) \equiv X_i(t)/X_{tot}(t)$. Then the total is always 1, since $\sum R_i = 1$.

which we need to estimate the asset distribution. We will extend the TMHH model to introduce a volume of trade.

Extended Takayasu-Miura-Hirabayashi-Hamada Model

Each trader demands/offers a bid/ask price and a buy/sell order volume at every time. The bid and ask prices are given as in Eq.(3). We will extend the TMHH model to introduce a volume of trade.

We assume that a buy/sell order volume is proportional to the trader's assets.

$$\begin{aligned} \mu_{\mathrm{b}}(B_i) &= \kappa E_i(t-1)/p(t-1) \\ \mu_{\mathrm{s}}(S_i) &= \kappa V_i(t-1) \end{aligned} \tag{5}$$

where $\mu_{\mathrm{b}}(B_i)/\mu_{\mathrm{s}}(S_i)$ is a buy/sell order volume with the bid/ask price of the ith trader. κ is a constant. We choose $0 < \kappa \leq 1$. This means that any trader can not demand/offer a buy/sell order volume beyond of the trader's assets.

Instead of Eq.(3) in the original TMHH model, we determine the stock price and volume of trade using the following cumulative demand and offer functions [2],

$$\begin{aligned} D(q) &= \textstyle\sum_{B_j \geq q} \mu_{\mathrm{b}}(B_i) \\ O(q) &= \textstyle\sum_{S_j \leq q} \mu_{\mathrm{s}}(S_i) \end{aligned} \tag{6}$$

The balance between the demand and offer determines the stock price and volume of trade. (1) If $D(p*) = O(p*)$ for a unique $p*$, then the stock price is $p(t) = p*$. If $D(p*) = O(p*)$ is unique, the volume of trade is $D(p*) = O(p*)$. If $D(p*) = O(p*)$ is not unique, the volume of trade is the maximum of $D(p*) = O(p*)$. (2)If $D(p*) = O(p*)$ for all $p*$ in $[p_{\mathrm{min}}, p_{\mathrm{max}}]$, then the stock price is $p(t) = (p_{\mathrm{min}} + p_{\mathrm{max}})/2$. The volume of trade is $D(p*) = O(p*)$. To satisfy the balance with respect to volume, we need to reduce some buy/sell order volumes.

Numerical Simulations

We have simulated time evolution of the stock market and obtained the asset distribution for long times. We chose the following parameters and initial values: $N = 100 \sim 300$, $\Lambda = 1000$, $p_0 = 3000$, $V_{tot} = 10000$, $E_{tot} = p_0 V_{tot}$, $\kappa = 0.5$, $c_i = 0$. The initial bid price of the ith trader $B_i(t)$ is given as random number within $(p_0 - \Lambda, p_0)$. The initial ask price of the ith trader is given as $S_i(t) = B_i(t) + \mathrm{L}_i$. Each trader has equal initial assets.

Each trader is characterized by two parameters. One is the difference between bid and ask prices, L_i. The other is the change of bid and ask price, $|a_i|$. The period of trade is approximately $T_i \sim 2\mathrm{L}_i/|a_i(t)|$. $|a_i|$ characterizes the weakness of traders. A smaller $|a_i|$ means a stronger trader.

(i)Common $L_i = \Lambda$ and $|a_i| = A$

In this case, the traders are almost identical. The period of trade is almost the same. There are no strong/weak traders.

The mechanism of stability works in this financial market. A trader who orders a greater volume has a tendency to lose/gain the asset. Each trader gains and loses assets under the mechanism of stability. We expect the assets to be distributed as a stationary Gaussian.

As is illustrated in Fig.1, we have confirmed our expectation.

(ii)Common $|a_i| = A$, randomly distributed L_i

We choose L_i as a random number within $[L_{\min}, L_{\max}]$. Then the period of the ith trader is approximately given as $T_i \sim 2L_i/A$.

In this case, there are still no strong/weak traders. Therefore, each trader gains and loses assets under the mechanism of stability. The assets are distributed as the stationary Gaussian shown in Fig.2. There is almost no correlation between the asset and period.

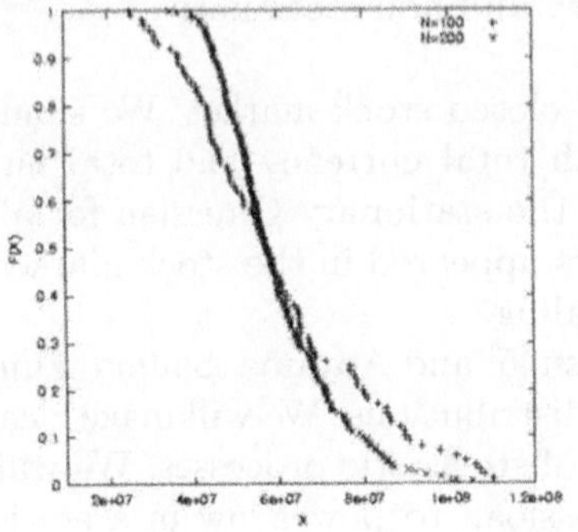

Fig. 1. Accumulated asset distribution and the error function with $\Lambda = 1000$ and $A = 30$.

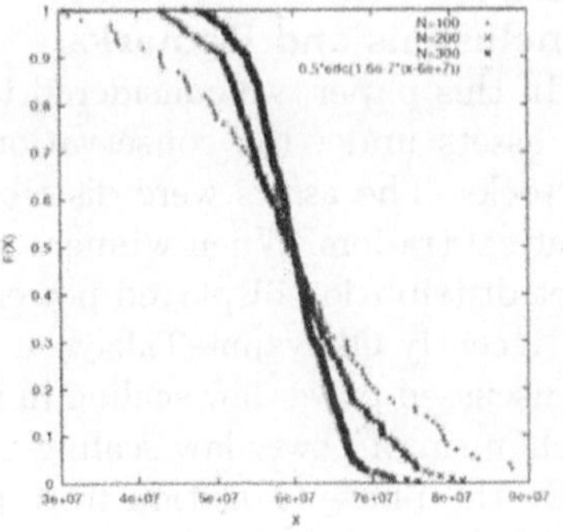

Fig. 2. Accumulated asset distribution with $L_{\min}$=1000 and $L_{\max}$=2000.

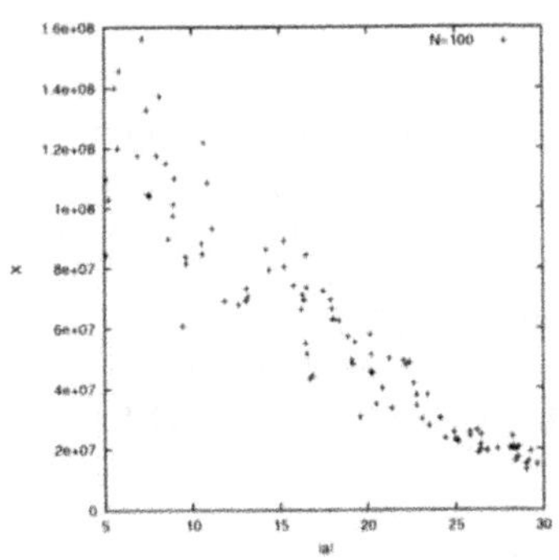

Fig. 3. a vs X with $|a_{\min}|$=5, $|a_{\max}|$=30, Λ=1000.

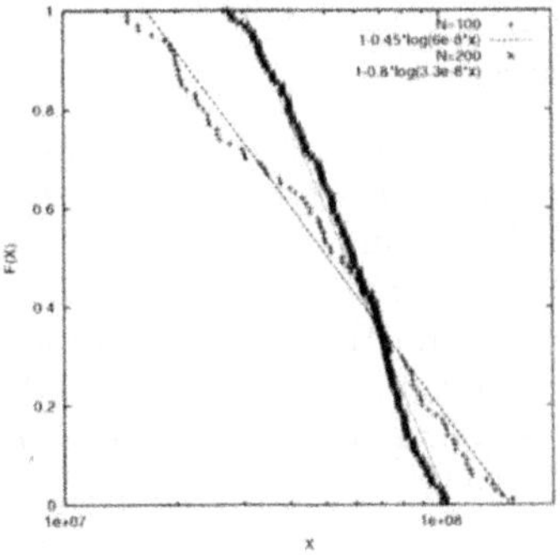

Fig. 4. Accumulated asset distribution. X-axis is log scale. $|a_{\min}|$=5, $|a_{\max}|$=30, Λ=1000.

(iii)Common $L_i = \Lambda$, randomly distributed $|a_i|$

We choose randomly distributed $|a_i|$ within the interval $[a_{\min}, a_{\max}]$. Then different behaviors appear in the stock market as is illustrated in Fig.4. Since $|a_i|$ characterizes the weakness of the ith trader (Fig.3), differences of $|a_i|$ make superiors and inferiors in the stock market. Beyond the stability mechanism winners and losers appear. This is an essential difference from a gas system.

In Fig.4, the cumulative assets distribution is fit as

$$F(X) = \int_{\infty}^{X} P(X')dX' \propto 1 - \alpha \log(\beta X). \tag{7}$$

The asset distribution displays $1/X$ scaling. This means that the asset distribution is on the border between stationary and non-stationary.

Conclusions and Remarks

In this paper, we considered the ideal closed stock market. We simulated the assets under the conservation of both total currency and total number of stocks. The assets were distributed as the stationary Gaussian for almost identical traders. When winners and losers appeared in the stock market, the asset distribution displayed power law scaling.

Recently Okuyama–Takayasu–Takayasu [5] and Aoyama–Souma–Fujiwara [6] discussed power law scaling in income distributions. We will make clear the mechanism of power law scaling in terms of stochastic processes. We will also study the phase transition from the Gaussioan to power-law in a statistical mechanical point of view.

Acknowledgements

We are grateful to H. Yoon, M. Tanaka-Yamawaki, H. Takayasu and K. Nelson for useful discussions. We thank Y. Kondo and H. Matsumoto for their careful checking of program of the model. This work was supported by JSPS.

References

1. V. Pareto. *Le Cours d'Économie Politique,.* Macmillan, London, (1897).
2. J. Voit. *The Statistical Mechanics of Financial Markets.* Springer, (2001).
3. H. Takayasu, H. Miura, H. Hirabayashi, and K. Hamada. *Physica* **A184**, (1992)127-134.
4. A.Sato and H.Takayasu. *Physica* **A250**, (1998)231-252.
5. K. Okuyama, M. Takayasu, and H. Takayasu. *Physica***A269**, (1999)125-131.
6. H. Aoyama, W. Souma, and Y. Fujiwara. *Mathematical Science* **472** *(Japanese)*, (2002)44-50.

Market Simulation Displaying Multifractality

Kazuko Yamasaki and Kenneth J. Mackin

Tokyo University of Information Sciences,1200-2,Yatou town,Chiba,Japan
yamasaki@rsch.tuis.ac.jp

Summary. We proposed a market simulation model (micro model) which displays multifractality and reproduces many important stylized facts of speculative markets. From this model we analytically extracted the MMAR model (Multifractal Model of Asset Returns)[3] for the macroscopic limit.

1.1 Introduction

Current Studies by Market Simulations
Many studies have tried to reproduce the stylized facts of speculative markets by means of market simulation with the aim of understanding the true nature of speculative markets. Stylized facts studied include (1)fat-tail, (2)long-memory of nonlinear function of returns, (3)short-memory of raw returns and (4)the time scale invariance property. In addition, recently many empirical investigations have shown the multifractality of time series of speculative markets [1]. These stylized facts are very characteristic and universal so that we expected if the stylized facts could be reproduced by a market simulation, we could narrow the cause of these facts in order to analyze the cause and effect. Recently many market simulations have succeeded in reproducing some of the facts, but these reports have given different or even contradictory explanations. The different results between reports comes from the nature of multi-agent simulations, which includes a lot of freedom. Though we believe the fractality (or multifractality) has been one of the most important feature of the speculative markets since first mentioned by Mandelbrot, most market simulations have not examine this aspect. This research is based on the following point of view.

Extraction of Macro Model from Micro Model
In this research, we define the model which includes individual variables for each agent "the micro model". (e.g. Each trader's parameters used in buy-sell decisions.) We define the model which is described only with macro variables such as price and volume "the macro model". We belive such extraction is

important for understanding of the mechanism of speculative market, as it was first emphasized by Takayasu et. al.[4]. This also makes the correspondence between the result and the cause clear. The following table compares 3 previous macro model stochastic equations.

	GARCH	FBM($H \neq 0.5$)	MMAR
(2)long memory	no	yes	yes
(3)short memory	yes	no	yes
(4)fractality	no	yes	yes

As seen in the table, The GARCH model and FBM (Fractional Brownian Motion) model can reproduce only a part of the stylized facts. The MMAR model [3], proposed by B. Mandelbrot can reproduce most of the main stylized facts. In addition, the model can show both long-memory and Martingale property simultaneously.

MMAR Model

MMAR model can be described as follows.

$$\log P(t) - \log P(0) = B_H[\theta(t)] \tag{1.1}$$

Here, $B_H(t)$ is a fractional Brownian motion with self-affin index H, and $\theta(t)$ is a stochastic trading time which is a cumulative distribution function of multifractal measure. (The trading time is counted every time a trade occurs.) An example of multifractal measure is

$$\mu = \prod_{k=0}^{K} M_k \tag{1.2}$$

Here, $M_k = M(\eta_1, \eta_2, ...\eta_k)$ is a stochastic variable which changes with the kth time scale. $B_H(t)$ and $\theta(t)$ are assumed to be independent. The meaning of this model is that the price changes under fractional Brownian motion along with the trading time, but the trading time shows multifractality.

1.2 Proposed Market Model

In this model, we aimed at analytically extracting the macro model from the micro model sacrificing some reality.

Time Scale

In our daily lives we commonly apply many layers of time scales.

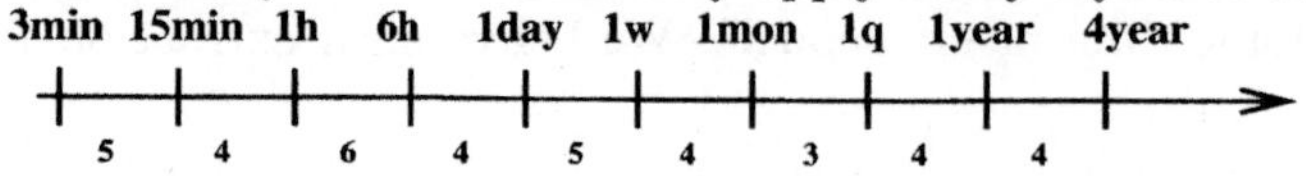

We supposed that the role of many layers needed for multifractality was carried by the different information in each layer. Each trader decides his judgment based on such information. In this model, all the information comes from external sources (e.g. a macro index such as GDP). This is because price and trading time are independent in the MMAR model, and if a trader's judgment decides the trading time, he cannot refer to the prices for his judgment.

Trader Judgment

M_k is the strength of information for source k. Generally if a trader's judgment consists of the combination of the logical products (and) and the logical sums (or) of several sources of information
((M_1 and M_2) or M_3) and (M_4 or M_5) and M_6
this can be rewritten in the form
(M_1 and M_2 and M_4 and M_6) or
(M_3 and M_4 and M_6) or ... etc.
We supposed a trader makes an order in single shares with a probability in proportion to the strength of his reference sources of information. Here, we do not distinguish between buying order and selling order. The strength of the source of information M_k is a stochastic variable, and m_k is its sample value. We define the probability of a trader making an order, as follows.

$$\mu_i(t) = \sum_{j=1}^{l_i} M_1 M_{02} M_{03} M_{04} M_5 M_{06} M_7 M_8 .... M_K = \sum_{j=1}^{l_i} \prod_{k=0}^{K} M_{0k}^{1-hi,k} M_k^{h_{i,k}}$$

$$h_i = (h_{i,1}, h_{i,2}, ...) = (0,0,0,1,0,0,1,0...) \qquad \sum_{k=1}^{K} h_{i,k} = Kz \tag{1.3}$$

Here, K is the number of the layers of time scale and these layers respectively include each source of information that is available for traders. l_i is the number of lines including only logical products. For simplicity, we set $l_i = 1$. A trader uses the information of Kz sources, because he uses only z number of sources whose corresponding element of his h is equal to 1. M_{0k} is a fixed stochastic variable and is always takes $m_0 = 0.5$. M_k takes either $m_1 = 0.9$ or $m_2 = 0.1$ in random order, at the intervals of $\frac{T}{2K}$

Market Price

We suppose that traders make an order by the equation (3) and all orders are traded. Then the trading frequency and volume of the market is in proportion to following equation.

$$\mu(t) = \frac{1}{N} \sum_{i=1}^{N} \mu_i(t) \tag{1.4}$$

Here, N is the number of traders. Like the MMAR model, every unit of the trading time is set every time a trade is completed. In this case the trading time is proportional to trading frequency. The price is assumed to behave in fractional Brownian motion (FBM) along this trading time. Here, $H = 0.5$ for which FBM is normal Brownian motion.

Extraction of the Macro Model

Using the trading frequency in an time interval (volume) $\mu(t)$, Hoolder exponents, scaling functions and multifractal spectrums are calculated as follows.

$$\mu(t) = \frac{1}{N}\sum_{i=1}^{N}\prod_{k=0}^{K} M_{0k}^{1-h_{i,k}} M_{k}^{h_{i,k}} = \frac{1}{N}\sum_{i=1}^{N}\prod_{k=0}^{K}\sum_{h_k=0}^{1} \delta(h_{i,k}-h_k) M_{0k}^{1-h_k} M_{k}^{h_k}$$
$$= \sum_{h_1=0}^{1}\sum_{h_2=0}^{1}\sum_{h_3=0}^{1} \rho(h_1,h_2,h_3,...)\prod_{k=0}^{K} M_{0k}^{1-h_k} M_{k}^{h_k}$$
$$\rho(h_1,h_2,h_3,...) = \frac{1}{N}\sum_{i=1}^{N}\prod_{k=0}^{K}\delta(h_{i,k}-h_k) \underset{N\to\infty}{\sim} \prod_{k=0}^{K}\frac{1}{N}\sum_{i=1}^{N}\delta(h_{i,k}-h_k) = \prod_{k=0}^{K}\rho(h_k)$$
$$\rho(1) = \frac{1}{N}\sum_{i=1}^{N}\delta(h_{i,k}-1) = z \qquad \rho(0) = \frac{1}{N}\sum_{i=1}^{N}\delta(h_{i,k}) = 1-z$$
$$\mu(t) = \prod_{k=0}^{K}\sum_{h_k=0} \rho(h_k) M_{0k}^{1-h_k} M_{k}^{h_k} = \prod_{k=0}^{K}(M_{0k}(1-z)+M_k z)$$
$$\mu_n = (m_0(1-z)+m_1 z)^n (m_0(1-z)+m_2 z)^{(K-n)} \quad N_n = {}_K C_n \tag{1.5}$$

Hoolder exponents

$$\alpha = \frac{\log(\mu_n)}{\log 2^{-K}} = -\frac{n}{K}\alpha_1 - (1-\frac{n}{K})\alpha_2$$
$$\alpha_1 = \log_2(m_0(1-z)+m_1 z) \qquad \alpha_2 = \log_2(m_0(1-z)+m_2 z) \tag{1.6}$$

scaling functions($N >> 1$,$K >> 1$)

$$\tau_\theta(q) = -\log_2[(m_0(1-z)+m_1 z)^q + (m_0(1-z)+m_2 z)^q] \quad \text{(frequency)}$$
$$\tau(q) = H(\tau_\theta(q)+1)-1 \quad \text{(return)} \tag{1.7}$$

multifractal spectrum($N >> 1$,$K >> 1$)

$$f_\theta(\alpha) = \frac{\log(N_n)}{\log(2^K)} = -\frac{\alpha-\alpha_2}{\alpha_1-\alpha_2}\log_2(\frac{\alpha-\alpha_2}{\alpha_1-\alpha_2}) - \frac{\alpha_1-\alpha}{\alpha_1-\alpha_2}\log_2(\frac{\alpha_1-\alpha}{\alpha_1-\alpha_2}) \quad \text{(frequency)}$$
$$f(\alpha) = f_\theta(\alpha/H) \quad \text{(return)} \tag{1.8}$$

1.3 Simulation

We show the results of the micro model and the macro model. There are some studies that state normal moment estimation lacks accuracy, so we adopted the MF_DFA (MultiFractal Detrended Fluctuation Analysis) method [2] for multifractal analysis. The parameters used were $N = 1000, K = 50, T = 10000, m_1 = 0.9, m_2 = 0.1, z = 0.1$. We generated time series in the length of 10^7.

Fig. 1.1 shows the stylized facts of speculative markets. Fig. 1.2 (top) show qth moments of frequency (left) and return (right) in $log - log$ scale. They are located along the straight lines, i.e. they show fractal properties. Fig. 1.2 (middle) show scaling functions of frequency (left) and return (right). They show the multifractal property stronger in the market simulation than in the macro model. Fig.1.2 (bottom) show multifractal spectrums. In this case, it appears the difference between random walk and the macro model is small, because in the left and right side of the graph, large values of $|q|$ dominate. The result of the market simulation clearly shows multifractal property.

1.4 Conclusion

We Proposed an example of a market simulation (micro model) that showed multifractal properties. The characteristic aspect of our model is that each trader makes his decision by the logical products of a few probabilities, which represent each time scale respectively and shows the appropriateness or accuracy of a source of information. From this market simulation model, we can extract the MMAR model as a macroscopic stochastic equation for the limit of $N >> 1, K >> 1$. This simple simulation can make clear what is the cause of each stylized fact, and which facts come from the same cause.

1. Fat-tail depends on the kind of multifractal cascade of trading time.
2. Long-memory is caused by multifractal cascades.
3. Short-memory is caused by the Brownian motion of price along the trading time.
4. Multifractal property is caused by multifractal cascades.
5. The layers of a time scale originates from the time scale layers used in our daily lives.
6. The multifractal cascade originates from each trader's strategy through sources of information representing each time scale.

References

1. M. Ausloos and I. Ivanova. *Comput. Phys. Commun*, 147.
2. J.W. Kantelhardt, S. Zschiegner, E. Koscielny-Bunde, S. Havlin, A. Bunde, and H.E. Stanley. Multifractal detrended fluctuation analysis of nonstationary time series. *Physica A*, 316.
3. B. Mandelbrot, A. Fisher, and L. Calvet. A multifractal model of asset returns. Yale university working paper.
4. A. Sato and H. Takayasu. Dynamic numerical models of stock market price - from microspic determinism to macroscopic randomness. *Physica A*, 250.

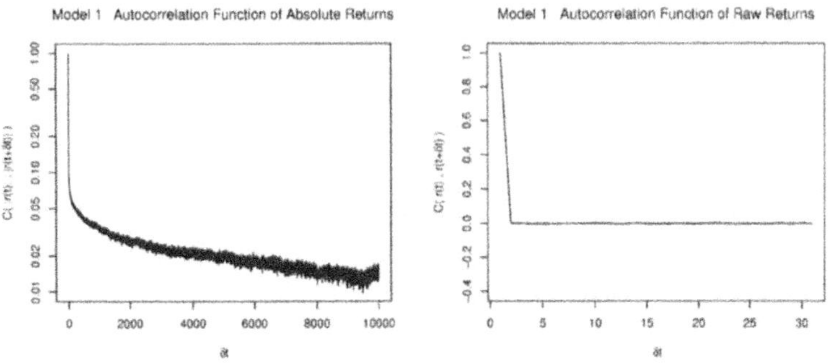

Fig. 1.1. stylized facts(market simulation)

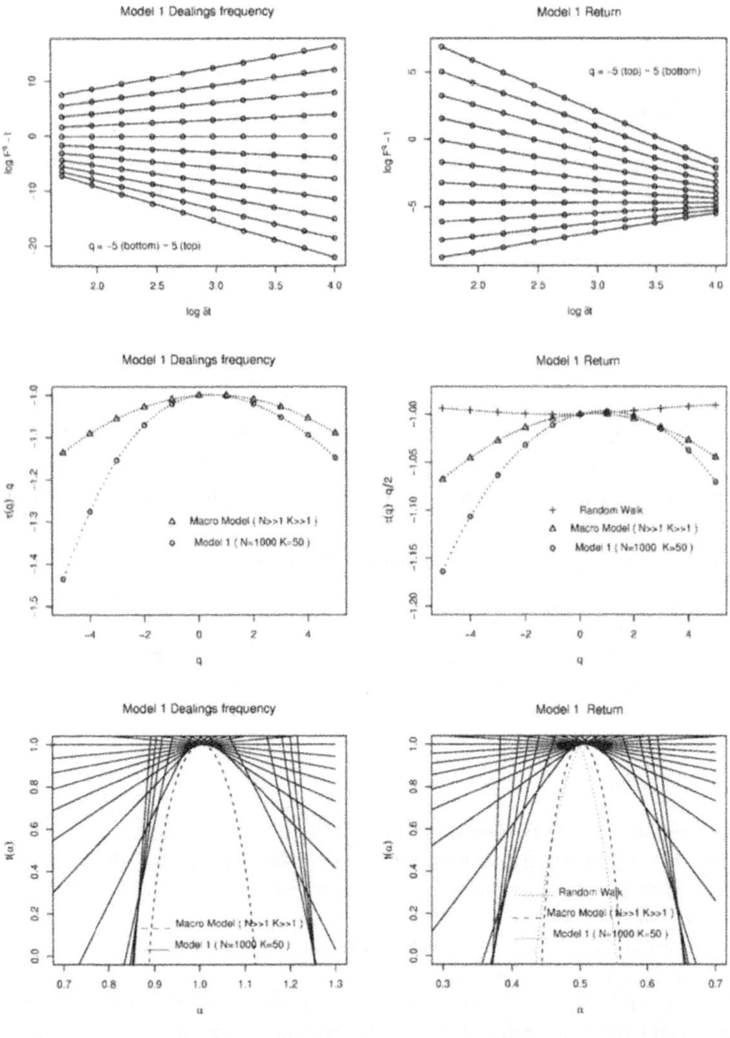

Fig. 1.2. Multifractal Property
The comparison of market simulation, macro model and random walk.

Random Graph Herding Model – An Application for Emerging Country Currency Markets –

Masahiro Fukuhara[1] and Yasufumi Saruwatari[2]

[1] Graduate School of Business Sciences, The University of Tsukuba, Tokyo, 3-29-1 Otsuka, Bunkyo-ku, Tokyo 112-0012, Japan
fukuhara@gssm.otsuka.tsukuba.ac.jp

[2] Institute of Policy and Planning Sciences, The University of Tsukuba, Tokyo, 3-29-1 Otsuka, Bunkyo-ku, Tokyo 112-0012, Japan
saru@gssm.otsuka.tsukuba.ac.jp

Summary. For analyzing currency crisis condition, we create a new model, which explains emerging country currency market and behavior of its investors. We expand a pure strategic game between speculators in a standard game theory to the overall market effect among speculators by using the random graph application. We find that three factors are critical for the herding in emerging country currency market, which are local currency market conditions, how much the speculators communicate each other, and what fraction of observed speculator offsets their position.
Key words. Herding, Crisis, Currency, Random graph

1 Introduction

After the Asian currency crises, many financial institutions and regulators have developed alarm systems for identifying currency crisis, especially for emerging country currency markets (ECCM). In addition to exogenous factors such as unexpected monetary shocks, it is natural to consider there exists an important factor for understanding and predicting currency crisis: those are endogenous factors. Endogenous factors are generated through the interaction of agents in ECCM. An agent with a different view who bets against the wind due to its own information and its strategic behavior creates a new trend if others follow him. Herding in ECCM is the typical example [1, 5], since it could arise when individuals face similar investment decisions under uncertainty and have private but imperfect information about the correct course of action.

The endogenous factors are influenced by market structure and its specific characteristics. One of well-known characteristics in a currency market is the existence of forward rate puzzle, that actual change in a spot currency market is poorly predicted by a forward rate [3]. Due to this, speculators are willing to take long positions in high interest rate currency usually belonging to emerging countries against hedgers since they know that forward rate puzzle exists, empirically. They, however, are ready to unwind the position when there is reverse back movement. In this paper, by employing the strategic game theory among speculators, we will show that there is a critical endogenous factor for currency crisis.

Our pure strategic game depicts a local relationship between speculators in a currency market. Clearly, the currency market is multi-dealer markets, not a centralized market. We will show the overall market effect caused by speculator's behavior, which is characterized by our pure strategic game, by using an application of the random graph [4]. It explains the herding by a network with interacting agents, whose decisions are determined by the actions of their neighbors based on a threshold rule, while there is no consideration of agent's rationality. Our model captures the overall effect of the endogenous factors by considering both speculators behavior and overall market. It corresponds to showing the possibility of crisis in ECCM from endogenous factors.

By employing our pure strategic game between speculators and our random graph model among speculators, it is possible to calculate the probability of currency crisis caused by endogenous factors explicitly.

2 Speculator behavior – Local model –

2.1 Model building

We assume that the market is in equilibrium among speculators, hedgers, and other participants at time $t-1$, and that only speculators exchange information and initiate the crash. This assumption is not strong, since speculators actually exchange information for getting private information and they know that some of market participants have private information in ECCM. In addition, hedgers and other participants have no strategic behavior but just trade at the market exchange rate in general.

Now, a speculator A asks z other speculators among the set N of speculators about their views in ECCM at time t. Consider the case that A realizes that x of z speculators offset their long position at time $t-1$ for some reasons such as a large realized loss in another asset. Suppose that there is uncertainty of its information [2]. Then, A starts to consider offsetting their position while A knows that continuing long position brings A benefits, unless some negative public information for the currency is announced as an exogenous shocks. Firstly, we will show the condition for deciding changing A's position through its payoff structure in local relationship.

Suppose that speculator A having one unit long forward position of emerging country currency Q against one unit of safe currency such as the US dollar at the forward rate $f_{t-1,t+1}$ at time $t-1$ and settled at $t+1$. At time

t, speculators have exchanged private information or rumors about Q. Each speculator is better off if no one unwinds the position since forward rate puzzle is assumed to exist. Therefore, the inequality $f_{t-1,t+1} > E_t^{A,W}(S_{t+1})$ is hold, where $E^{A,W}(S_{t+1})$ is an expected spot rate at $t+1$ of Q against unit value of safe currency at time $t+1$ assuming the state W by A, and W denotes the state that speculator A believes the existence of forward rate puzzle at time t. They do not focus on S_t, spot rate at t, unless it exists crises between $t-1$ and t for emerging country currency Q, due to the belief of forward rate puzzle in a certain period of time. On the other hand, A is afraid that reversal trade has started and forward rate puzzle has collapsed between time t and $t+1$. In this case, $f_{t-1,t+1} < E^{A,B}(S_{t+1})$ holds, where B is the state that speculator A believes the collapse of forward rate puzzle. Also, A know that if A starts to unwind the position, it might create market crash and therefore $S_t < E^{A,B}(S_{t+1})$. The last case is: A quits the market. It is shown as $S_t > E^{A,W}(S_{t+1})$.

Remark that the pure game and "Bayesian" model are not applicable directly in showing the behavior of ECCM since each speculator doesn't see the behavior of all other speculators and these models do not take into account the importance of private information. Actually, each speculator can see only limited other speculators behavior and exchange private information.

A sub-game, which explains local relationship among z rather than all participants in the market, is constructed. Let z be the average number of speculators that each speculator access, and s^A the variable showing that speculator A is in "hold position." Clearly, s^A is 1 (resp. 0) if A takes hold position (resp. offset position). Then, the payoff function of A is defined as follows:

$$\max\{(1-q)(s^A(f_{t-1,t+1} - E_t^{A,W}(S_{t+1})) + (1-s^A)(f_{t-1,t+1} - S_t)) \\ +q(s^A(f_{t-1,t+1} - E_t^{A,B}(S_{t+1})) + (1-s^A)(f_{t-1,t+1} - S_t))\}$$

where $q = c\frac{x}{z} = \frac{x}{z}$, c is the confidence level of information and assumed to be 1 here (i.e. 100% confidence), and x is the number of neighbors offsetting the position. Note that it means A makes decision whether to hold their long position or not taken at $t-1$ after observing q. Naturally, we can assume $E_t^{A,W}(S_{t+1}) < S_t < E_t^{A,B}(S_{t+1})$.

2.2 Result

We assume that all speculators are homogeneous, and therefore each speculator plays the same game, has same views on a currency market, and has the same strategy and payoffs of the game.

From the relationship, the decision of speculator A can be characterized by the following conditions:

$$s^A = 1, \quad \text{if } (1-q)E_t^{A,W}(S_{t+1}) + qE_t^{A,B}(S_{t+1}) > S_t \\ s^A = 0, \quad \text{if } (1-q)E_t^{A,W}(S_{t+1}) + qE_t^{A,B}(S_{t+1}) \leq S_t$$

Thus, the following inequality can be derived.

$$q = \frac{x}{z} \geq \frac{(S_t - E_t^{A,W}(S_{t+1}))}{E_t^{A,B}(S_{t+1}) - E_t^{A,W}(S_{t+1})}$$

We call it as the currency crisis condition. The above shows that each speculator starts to offset its long position, when

- The larger expected positive return, the more likely to hold their position.
- The larger difference of expected return in good and bad state (the more uncertainty), the more likely to offset their position.
- The more skew for good performance, the more likely to hold their position.

The above model explains only the local relationship among limited numbers of speculators while many herding models [1, 5] only analyze its effects since they deal with basically centralized investor market. However, those models are not sufficient to analyze de-centralized investor market such as the currency markets, which are made by unobserved number of groups and we need take into account these inter-action explicitly. In the next section, we investigate its interaction and the overall effect, and construct a global model.

3 Market overall effect – Global model –

3.1 Model building

Even if some speculators start to offset their long position and its information is propagated to other speculators, it may not create herdings in currency markets. Actually, there are many small local endogenous shocks, but there are rarely real endogenous overall market shocks, which are in fact herdings in currency markets.

We suppose that the condition of information sharing among speculators is the key for the possible overall market herding, and when a speculator has observed a large fraction of other speculators offsetting their positions over information exchanged with them. We will see its condition and calculate the probability of the overall market herding from a speculator behavior through the analysis of our model.

Watts [4] shows information herding relationship by employing the theory of random graphs. In its model, binary decision, neighbors of an origin, and network play important roles.

By applying Watts model to our currency herding model, firstly, we construct a complete network with the set N of vertices, each vertex denotes a speculator. In addition, for each edge in the network, a probability p_k is associated. The probability p_k of an edge between vertices v_i and v_j stands for the possibility of exchanging information between speculators showing vertices v_i and v_j. Let z be the average number of vertices (speculators) that each vertex may exchange information. Suppose that all speculators initially hold long position (denoted by state 0), and a speculator (called "seed") starts to offset its position at $t-1$. Under some certain condition, its movement may grow. We define the movement as overall market herding in a currency market. It means that the state of almost all vertices are turning to another state, denoted by state 1.

The model depicts its dynamics by using random graph theory, where the generating function of vertices simplifies the problem, because we can extract all the moments of the degree distribution of vulnerable vertices.

By applying some technical calculations, "herding condition" can be derived as follows:

$$G_0''(1) = \sum_k k(k-1)\rho_k p_k = z$$

where G_0 is a generating function of vertex degree,

$$\rho_k = \begin{cases} 1 & \text{if } k = 0 \\ F(\frac{1}{k}) & \text{if } k > 0, \end{cases}$$

$F(\phi) = \int_0^\phi f(\varphi)d\varphi$, and $f(\varphi)$ is a distribution for assigning a threshold and normalized such as $\int_0^1 f(\varphi) = 1$.

When $G_0''(1) < z$, all vulnerable clusters are small and therefore no herding has made in a currency market.

3.2 Result

By using the model, we can analyze the dynamics of herdings. Here, for simplicity, we assume that all speculators have observed the same fraction of offsetting speculators, ϕ_c. That is:

$$f(\phi) = \delta(\phi - \phi_c),$$

where δ is a delta function which connects discrete fraction to continuous function. Also, any pair of speculators is connected with probability $p = \frac{z}{|N|}$. The two conditions indicate that p_k is "Poisson distribution," $p_k = \frac{e^{-z}z^k}{k!}$. In this simplified case, the herding condition reduces to:

$$zQ(K_c - 1, z) = 1$$

where $K_c = \frac{1}{\phi_c}$ and $Q(a, x)$ is the incomplete gamma function.

Table 1 shows the probability of herdings, showing %, which are calculated by the above condition. Note that z denotes the average number of speculators with whom each speculator can exchange information, and "threshold" stands for q, the currency crisis condition.

Table 1. Probability of herding (%)

	z													
threshold	0.5	1.0	1.5	2.0	2.5	3.0	3.5	4.0	4.5	5.0	5.5	6.0	6.5	7.0
0.05	0	0	58	80	89	94	97	98	99	99	100	100	100	100
0.10	0	0	58	80	89	94	96	97	97	96	95	95	90	87
0.15	0	0	56	77	85	89	89	89	81	75	68	58	0	0
0.20	0	0	48	67	73	71	65	50	34	15	0	0	0	0
0.25	0	0	0	0	0	0	0	0	0	0	0	0	0	0

If this condition is met, sellers dominate the market through herding, and it may create market crash in the worst case. For example, assume that speculator A exchanges private information for his/her long position in ECCM with 5 other speculators and realizes one of them offsets their long position between $t-1$ and t. Based on his/her prediction, $\frac{(S_t - E_t^{A,W}(S_{t+1}))}{E_t^{A,B}(S_{t+1}) - E_t^{A,W}(S_{t+1})}$ becomes 0.18. In this case, the first criteria that observed fraction $0.2 = \frac{1}{5}$ exceeds his/her offset condition 0.18 is met. Next, we check whether herding occurs or not in the above table. Since we suppose homogeneous agents, z is 5 and $\frac{x}{z} = 0.2$. Therefore, the probability of herding is 15% in the example.

4 Conclusion

Based on our ECCM model, we investigated three important factors for the condition of currency crisis: local currency market condition in a pure strategic game, how much speculators communicate among their neighbors, and what fraction of observed speculators are offsetting the position.

References

1. S. Bikhchandani and S. Sharma, *Herd Behavior in Financial Markets: A Review,* IMF Working Paper WP/00/48, March 2000.
2. H.H. Cao and D. Hirshleifer, *Conversation, Observational Learning, and Informational Cascades,* http://www.cob.ohio-state.edu/fin/dice/papers/2001/2001-5.pdf, November 13, 2000.
3. C. Engel, *The Forward Discount Anomaly and the Risk Premium: A Survey of Recent Evidence,* Journal of Empirical Finance, Vol.3, pp.123–192, 1996.
4. D.J. Watts, *A Simple Model of Fads and Cascading Failures,* Santa Fe Working Paper, http://www.santafe.edu/sfi/publications/Working-Papers/00-12-062.pdf, December 2000.
5. I. Welch, *Sequential sales, learning, and cascades,* The Journal of Finance, Vol.XlVII, No.2, June 1992.

Multivariable Modeling on Complex Behavior of a Foreign Exchange Market

Tomoya SUZUKI[1], Tohru IKEGUCHI[2] and Masuo SUZUKI[1]

[1] Graduate School of Science, Tokyo University of Science,
1-3 Kagurazaka, Shinjuku-ku, Tokyo, 162-8601 JAPAN
j1202705@ed.kagu.tus.ac.jp

[2] Graduate School of Science and Engineering, Saitama University,
225 Shimo-Ohkubo, Sakura-ku, Saitama, 338-8570 JAPAN

Summary. We discover a remarkable property of a foreign exchange market: when the spreads, the difference between ask and bid prices, become large, the dealing time intervals become short and the price movements become strong. To discuss an interaction of these variables, we propose a model on complex behavior of the foreign exchange market. Then, we confirm the plausibility of the proposed model by comparing the statistical properties of the real market and the model. Finally, we discuss why there exists the interaction on the basis of our model, and clarify the mechanism for producing complex behavior with the interaction.

Key words: foreign exchange market, spreads, dealing time intervals, middle price, raster plots, peri-stimulus time histograms

1 Introduction

The present paper mainly focuses on the relation between the dealing time intervals and the spreads, the difference of bid and ask prices, which could affect the price movements for the following reason. Since dealers generally wait for an opportunity of getting large earnings, the dealing time interval which corresponds to dealing timings and the spread may provide important information to dealers' mind getting earnings and hedging risks and directly link to dealers' motivation for getting a deal. Thus, these variables could be essential factors for modeling exchange markets.

To confirm the interaction among these variables, we introduce novel analysis tools which are based on raster plots[1] and peri-stimulus time histograms (PSTH)[2]. These methods are popular in the field of neurophysiology and are suitable to observe ensemble behavior. Namely, these methods are useful for visually understanding the complex behavior of multivariables in foreign exchange markets. From analysis results, we see that when the spreads become large suddenly, the dealing time intervals become short and the price movements become large. Since the expansion of the spread means that ask and bid prices are separated from the middle price, it is natural to consider that the dealer tries to sell at higher prices and to buy at lower prices. Such a bull quotation is anticipated to lead the situation that dealing time intervals become longer since it is not so easy to find a dealing partner. However, our analyses show completely opposite tendencies.

In order to understand the mechanism of producing such unexpected behavior in a foreign exchange market, we propose a plausible model of explaining the behavior in the foreign exchange market. Then, from the viewpoint of the proposed model, we clarify the reasons why there exists the singular interaction among these three variables in the real market.

2 Analyzing Real Dealing Data

For analyses, we use the time series of tick data between the U.S. dollar and the Swiss franc observed in an interbank market[3]. The tick data set is recorded from January 1986 to April 1991 (total 1,322 days), and the total number of the data points is 282,956. Generally, a tick data set has the following intrinsic aspects. There is a sort of discontinuity on dairy tick data, because banks close at nights, weekends and holidays. Then the first and the last dealing times are different from each other day by day. The dealing-time stamps are recorded at actual timings when every deal occurs.

Now, we denote the temporal index by n, which indicates a discrete time and increases one by one whenever each dealing occurs. Moreover, we denote each middle price, namely a mean value of bid and ask prices, by $P(n)$ [dollar] and their temporal differences by $dP(n)$ $(= P(n) - P(n-1))$, respectively. We also denote the spread, the difference between bid and ask prices, by $S(n)$ [dollar] and the dealing time interval by $\tau(n)$ [sec].

First, we analyze the interaction of three variables by drawing raster plots (RPs)[1] and peri-stimulus time histograms (PSTHs)[2]. These methods are often utilized for analyzing neural spikes in the field of neurophysiology, and are good schemes to express spike timings visually. In the case of analyzing neural spikes, every observed datum is plotted in horizontal lines with external stimuli. However, for analyzing financial indices, it is not easy to consider the existence of external stimuli explicitly. Then, we propose a modified method in order to apply these techniques to analyze interbank data.

For drawing RPs, we consider each day as each trial, and occurrences of actual dealings as spikes. Since it is very rare to observe the case in which spreads become very large, we assume that a particular time at which spreads become very large is equal to the time at which external stimuli are applied in order to investigate alteration in the market's behavior by the external stimuli.

For evaluating PSTHs, we have to consider how to treat the variation of opening and closing times in each day. In the i-th temporal bin ($i = 1, 2, \cdots, B$), deals are denoted by $s_i(k)(k = 1, \cdots, N_i)$. When there are d_i days in which the start of dealing is later and the finish of dealing is faster than the median of the corresponding temporal bins, the average of the dairy dealing in the i-th temporal bin is calculated by $H_i = N_i/(D - d_i)$, where D is the number of total days. Moreover, in order to examine an ensemble behavior of the movement of middle prices, the difference of the middle price in the dealing $s_i(k)$ is used for calculating the following histogram, $h_i = \frac{1}{N_i}\sum_{k=1}^{N_i} |\Delta P_{s_i(k)}|$, where $\Delta P_{s_i(k)} = P_{s_i(k)} - P_{r_i(k)}$, and $P_{r_i(k)}$ is a one-step previous price of $P_{s_i(k)}$. Then, we divide the total dealing length from 9:00 a.m. to 5:00 p.m. into four sections (each has two hours long) and classify all tick data into each section on the basis of the dealing time.

Calculating RPs and PSTHs by the following two methods (i) and (ii), we compare both of these results.

(i) We randomly select 15 dealings, in which the actual dealings occur nearly at the median of each section, and we can temporarily identify their dealing time as the time of external stimuli. Then, their deals are aligned with the vertical line at $t = 0$ on the horizontal axis in RPs. Here, t is continuous time.
(ii) We select top 15 dealing data sets whose spreads become much larger from each temporal section. Stimuli are considered to be the same as those in (i).

If the movement of spreads has no relation to the dealer's action, we expect that no significant difference appears between the results obtained by the above two methods.

Figure 1 shows the results obtained by the above analyses. The temporal section is from 11:00 a.m. to 1:00 p.m, $B = 40$ (the duration of the temporal bin is about 1,000 [sec]) and $D = 15$. In Fig.1(a), the histogram H_i shows a kind of temporal rhythm (a shape of the letter "M") and the histogram h_i is almost flat (no particular rhythm). The M type rhythm shows a daily trend of the dealing time intervals which depend on the number of dealers who participate in the market. Since there are few dealers in early morning, lunch time and evening, the dealing time intervals tend to become larger in such time regions. Thus, H_i becomes shorter and the M type rhythm appears. On the other hand, from Fig.1(b), we find that many dealings appear in temporal bins near $t = 0$ and the large amount of dealings destroys the rhythm of the shape "M," then the dealing time interval becomes shorter. Moreover, the histogram h_i also shows that the movement of middle prices has a peak in the temporal bins at $t = 0$, which means that the expansion of the spreads makes the movement of middle prices larger.

The term *expansion of the spread* means that the bid and ask prices separate from the middle price, which indicates that dealers try to buy at lower prices and to sell at higher prices. It is very singular, because in spite of such bull quotations, this dealing time intervals become shorter as introduced

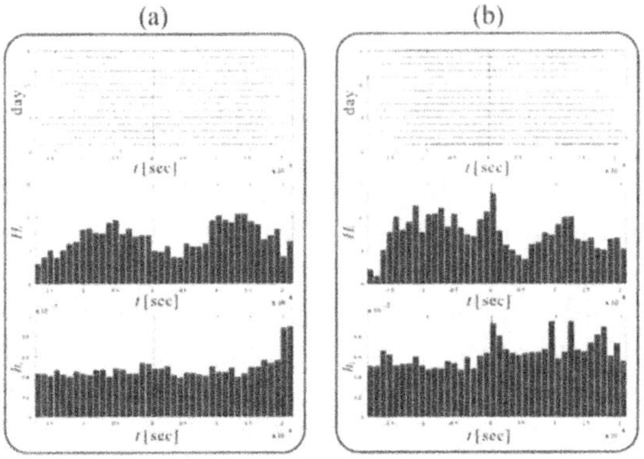

Fig. 1. Top plots are modified RPs, and middle and bottom plots are modified PSTHs. Figures (a) and (b) correspond to the methods (i) and (ii), respectively.

in Sec.1. Such tendencies are also observed if we use other sections (9:00 a.m.∼11:00 a.m., 1:00 p.m.∼3:00 p.m. and 3:00 p.m.∼5:00 p.m.) as the central time.

Moreover, from Refs.[4, 5], we obtain the result that prediction accuracy of middle-price movements is better by using these three variables than using only the middle price. The existence of the interaction among three variables is also supported by this improvement on prediction accuracy.

3 A Model for a Foreign Exchange Market

In this section, we propose two models of the interbank exchange dealing, which incorporate the following three variables, dealing time intervals, spreads and middle-price movements, in order to discuss the singular interaction among these variables.

Model 1: a mechanism for deciding bid and ask prices by each dealer
We calculate a distribution of possible future prices according to past middle-price movements during last p terms, $\{P(n-p+1), \cdots, P(n)\}$. The distribution is based on a geometric Brownian motion[6, 7], and its standard deviation temporarily changes according to the past movements. Moreover, though its mean value also changes, it is almost equal to $P(n)$[8]. Since a future price is decided by a dealer's quotation, possible future prices correspond to predicted future prices by all dealers. Since each dealer does not know the predicted future prices by the others, the dealer predicts them and evaluates the expected gain and the risk by prediction errors. Each dealer estimates the expected return by the expected gain and risk and decides best bid and ask prices, that is a best spread $\dot{S}(n+1)$, by maximizing the expected return.

As one of the characteristics of Model 1, when past middle prices fluctuate heavily, the standard deviation of the distribution of possible future prices becomes large. Namely, the dealers' quotations are distributed widely. In this case, even if a dealer quotes a bull quotation, he can find a dealing partner. Thus, a best spread tends to be large.

Model 2: describing a dealing process for deciding dealing time intervals First, we select a dealer A who can really get a deal by referring to a real future price $P(n+1)$. In this model, we mainly consider how we can model the behavior of spreads and dealing time intervals. Then, we use real future prices for the present purpose. However, even if we do not have real prices, we can substitute the ARCH model[9] or the GARCH model[10] for real prices[8]. In Model 1, we know this dealer's quotation and the distribution of predicted future prices by the other dealers. Next, we randomly select a dealing partner of the dealer A and iterate this random selection until a dealing really occurs, that is the predicted price of a dealing partner is larger than the ask price or is smaller than the bid price of the dealer A. Once it occurs, we record the number of iterations as the decided dealing time interval $\tau(n+1)$. By iterating the above processes, we are able to decide both of the spreads and the dealing time intervals.

In order to confirm the plausibility of our models, we analyze whether or not the data obtained from our models have similar properties to the real data[8]. By this estimation, we confirm that the data obtained by the model have almost the same statistical properties as the real data. In particular, each power spectrum and each cumulative distribution function have the same power laws of same slopes. Moreover, by calculating RPs and PSTHs, the same singular interaction is confirmed from the data obtained by the our model as well[8]. From these results, we can consider that our model has very good plausibility for explaining a producing mechanism of complex behavior observed in real markets.

4 Discussion on the singularity of h and H

Once the spread becomes larger, the middle-price movement is large, that is, the singularity on h appears. In Model 1, since a best spread $\dot{S}(n+1)$ depends on the past middle-price movements for last p terms, $\{P(n-p+1),\cdots,P(n)\}$, only the last price movement $|dP(n)|$ makes $\dot{S}(n+1)$ larger than $\dot{S}(n)$. Moreover, by analyzing real data, we can confirm that $|dP|$ has a slight temporal dependency[8]. Thus, when $S(n+1)$ becomes larger, since the fact that the last price movement $|dP(n)|$ is large affects to the temporal dependency of $|dP(n+1)|$, then $|dP(n+1)|$ becomes large. By above discussion, we can explain the reason why the singularity on h appears.

In Model 2, when $|dP(n+1)|$ is large, the dealer A, who can really get a deal, puts his own middle price near to the tail of the distribution of the predicted future price by all dealers as shown in Fig.2(a). Figure 2(b) is the situation when the $|dP(n+1)|$ is small. If $|dP(n+1)|$ is large (Fig.2(a)) rather than when the $|dP(n+1)|$ is small (Fig.2(b)), there exists more dealing

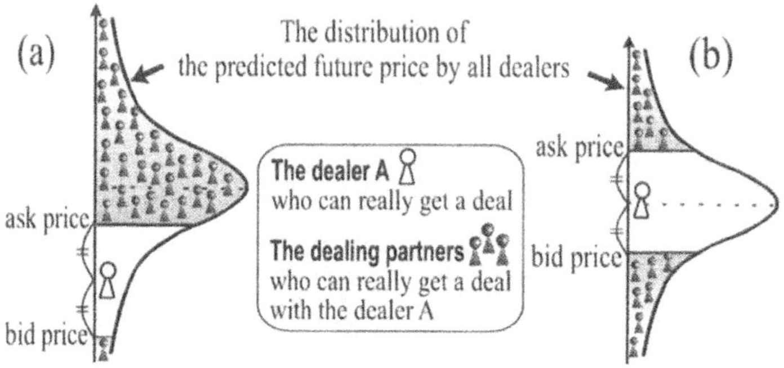

Fig. 2. Relations between a dealer and its dealing partners. The mean value of each distribution is almost equal to $P(n)$.

partners who can really get a deal with the dealer A[8]. Therefore, in the situation as Fig.2(a), it is easy to find a dealing partner randomly in Model 2 for a few numbers of iteration. Thus, as $|dP(n+1)|$ is large, the dealing time interval $\tau(n+1)$ becomes shorter. Thus, we can explain the reason why the singularity on H appears.

5 Conclusions

By using modified RPs and PSTHs, we discover the singular interaction among the three important variables, the middle-price movements, the dealing time intervals and the spreads, in a real foreign exchange market. In order to confirm the mechanism of such interaction among these variables, we introduce a plausible model with multivariables. We also mention why there exists the singular interaction on the basis of our model.

References

1. Gerstein G. L., et al., IEEE Trans., SMC-13, 668-676 (1983).
2. J. Krueger, Rev. Physiol. Biochem. Pharmac. **98**, 177-233 (1983).
3. A. S. Weigend, et al. eds., "Time Series Prediction," Addison Wesley (1993).
4. T. Suzuki, T. Ikeguchi and M. Suzuki, Proceedings of 2001 International Symposium on Nonlinear Theory and its Applications, **1**, 287-290 (2001).
5. T. Suzuki, T. Ikeguchi and M. Suzuki, "Multivariable Nonlinear Analysis of Foreign Exchange Rates," Physica A (2003), in press.
6. F. Black and M. Scholes, J. Polit. Econ., **81**, 637-654 (1973).
7. R. C. Merton, Bell J. Econ. Management Sci., **4**, 141-183 (1973).
8. T. Suzuki, T. Ikeguchi and M. Suzuki, "Modeling on complex behavior of inter-bank exchange markets," submitted to possible publication (2003).
9. R. F. Engle, Econometrica, **50**, 987-1002 (1982).
10. T. Bollerslev, J. Econometrics, **31**, 307-327 (1986).

Gibbs measure and Markov chain modeling for stock markets

Jun-ichi Maskawa

Department of Management Information, Fukuyama Heisei University, Fukuyama, Hiroshima 720-0001, Japan maskawa@heisei-u.ac.jp

Summary. We reviewed the recent work on Gibbs measure (statistical physics model) describing the collective price jumps in stock markets. We started with the study of a multivariate Markov chain model as a stochastic model of the price changes of portfolios in the framework of the mean field approximation. The time series of price changes were coded into the sequences of up and down spins according to their signs. As the stationary state of the Markov chain, Gibbs measure was naturally derived, which formally coincides with spin glass model of disordered magnetic systems. The linear response of the system to external fields was examined to prove the fluctuation response theorem. Finally, the analysis of actual portfolios based on this model was briefly summarized.

1 Introduction

The cross-correlation of the price fluctuation between different stock issues has a practical importance as a main material for risk management (Markowitz 1991). It is also a distance between individual companies in the subdominant ultrametric space of companies (Mantegna 1999, Mantegna and Stanley 2000), and we can recognize a business sector by the eigenvector of the cross-correlation matrix filtered by using random matrix theory (Gopikrishnan et al. 2001).

In this paper, we reviewed the recent work on Gibbs measure (statistical physics model) describing the collective price jumps in stock markets. We study a multivariate Markov chain model as a stochastic model of the price changes of portfolios (Maskawa 2003). In the framework of the mean field approximation (two body approximation), Gibbs measure is naturally derived as the stationary state of the Markov chain, which is written only in terms of cross-correlation. Through the analysis of actual portfolios, it has been shown that the synchronization of the direction of the price changes is well described by the model (Maskawa 2002a, 2002b).

2 Markov chain model

We study the discrete stochastic dynamics of collective price changes of portfolio in the framework of multivariate Markov chain model (Maskawa 2003). We concentrate on the direction of the stock price change, that is, the sign of the price change $\sigma_i^t = \pm 1$ ($i = 1$ to portfolio size N, $t = 0$ to observation time T) . The joint probability $p(\sigma_1^t, \sigma t_2, \ldots, \sigma t_N)$ and the transition probability $p(\sigma_1^{t+1}\sigma_2^{t+1}, \ldots, \sigma_N^{t+1}|\sigma_1^t\sigma_2^t, \ldots, \sigma_N^t)$ for a given portfolio are obtained as relative frequencies. Here we assume the stationarity of the process and, for simplicity, the time reversal symmetry, that is, $p(\sigma_1^t, \sigma t_2, \ldots, \sigma_N^t) \simeq p(-\sigma t_1, -\sigma t_2, \ldots, -\sigma_N^t)$ and $p(\sigma_1^{t+1}\sigma_2^{t+1}, \ldots, \sigma_N^{t+1}|\sigma_1^t\sigma_2^t, \ldots, \sigma_N^t) \simeq p(-\sigma_1^{t+1} - \sigma_2^{t+1}, \ldots, -\sigma_N^t| - \sigma_1^t - \sigma_2^t, \ldots, -\sigma_N^t)$.

We start with small portfolios consisting of two stock issues. The generalization of our model to arbitrary size of portfolio will be constructed later by recurrence relations. Under the assumption of the time reversal symmetry, there are 6 independent elements in the transition matrix, those are, in our notation,

$$\begin{array}{lll} p(-1-1|-1-1) = a & p(-11|-1-1) = e & p(1-1|-1-1) = f \\ p(-1-1|-11) = b & p(-11|-11) = d & p(-1-1|1-1) = c. \end{array} \quad (1)$$

The stationary state is expressed in terms of the elements of transition matrix as

$$\begin{array}{ll} p(-1-1) = p(11) & = (b+c)/2(b+c+e+f) \\ p(-11) \quad = p(1-1) & = (e+f)/2(b+c+e+f). \end{array} \quad (2)$$

The averages $< \sigma_1 >=< \sigma_2 >= 0$, and the joint probability (2) has an expression in terms of the covariance $C_{12} =< \sigma_1\sigma_2 >= (b+c-e-f)/(b+c+e+f)$ as

$$p(\sigma_1\sigma_2) = (1 + C_{12}\sigma_1\sigma_2)/4 \simeq e^{C_{12}\sigma_1\sigma_2}/2(e^{C_{12}} + e^{-C_{12}}) \quad (3)$$

up to a second order in C_{12}.

The most general form for the different-time joint probability is

$$\begin{aligned} p(\sigma_1^t, \sigma_2^t, \sigma_1^{t+1}, \sigma_2^{t+1}) &= A(1 + a_1\sigma_1^t + a_2\sigma_1^{t+1} + a_3\sigma_2^t + a_4\sigma_2^{t+1} \\ &+ a_{12}\sigma_1^t\sigma_2^t + a'_{12}\sigma_1^{t+1}\sigma_2^{t+1} + a_{11}\sigma_1^t\sigma_1^{t+1} + a_{22}\sigma_2^t\sigma_2^{t+1} + a_{12}\sigma_1^t\sigma_2^{t+1} + a_{21}\sigma_2^t\sigma_1^{t+1} \\ &+ b_1\sigma_1^t\sigma_2^t\sigma_1^{t+1} + b_2\sigma_1^t\sigma_2^t\sigma_2^{t+1} + b_3\sigma_1^t\sigma_1^{t+1}\sigma_2^{t+1} + b_4\sigma_2^t\sigma_1^{t+1}\sigma_2^{t+1} \\ &+ c_{12}\sigma_1^t\sigma_2^t\sigma_1^{t+1}\sigma_2^{t+1}). \end{aligned} \quad (4)$$

By the assumption of the time reversal symmetry, we can eliminate the constants a_i and b_i ($i = 1 - 4$). Rearranging the terms in the equation (4), we have another useful form

$$\begin{aligned} p(\sigma_1^t, \sigma_2^t, \sigma_1^{t+1}, \sigma_2^{t+1}) = 1/Z^{(2)} \exp(& v_{12}\sigma_1^t\sigma_2^t + v'_{12}\sigma_1^{t+1}\sigma_2^{t+1} \\ &+ w_{11}\sigma_1^t\sigma_1^{t+1} + w_{22}\sigma_2^t\sigma_2^{t+1} + w_{12}\sigma_1^t\sigma_2^{t+1} + w_{21}\sigma_2^t\sigma_1^{t+1} \\ &+ u_{12}\sigma_1^t\sigma_2^t\sigma_1^{t+1}\sigma_2^{t+1}). \end{aligned} \quad (5)$$

Using the definition of transition probability,

$$p(\sigma_1^{t+1},\sigma_2^{t+1}|\sigma_1^t,\sigma_2^t) = \frac{p(\sigma_1^t,\sigma_2^t,\sigma_1^{t+1},\sigma_2^{t+1})}{\sum_{\sigma_1^{t+1},\sigma_2^{t+1}} p(\sigma_1^t,\sigma_2^t,\sigma_1^{t+1},\sigma_2^{t+1})}, \tag{6}$$

we obtain the relations

$$e^{v'_{12}} = \sqrt[8]{\frac{abc(1-a-e-f)}{def(1-b-c-d)}},\ e^{w_{11}} = \sqrt[8]{\frac{abde}{cf(1-b-c-d)(1-a-e-f)}}$$

$$e^{w_{22}} = \sqrt[8]{\frac{acdf}{be(1-b-c-d)(1-a-e-f)}},\ e^{w_{12}} = \sqrt[8]{\frac{abf(1-b-c-d)}{cde(1-a-e-f)}}$$

$$e^{w_{21}} = \sqrt[8]{\frac{ace(1-b-c-d)}{bdf(1-a-e-f)}},\ e^{u_{12}} = \sqrt[8]{\frac{ad(1-b-c-d)(1-a-e-f)}{bcef}}. \tag{7}$$

From the consistency condition

$$p(\sigma_1^t,\sigma_2^t) = \sum_{\sigma_1^{t+1},\sigma_2^{t+1}} p(\sigma_1^t,\sigma_2^t,\sigma_1^{t+1},\sigma_2^{t+1}), \tag{8}$$

we have

$$e^{v_{12}} = \sqrt{\frac{b+c}{e+f}}\sqrt[8]{\frac{aef(1-a-e-f)}{bcd(1-b-c-d)}}. \tag{9}$$

Under the assumption of stationarity, the equality $e^{v_{12}} = e^{v'_{12}}$ must be fulfilled, which means the equation $(e+f)/(b+c) = \sqrt{ef/bc}$.

We generalize our model to portfolios with arbitrary size N. Here we adopt a mean field approximation. For N-issues model, our ansatz for the joint probability is the following form,

$$p(\sigma_1,\cdots,\sigma_N) = 1/Z^{(N)} \times \prod_{1\le i<j\le N} e^{J_{ij}^{(N)}\sigma_i\sigma_j}., \tag{10}$$

Using the identity

$$p(\sigma_1,\cdots,\sigma_n) = 1/Z^{(n+1)} \times \sum_{\sigma_{n+1}} \prod_{1\le i<j\le n+1} e^{J_{ij}^{(n+1)}\sigma_i\sigma_j} \tag{11}$$

we have recurrence relations as the consistency equations

$$J_{ij}^{(n+1)} = J_{ij}^{(n)} - C_{i(n+1)}C_{j(n+1)} \tag{12}$$

up to a second order in C_{ij}. Solving the equation (12), we have the expressions for the renormalized coefficients

$$J_{ij}^{(N)} = C_{ij} - \sum_{1\le k\le N, k\ne i, k\ne j} C_{ik}C_{jk} \simeq -\mathrm{C}_{ij}^{-1}, \tag{13}$$

and the expression for the partition function

$$Z^{(N)} = 2^{N-1}(2 + \sum_{i<j} C_{ij}^2). \tag{14}$$

up to a second order in C_{ij}, where $\mathsf{C} = [C_{ij}]$ is a $N \times N$ covariance matrix. Therefore we found that Gibbs measure was naturally derived as the stationary state of the Markov chain in the framework of mean field approximation, which formally coincides with spin glass model of disordered magnetic systems (Sherrington and Kirkpatrick 1975).

In order to derive the model of transition probability,

$$p(\sigma_1^{t+1}, \cdots, \sigma_N^{t+1} | \sigma_1^t, \cdots, \sigma_N^t) = \frac{p(\sigma_1^t, \cdots, \sigma_N^t, \sigma_1^{t+1}, \cdots, \sigma_N^{t+1})}{p(\sigma_1^t, \cdots, \sigma_N^t)} \tag{15}$$

we write the different-time joint probability in the following form,

$$\begin{aligned} p(\sigma_1^t, \cdots, \sigma_N^t \;,\; \sigma_1^{t+1}, \cdots, \sigma_N^{t+1}) = 1/\mathsf{Z}^{(N)} \times \prod_{1 \le i < j \le N} & \exp(v_{ij}^N (\sigma_i^t \sigma_j^t + \sigma_i^{t+1} \sigma_j^{t+1}) \\ & + \frac{1}{N-1}(w_{ii}^N \sigma_i^t \sigma_i^{t+1} + w_{jj}^N \sigma_j^t \sigma_j^{t+1}) + w_{ij}^N \sigma_i^t \sigma_j^{t+1} + w_{ji}^N \sigma_j^t \sigma_i^{t+1} \\ & + u_{ij}^N \sigma_i^t \sigma_j^t \sigma_i^{t+1} \sigma_j^{t+1}). \end{aligned} \tag{16}$$

Using the identity

$$p(\sigma_1^t, \cdots, \sigma_n^t, \sigma_1^{t+1}, \cdots, \sigma_n^{t+1}) = \sum_{\sigma_{n+1}^t, \sigma_{n+1}^{t+1}} p(\sigma_1^t, \cdots, \sigma_{n+1}^t, \sigma_1^{t+1}, \cdots, \sigma_{n+1}^{t+1}) \tag{17}$$

we have general formulas,

$$\begin{aligned} v_{ij}^N &= v_{ij} - \sum_{1 \le k \le N, k \ne i, k \ne j} (v_{ik} v_{jk} + w_{ik} w_{jk}) \\ w_{ii}^N &= w_{ii} - \sum_{1 \le k \le N, k \ne i, k \ne j} ((w_{ik} + w_{ki}) v_{ik}) \\ w_{jj}^N &= w_{jj} - \sum_{1 \le k \le N, k \ne i, k \ne j} ((w_{jk} + w_{kj}) v_{jk}) \\ w_{ij}^N &= w_{ij} - \sum_{1 \le k \le N, k \ne i, k \ne j} (w_{ik} v_{jk} + w_{kj} v_{ik}) \\ w_{ji}^N &= w_{ji} - \sum_{1 \le k \le N, k \ne i, k \ne j} (w_{jk} v_{ik} - w_{ki} v_{jk}) \\ u_{ij}^N &= u_{ij} - \sum_{1 \le k \le N, k \ne i, k \ne j} u_{ik} u_{jk} \quad (1 \le i < j \le N), \end{aligned} \tag{18}$$

and

$$Z^{(N)} = 2^{2N-1}(2 + \sum_{i<j}(2v_{ij}^2 + \frac{1}{(N-1)^2}(w_{ii}^2 + w_{jj}^2) + w_{ij}^2 + w_{ji}^2 + u_{ij}^2)). \tag{19}$$

Our modeling of the transition matrix extremely reduce the number of independent parameters. That is very important for practical prediction, because the number of the independent parameters $2^{2N} - 2^N$ for naive Markov chain model immediately becomes an astronomical number when we consider large portfolios.

3 Linear response

From the equation (10), let us consider a N-issues portfolio as a system following the statistical mechanics whose Hamiltonian is given by the expression

$$H[\sigma, h] = -\sum_{<i,j>} J_{ij}\sigma_i\sigma_j - \sum_i h_i\sigma_i, \tag{20}$$

where the the interaction coefficients J_{ij} (index N is omitted) are given by the equation (13). The second term in the right hand side is added to examine the response of the system to external fields.

Here we will show that our system follows the fluctuation response theorem. Thouless-Anderson-Palmer (TAP) equation is given by

$$m_i = \tanh(\sum_j J_{ij}m_j + h_i - \sum_j J_{ij}^2(1 - m_j^2)m_i). \tag{21}$$

as the consistency condition of the mean field $m_i =< \sigma_i >$ (Thouless et al. 1977). The partial differentiation of TAP equation (21) with respect to h_i yields the equation

$$\sum_k A_{ik}[\mathbf{m}]\chi_{kj} = \delta_{ij} \tag{22}$$

where $\chi_{ij} = \partial m_i / \partial h_j$ is susceptibility and

$$A_{ij}[\mathbf{m}] = -J_{ij} - 2J_{ij}^2 m_i m_j + \delta_{ij}[\sum_k J_{ik}^2(1 - m_k^2) + \frac{1}{1 - m_i^2}]. \tag{23}$$

The fluctuation-response theorem relates the susceptibility χ_{ij} and the covariance $C_{ij} =< (\sigma_i - m_i)(\sigma_j - m_j) >$ as

$$\chi_{ij} = C_{ij}. \tag{24}$$

The equation $J_{ij} = (-1 + \sqrt{1 - 8C_{ij}^{-1}m_i m_j})/4m_i m_j$ fulfill the equation (24). In the case with $m_i = 0$, we have the equation $J_{ij} = -C_{ij}^{-1}$ again.

4 Data analysis

In this section, the analysis of actual portfolios based on this model is briefly summarized. The time series of price changes are coded into the sequences of up and down spins. The covariance C_{ij} for $N(N-1)/2$ pairs of i and j are derived from the coded data. Then the interaction coefficients J_{ij} are calculated by the equation (13). The energy spectra of the system (portfolio energy) are defined as the eigenvalues of the Hamiltonian $H[\sigma, 0]$.

•The probability density of portfolio energy for each dataset is empirically obtained from the relative frequency during the observation times as $p_E(E) = P(E - \Delta E/2 \leq H[\sigma, 0] \leq E + \Delta E/2)/\Delta E$. In our formalism that is given by the equation $p(E) = n(E)e^{-E}/Z$ using the density of states $n(E)$ and the partition function Z. The empirical measure $p_E(E)n(0)/p_E(0)n(E)$ can be compared with Gibbs measure as the theoretical prediction.

•Monte Carlo simulation of the probability distribution of the system magnetization $m = (1/N)\sum_{i=1}^{N}\sigma_i$, which describes the degree of the alignment of price changes in portfolio, can be compared with empirical result.

Those two items are good tests of our model. We have investigated several datasets of Dow-Jones portfolio consisting of N=30 stock issues in the framework of the model for several sampling intervals. The price changes sampled at 1 and 10-minutes intervals are well described by the model (Maskawa 2002a). On the other hand, some deviation of daily data from theoretical line was observed. We investigated the theory in more detail by the mode analysis of TAP equation (21), and explained this deviation in the framework of the theory (Maskawa 2002b).

References

1. Gopikrishnan P, Rosenow B, Plerou V, Stanley H E (2001) Quantifying and interpreting collective behavior in financial markets. Phys. Rev. E 64:035106(R)
2. Mantegna R N (1999) Hierarchical structure in financial markets. Eur. Phys. J. B 11:193-197
3. Mantegna R N, Stanley H E (2000) An Introduction to Econophysics. Cambridge University Press, Cambridge
4. Markowitz H M (1991) Portofolio Selection. Blackwell, Oxford
5. Maskawa J (2002a) Spin-Glass like network model for stock market. In: Takatasu H (Ed) Empirical Science of Financial Fluctuation. Springer-Verlag, Tokyo
6. Maskawa J (2002b) Ordered phase and non-equilibrium fluctuation in stock market. Physica A 311:563
7. Maskawa J (2003) Multivariate Markov chain modeling for stock markets. Physica A in press
8. Sherrington D,Kirkpatrick S (1975) Solvable model of a spin glass. Phys. Rev. Lett. 35:1972
9. Thouless D J,Anderson P W, Palmer R G (1977) Solution of solvable model of a spin glass. Phil. Mag. 35:593-601

II. Other Topics

II-1 Income Distribution

Entropy in the Economy

Philip K. Rawlings, David Reguera, and Howard Reiss

UCLA, Department of Chemistry and Biochemistry, 607 Charles E. Young Drive East, Los Angeles, 90095 CA, USA

Summary. Based on a few simple assumptions, a derivation of the general thermodynamic behavior for an economic system is presented. The concepts of economic temperature and entropy are explained and an expression which is isomorphic with the fundamental equation of thermodynamics is obtained. As a practical application, the Pareto distribution of incomes is shown to be a consequence of assuming that a quasi-equilibrium economic system obeys a particular conservation law.

Key words. Economic entropy, Economic temperature, Thermodynamics, Income distribution

Equilibrium distribution of a conserved quantity and the fundamental equation of an equilibrated economic agent

The authors of this paper do not claim that an economy can achieve equilibrium. However, it seems possible that over a limited period of time a real economy might behave as though it were quasi-equilibrated. For this situation, a hypothetical equilibrated economy could serve as a useful reference state from which some features (either qualitative or semi-quantitative) of the real economy could be deduced. With this in mind, the present paper will develop certain properties of an equilibrated economy. An important aspect of this development will be the discovery of entropy in the economy and its relation to other economic variables. Of course, this is not the first time that entropy has been examined within this context. However, in previous studies, entropy has been defined at the outset, and an information theoretic approach involving the maximum entropy principle has been used to determine the equilibrium distributions of certain economic quantities assumed to be conserved. One of the most thorough examples of this approach is due to Montroll and his coworkers (Montroll and Shlesinger 1983).

We find it more instructive to follow the conventional approach of physics in which entropy is not defined at the outset but, instead, first appears as a quantity in a differential expression which, in physics, would correspond to the fundamental equation of thermodynamics, or the combined first and second laws. Many would argue that the two approaches are the same and that statistical thermodynamics can be based equally on either of them. We are not prepared to argue this question, and to a large degree concur, even though there is still some controversy over this point. Nevertheless we favor the more conventional physics approach.

Choosing this approach we investigate the distribution of a conserved economic quantity U over a large set of agents competing for shares of U. We imagine U to be quantized so as to be available in packages of size u_i where i denotes the i *th* size. In a particular distribution there will be n_i agents having a share u_i of U. Thus

$$U = \sum_i n_i u_i \,. \qquad (1)$$

The total number of agents will also be fixed and this number will be denoted by N. Thus

$$N = \sum_i n_i \,. \qquad (2)$$

What we are after is the distribution $\{n_i\}$ corresponding to equilibrium in the competition among agents. We note that, corresponding to a particular distribution, there will be

$$\Omega = \frac{N!}{\prod_i n_i!} \qquad (3)$$

ways of distributing shares of U among the N distinct agents.

Now $\{n_i\}$ is already subject to the constraints represented by eqs(1) and (2), but it may be subject to $c-2$ additional (economic) constraints, not explicitly indicated. Thus Ω will be subject to a total of c constraints. Besides eqs(1) and (2), these will be denoted by $x_3, x_4 \ldots\ldots, x_c$. We now make the assumption that, independent of the distribution, every arrangement of shares of U among the N agents is equally probable, so that the probability of distribution is proportional to the Ω for that distribution. (Notice that this assumption implies that the N agents are identical, at least in their abilities to compete for a share of U. It is also possible to develop a theory for cases in which the agents are not identical. A brief comment on this issue appears below.) In the usual manner, we then approximate the average or equilibrium distribution by the most probable distribution. In so doing, we recognize that these additional constraints determine the set of quanta u_i, i.e.

$$u_i = u_i(x_3, x_4 \ldots\ldots, x_c) \,. \qquad (4)$$

In other words the spectrum of u states is determined by the remaining constraints. By holding them constant we are able to maintain a constant spectrum during the variational procedure that leads to the most probable distribution. It is then clear that the most probable distribution is the most probable one subject to particular values of $x_3, x_4 ...., x_c$.

The equilibrium distribution is then

$$n_i(N, \beta, x_3, x_4 .....x_c) = N \frac{\exp\{-\beta u_i(x_3, x_4 .....x_c)\}}{Q(\beta, x_3, x_4 .....x_c)} \tag{5}$$

where

$$Q(\beta, x_3, x_4 ...., x_c) = \sum_i \exp\{-\beta u_i(x_3, x_4 ...., x_c)\} \tag{6}$$

and where β is related to the undetermined multiplier associated with eq(1). The probability p_i that an agent obtains a share of U equal to u_i is then clearly given by

$$p_i = \frac{n_i}{N} = \frac{\exp\{-\beta u_i\}}{Q} \tag{7}$$

where we do not show the functional dependencies of n_i and u_i.

(Here, it is worth noting that one way for agents to possess nonidentical abilities to acquire shares of U would be to have ω different, equally effective strategies to capture such shares. For that agent, such a "degeneracy" would lead to the replacement of eq(7) by a similar equation with ω multiplying the exponential in eq(7).)

Now, the equilibrium average share of U obtained by an agent is given by

$$\langle u \rangle = \sum_i p_i u_i \tag{8}$$

and the differential of $\langle u \rangle$ is then

$$d\langle u \rangle = \sum_i u_i dp_i + \sum_i p_i \sum_{j=3}^{j=c} \left(\frac{\partial u_i}{\partial x_j} \right)_{x_k \neq x_j} dx_j \; . \tag{9}$$

Now, eq(7) can be rewritten as

$$u_i = -\frac{1}{\beta}[\ln p_i + \ln Q] \tag{10}$$

and when this equation is substituted into eq(9) we obtain

$$d\langle u\rangle = \frac{1}{\beta} d\left[-\sum_i p_i \ln p_i\right] + \sum_{j=3}^{j=c}\sum_i p_i \left(\frac{\partial u_i}{\partial x_j}\right)_{x_k \neq x_j} dx_j \tag{11}$$

where we have used the fact that Q does not depend on i and $\sum_i p_i = 1$. It is convenient to define

$$X_i^{(j)} = -\frac{\partial u_i}{\partial x_j}, \qquad X_j = \sum_i p_i X_i^{(j)}. \tag{12}$$

Clearly, X_j is the average value of the $X_i^{(j)}$. Substitution of eq(12) into eq(11) yields

$$d\langle u\rangle = \frac{1}{\beta} d\left[-\sum_i p_i \ln p_i\right] - \sum_{j=3}^{j=c} X_j dx_j \,. \tag{13}$$

The quantity in square brackets in eq(13) has the form of a Gibbs entropy S. We can define it as the Gibbs entropy of an agent. Thus

$$S = -\sum_i p_i \ln p_i \tag{14}$$

and eq(13) can be written as

$$d\langle u\rangle = \frac{1}{\beta} dS - \sum_{j=3}^{j=c} X_j dx_j \,. \tag{15}$$

It is interesting to note that economic entropy has arisen naturally in this equation, and that, if the constraints x_j remain constant, an increase in agent's average share of *U*, namely $\langle u\rangle$, can only be obtained through an increase of the economic entropy, i.e. an increase of the 'risk' for the agents. It is also not an accident that this equation resembles the fundamental equation of thermodynamics for a *closed* system able to perform c-2 kinds of work. In this context, $1/\beta$ would be a temperature, and the various x_j would be extensive quantities conjugate to the "forces" X_j. It should be remembered that both p_i and *S* are quantities that correspond to fixed values of the c-2 variables x_j. Thus *S* and the x_j are the mutually independent variables of state of the agent "system" and $d\langle u\rangle$ is an exact differential. This equation may be regarded as the fundamental economic equation of an equilibrated economic agent. Because no loss of logic will be generated if we replace $1/\beta$ with T and call it "economic temperature", we can write

$$d\langle u\rangle = TdS - \sum_{j=3}^{j=c} X_j dx_j \,. \tag{16}$$

Since eq(16) is isomorphic with the fundamental equation of thermodynamics, all the useful transformations of thermodynamics may be utilized, including Maxwell equations, Legendre transformations, etc. Note that with the application of a sufficient number of constraints x_j, the remaining unconstrained system could possibly indeed be distributed quasi-randomly.

Application: Income Distribution

The approach developed in the previous section is completely general and thus can be applied to a wide variety of situations of economic interest. For illustrative purposes, we focus on a particular example, namely the distribution of incomes. Such distributions usually resemble the behavior plotted in Fig. 1. Typically, high income earners, amounting to at most a few percent of the population, are distributed following a power law known as the Pareto distribution (Pareto 1897). The rest follow a different distribution (usually Boltzmann or log-normal) in terms of the income (Dragulesco and Yakovenko 2001, Souma 2001). It is reasonable to assume that the high income earners behave differently, and are decoupled from the others. There is cultural evidence of this sort of lack of interaction (Montroll and Badger 1974). For example, it is commonly observed that the substructure of society is invariant to the superstructure. The lot of peasants is frequently insensitive to a change of government, i.e. invariant to whether that government is an absolute monarchy, a republic, etc. (This of course does not prevent individuals from jumping from one group to the other). As a consequence, each group might be treated independently, and we follow this course.

On the other hand, it is evident that the exchange of money has different implications for individuals with differing incomes. Usually a transaction made by a wealthy individual can be affected by the fraction of his income involved in that transaction, rather than the absolute amount of income involved. In contrast an individual with low income has less discretion and may be forced to purchase essentials independent of the fraction of his income that an essential might represent. Bernoulli suggested that the effect of income fraction on transaction could be described by a "utility" function

$$B(z) = \ln\left(z/\bar{z}\right) \tag{17}$$

where z is income and $\bar{z}$ is the average income of an individual agent (Montroll and Badger 1974).

It is reasonable to assume that shared utility is a globally conserved quantity, a hypothesis that can be tested experimentally. As a consequence, if the system reaches equilibrium the income distribution will be given by eq(7), which in terms of the utility becomes

$$p(z) = \frac{\exp\{-\beta u(z)\}}{Q} \propto z^{-\beta} \tag{18}$$

where z takes the place of i in eq(7), an $u(z)$ is $B(z)$.

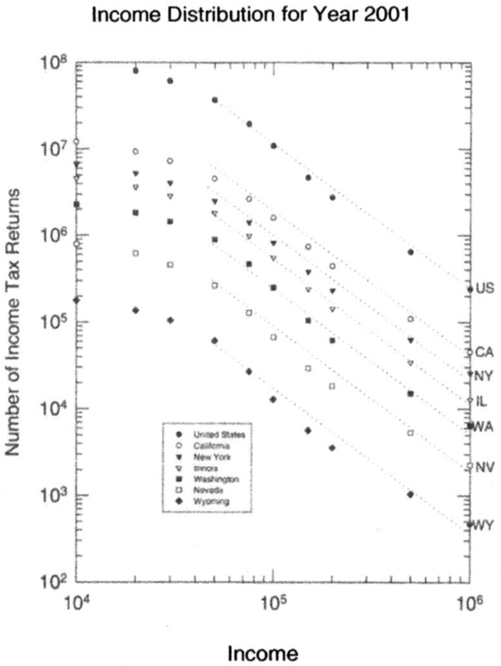

Fig.1 Distribution of individual income tax returns for the United States of America and for several individual states (CA, NY, IL, WA, NV and WY) in 2001. The lines are not fit to the data; they are all parallel to each other and serve as visual guides. As discussed in the text, the slope of the lines is related to the economic temperature.

Thus, if trading is based on the utility of income, we obtain Pareto's law where the exponent is related to the "economic temperature" defined in the previous section. It should be mentioned that Pareto distributions have also been obtained using different (non-equilibrium) models for the dynamics of income transactions (Montroll and Shlesinger 1983, Richmond and Salomon 2001).

The analogy with thermodynamics developed in the previous section suggests an "experimental" test of the reasonableness of the assumptions we have introduced. In thermodynamics, the "zeroth law" asserts that two systems in thermal contact with each other will have the same temperature at equilibrium. This also means that, at equilibrium, sub-systems of the same system will have the same temperature. Therefore, if the exponent of Pareto's law plays the same role as temperature in a thermodynamic system, it would imply that the exponents in different income distributions, which correspond to different economic systems in strong contact, should be the same.

Fig 1. plots the income distribution for both United States and several of its "sub-system" states. The plot shows that both the tails of income distributions of different states in the US, and that of the U.S. as a whole, are all power laws having practically the same exponent, i. e. the same *economic temperature*. This supports the hypothesis of a quasi-equilibrated system, and illustrates the potential usefulness of the thermodynamic analogy in the analysis of macroeconomic systems.

Concluding remarks

Entropy in economic systems is derived and analyzed using the physical approach of statistical thermodynamics. An example of the usefulness of the thermodynamic approach in discussing the behavior of economic systems is presented with reference to Pareto's law. The formulation of economic problems in terms of equilibrium statistical thermodynamic allows the use of many results derived in standard thermodynamics. Ideally, one could exploit these analogies and thereby translate many of the well-known results of statistical thermodynamics (stability, phase transitions, fluctuations…) to economic systems. Another interesting possibility is the potential development of an analogue of nonequilibrium thermodynamics, capable, in principle, of describing the time evolution of economic systems.

References

Dragulescu A and Yakovenko MV (2001). Physica A 299: 213-221; Eur. Phys. J. B 20:585-589; cond-mat/0211175.

Montoll EW and Badger W (1974). Introduction to Quantitative Aspects of Social Phenomena. Gordon and Breach, New York.

Montroll EW. and Shlesinger MF (1983). J. Stat. Phys. 32: 209-230.

Pareto V (1897). Cours d'Economie Politique, Lausanne.

Richmond P and Salomon S (2001). Int. J. Mod. Phys. C 12:333-343.

Souma W (2001). Fractals 9:463-470.

Investment strategy based on a company growth model

Takayuki Mizuno[1], Shoko Kurihara[1], Misako Takayasu[2], and Hideki Takayasu[3]

[1] Department of Physics, Faculty of Science and Engineering, Chuo University, Kasuga, Bunkyo-ku, Tokyo 112-8551, Japan
[2] Department of Complex Systems, Future University-Hakodate, 116-2 Kameda-Nakano, Hakodate 041-0803, Japan
[3] Sony Computer Science Laboratories, 3-14-13 Higashi-Gotanda, Shinagawa-ku, Tokyo, Japan

Summary. We first estimate the average growth of a company's annual income and its variance by using both real company data and a numerical model which we already introduced a couple of years ago. Investment strategies expecting for income growth is evaluated based on the numerical model. Our numerical simulation suggests the possibility that an investment strategy focusing on the medium-sized companies gives the best asset growth with relatively low risk.

Key words. Investment strategy, Income, Wealth.

1. Introduction

Studies on wealth distribution can be traced back more than 100 years. It is widely known that Pareto published the first report on the personal income distribution in 1897 [V. Pareto]. He showed that the probability density distribution of income follows a power law distribution in the high-income range. He claimed that a kind of social structure can be characterized by the power exponent because smaller value implies that the tendency of monopoly is stronger.

By analyzing a huge database of company incomes, our group has been investigating statistical properties of time evolution of company's income [T. Mizuno(1), T. Mizuno(2)]. In those papers, we showed that prediction of the average income in the future is possible although the income of each company has violent fluctuations. A country needs to implement many policies of company's income because company's income is closely related to the tax. In the present paper we estimate the average growth rate of a company by using a numerical model that is derived from the real data and discuss the performance of investment strategy for companies.

2. An empirical model for company's income growth

We have already introduced an empirical model describing the stochastic dynamics of company's income growth by the following equation [T. Mizuno(1), T. Mizuno(2)],

$$I_k(t+1) = \alpha(t) \cdot b(t)^{\frac{\sigma(I)}{\sigma_0}} \, I_k(t) + f(t) \tag{1}$$

where $I_k(t)$ is the income of company k at the t-th year, $b(t)$ and $f(t)$ are random noises. The factor $\alpha(t)$ is a stochastic variable taking either 1 or -1. The probability densities of these functions can be estimated from the company data. The power exponent $\sigma(I)/\sigma_0$ characterizes the size dependence of company growth statistics. This model is simple but is confirmed to reproduce known empirical laws of company's income.

3. The growth rate of company's income

For the purpose of discussing the investment strategy for income growth of American companies, we apply Eq.(1) with estimation the coefficient ($\alpha(t)$, $b(t)$, $f(t)$, $\sigma(I)/\sigma_0$) from a data set of income of American companies from 1989 to 1995 [T. Mizuno(2)][Moody's company data]. In order to find out an investment strategy we observe how a company of initial size I(t=0) grows in the next 5 years. We calculate the averaged growth rate of the cumulative income, that is defined by the ratio of the cumulative income normalized by the initial income, ΣI(t=5)/I(0), where ΣI(t=5)=I(t=1)+···+I(t=5). In Fig.1 the numerical values of ΣI(t=5)/I(0) estimated by Eq.(1) are compared with the values estimated directly from the real data of American companies (◆). From this figure we notice that the expectation of the cumulative income Σ I(t=5) is proportional to the initial income $I(0)$ for the companies whose initial income is larger than one million dollars/year. For the companies with initial incomes smaller than one million dollars/year the expectation value of cumulative income is nearly independent of the initial value of income. We can also estimate the income growth of a long term by using the numerical model. We show the case of $\Sigma I(t=15)$ by a line of USA_S(t=15) in Fig.1. Companies with initial incomes smaller than two million dollars/year also have the large average value of $\Sigma I(t=15)/I(0)$. These results imply that investing such smaller income companies will produce high growth on average.

Although a small company's growth rate is large on average, its variance is also large. In Fig.2 we plot the probability densities of growth rates for the term of 5 years ($\Sigma I(t=5)$) for three different initial income ranges. The probability density tails are very large for smaller companies as expected. In order to clarify the dependence on the initial income I(0) for this variance, we show the standard deviation of the probability density distribution of $\Sigma I(t=5)/I(0)$ in Fig.3. For companies with initial incomes smaller than one million dollars/year the standard deviation decreases rapidly in inverse proportion to initial income. For companies with initial incomes larger than one million dollars/year the standard deviation decreases slowly depending on initial income. Therefore, the risk caused by investment is small for large income companies.

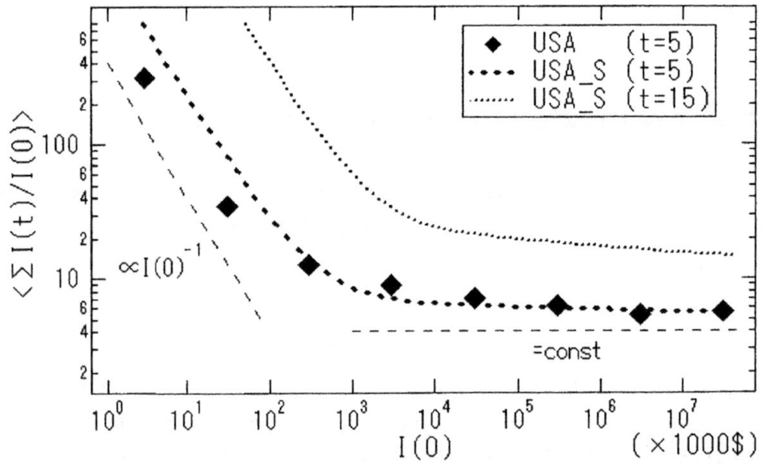

Fig.1 Averaged growth rate as a function of the initial income. Plots of USA are real data for 5 years. Plots of USA_S are results estimated by the numerical model.

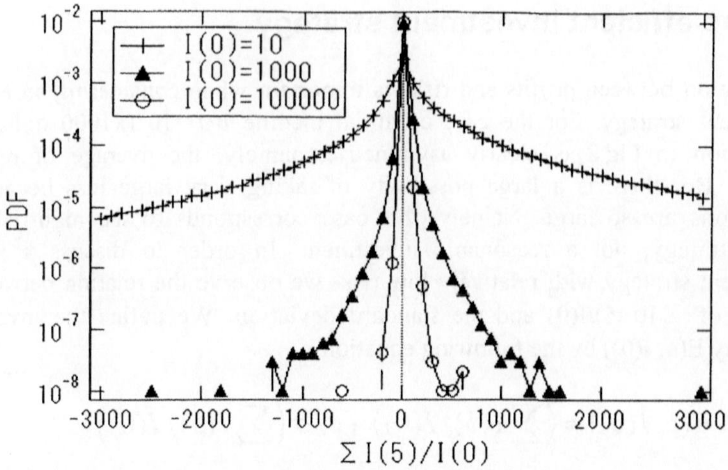

Fig.2 The probability density of growth rate for the term of 5 years (ΣI(t= 5)).

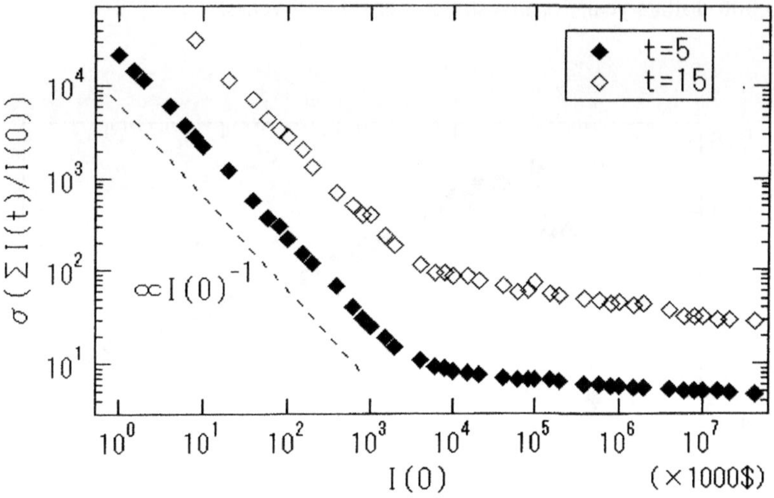

Fig.3 The standard deviation of the growth rate.

4. The efficient investment strategy

The relation between profits and risks is important when considering an efficient investment strategy. For the case of initial income I(0)=10 (x1000 dollars) the distribution in Fig.2 is clearly asymmetric, namely, the average of return is positive. But there is a large possibility of taking very large loss because the fluctuations are so large. Namely, this case corresponds to the high-risk-high-return strategy, not a reasonable investment. In order to discuss a rational investment strategy with relatively low risk, we observe the relation between the average of Σ I(t=5)/I(0) and the standard deviation. We define the investment efficiency E(c, I(0)) by the following equation,

$$E(c, I(0)) = \left\langle \sum I(5)/I(0) \right\rangle - c \cdot \sigma\left(\sum I(5)/I(0)\right), \qquad (2)$$

where, c is a constant showing the balances between profits and risks. We assume $c =$ 1/5 ~ 1/9 although this balance changes with contents of each investment. We show the relation between the investment efficiency E and the initial income I(0) in the case of $c = 1/8$ in Fig.4. In this case the best reasonable investment targets are the companies of size around the peak point of the investment efficiency E, that is, the companies with income 1,000,000 ~ 10,000,000 dollars/year for investment period of 5 years.

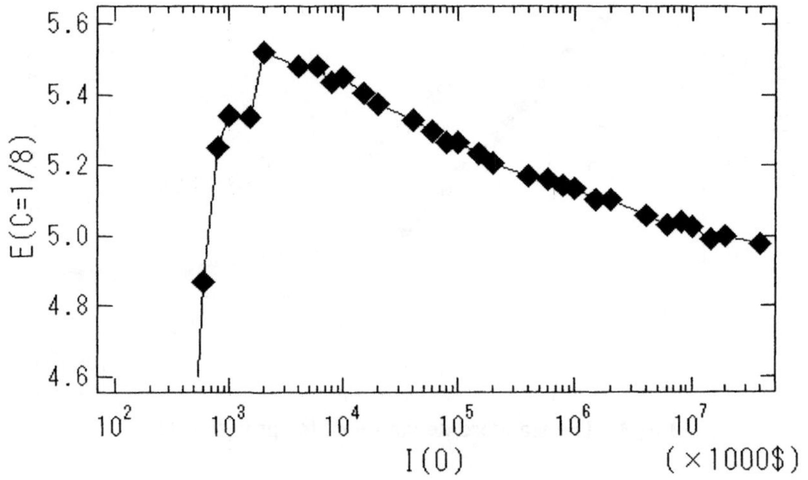

Fig.4 The investment efficiency E(c, I(0)).

5. Discussion

We discussed the expected profits and risks for various investment strategies by using the simple model of company income and showed that investment to the medium-sized companies is efficient with relatively low risk. Although we showed results only for companies in USA, similar results hold also for other countries such as Japan and England.

Turning our eyes to politics, Japanese government and banks tends to support large companies, however, judging from our results such investments can be categorized as too much low-risk-low-return strategy. Financial support to medium-sized companies will contribute to recovery of economical growth.

The correlation between the income and other indices (asset, etc.) [H. Takayasu] and the dynamics of other indices [M. H. R. Stanley] have been reported. We are going to use other indices of wealth in addition to income for the investment strategy in the future works. We expect that the investment strategy of high accuracy can be obtained by using these results.

Acknowledgement

We would like to thank Prof. Tohru Nakano and Prof. Mitsuo Kono for stimulus discussions.

References

V. Pareto, Le Cours d'Economie Politique, Macmillan, London, (1897).

T. Mizuno, M. Katori, H. Takayasu and M. Takayasu, in *Empirical Science of Financial Fluctuations – The Advent of Econophysics*, (Springer Verlag, Tokyo, 2002) 321-330.

T. Mizuno, M. Takayasu, H. Takayasu, preparing for publication.

Moody's company data 1989-1995. Big companies in U.S.A more than 10,000 companies.

H. Takayasu and M. Takayasu, *Econophysics – Toward Scientific Reconstruction of Economy* (in Japanese, Nikkei, Tokyo, 2001).

M. H. R. Stanley, et al, *Nature* 379 (1996) 804-806.

Growth and Fluctuations of Personal Income I

Yoshi Fujiwara[1], Hideaki Aoyama[2], and Wataru Souma[1]

[1] ATR Human Information Science Laboratories
Kyoto 619-0288, Japan
[2] Faculty of Integrated Human Studies, Kyoto University
Kyoto 606-8501, Japan

Summary. Pareto's law states that the distribution of personal income obeys a power-law in high-income range. Many researchers have proposed models over a century since its discovery. But the dynamical understanding of the phenomenon necessitates empirical study on growth of individual income. Here we uncovered several properties in the fluctuations of *individual* income growth, derived from income tax data including all taxpayers, more than 80,000, with high-income in Japan for two consecutive years, 1997 and 1998, in a relatively stable period of economics. In this work [8], we show (i) that the dynamics has approximate time-reversal symmetry in the period, (ii) that the distribution of growth rate is independent of income-tax in the past yearm, and (iii) that the Pareto's law breaks down in "bubble" collapse of japanese economy in 1990. It is shown in [3] how these phenomenological facts are consistent with Pareto law. These findings provide first direct observation of dynamical process of money flow in an economic network, and its relation to risky assets' price fluctuations of lands and stocks.

Key words. personal income, Pareto law, Zibrat law, growth and fluctuations

1 Introduction

High-income distribution follows a power-law: the probability $P_>(x)$ that a given individual has income equal to, or greater than x, obeys

$$P_>(x) \propto x^{-\mu}, \tag{1}$$

with a constant μ called the Pareto index. This phenomenon, now known as a classic example of fractals, has been observed [4,6,2,18,7] in many different countries, we found that the law holds with striking accuracy in the high-income region, with μ varying typically around 2 reflecting economic conditions [2,18,8] (see Fig. 1 (a)).

Understanding the origin of the law has importance in economics because of its linkage with consumption, business cycles, and other macro-economic activities. Many researchers, recently including those in non-equilibrium statistical physics, have proposed models for power-law [9,6,12,14,11,5,17]. Many kinds of proposed scenarios have predicted a power-law distribution as a static snapshot. However, in order to test models, it has been highly desirable to have direct observation of the dynamical process of growth and fluctuations of personal income. That is what we address in this work.

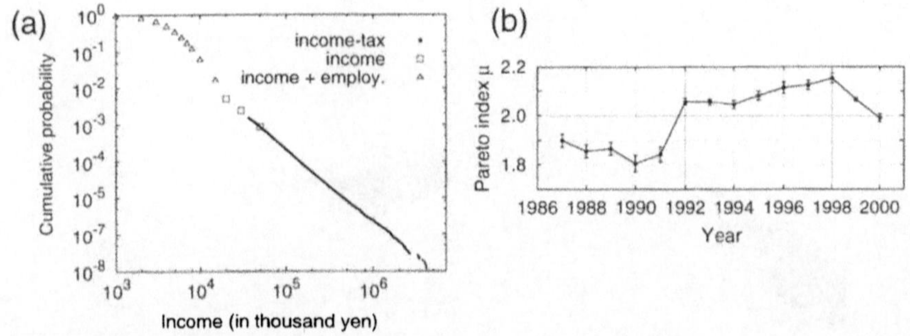

Fig. 1. (a) Cumulative probability distribution of personal income in the year 2000. Note that the dots are income tax data of about 80,000 taxpayers. (See [8] for the details). (b) Annual change of Pareto index μ from the year 1987 to 2000. The complete list of income tax data in each year was used for the estimation of index.

2 Growth and Fluctuations

For that purpose, we employ Japanese income tax data covering most of the power-law region in Fig. 1 (a). It is an exhaustive list of all taxpayers — full names, addresses and tax amounts — who paid 10 million yen (approx. 10 thousand euros/dollars) or more a year through tax offices of the Japanese National Tax Administration (NTA). Fig. 1 (b) plots Pareto indices, estimated from this income tax data, from 1987 to 2000. The abrupt change of μ in 1991 and 1992 is considered to coincide with the fact that Japanese economy experienced an abnormal rise and rapid fall of prices in the risky assets of lands and shares due to speculative investment, "bubble" and "burst". We examined a relatively stable period in the economy, namely 1997 and 1998. The complete datasets of 93,394 persons in 1997 and 84,571 in 1998 were used. Identification of individuals listed in both years was done (see [8] for the details). The number of such identified persons appearing in the two consecutive years was 52,902. The rest of the persons in 1997 and 1998 can thus be regarded as those disappearing from or novel in the list.

Scatter plot in Fig. 2 (a) shows each point representing a person who paid income tax of T_1 in 1997 and T_2 in 1998 (both in units of thousand yen). This represents the joint distribution $P_{12}(T_1, T_2)$. The plot is consistent with approximate *time-reversal symmetry* in the sense that the joint distribution is invariant under the exchange of values T_1 and T_2, i.e. $P(T_1, T_2) = P(T_2, T_1)$.

Our concern is the annual change of individual income-tax, or growth. Growth rate is defined as $R = T_2/T_1$. It is customary to use the logarithm of R, $r \equiv \log_{10} R$. We examine the probability density for the growth rate $P(r|T_1)$ on the condition that the income T_1 in the initial year is fixed (see Fig. 2 (b)). The result shows that the distribution for growth rate r is *statistically independent* of the initial value of T_1. In a mathematical notation, we

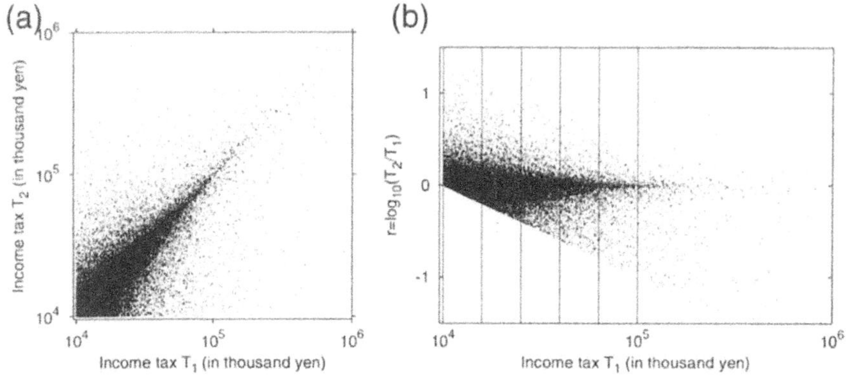

Fig. 2. (a) Scatter-plot of all individuals whose income tax exceeds 10 million yen in both 1997 and 1998. These points (52,902) were identified from the complete list of high-income taxpayers in 1997 (93,394) and in 1998 (84,571) (numbers in parentheses), with income taxes T_1 and T_2 in each year. (b) The same as (a) with vertical axis for $r = \log_{10}(T_2/T_1)$. The segments are bins for $T_1 \in [10^{4+0.2(n-1)}, 10^{4+0.2n}]$ $(n = 1, \cdots, 5)$.

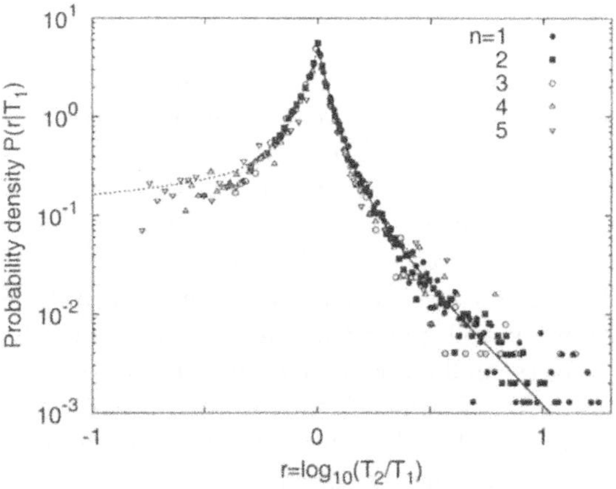

Fig. 3. Probability density $P(r|T_1)$ of growth rate $r \equiv \log_{10}(T_2/T_1)$ from 1997 to 1998. Note that due to the limit $T_1 > 10^4$ (in thousand yen), the data for large negative growth, $r < 4 - \log_{10} T_1$, are not available. Different bins of initial income-tax with equal size in logarithmic scale were taken as $T_1 \in [10^{4+0.2(n-1)}, 10^{4+0.2n}]$ $(n = 1, \cdots, 5)$ to plot probability densities separately for each such bin. The solid line in the portion of positive growth $(r > 0)$ is an analytic fit. The dashed line $(r < 0)$ on the other side is calculated by a scaling relation derived in [3].

found that $P_{1R}(T_1, R) = P_1(T_1)\, P_R(R)$, where P_{1R} is the joint distribution for T_1 and R, P_1 and P_R are the distributions for T_1 and R, respectively.

This "universal" distribution for the growth rate has a skewed and heavy-tailed shape with a peak at $R = 1$. How is such a functional form consistent with the approximate time-reversal symmetry shown in Fig. 2? And how these phenomenological facts are consistent with Pareto's law in Fig. 1? The answer to this question is given in [3].

3 Breakdown of Pareto's law

In economically unstable period such as "bubble" and "crash", personal income distribution deviates from power-law. Fig. 4 shows the cumulative distributions of income tax in the successive years 1991 and 1992. The year 1991 coincides with the peak of speculative bubble of land price (see also Fig. 1 (b)). One can observe that the 1991 data cannot be fitted by Pareto's law in the entire range of high income, while one year later the distribution went back to power-law.

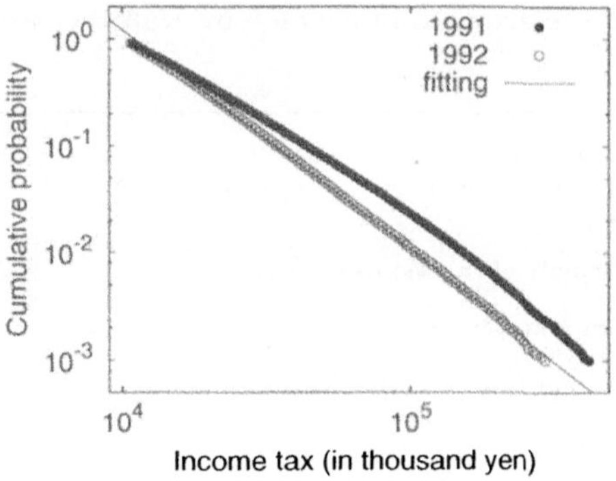

Fig. 4. Cumulative probability distributions of income tax in 1991 and 1992. The fitted line is for $\mu = 2.057 \pm 0.005$ (see [8] for estimation).

Sample survey by NTA on income earners gives data about income types. Picking those persons with total income exceeding 50 million yen (who are necessarily included in the exhaustive list described so far), it can be observed that in terms of the numbers chief income-sources are employment income, rental of real estate, and capital gains from lands and stock shares (Fig. 5 (a)), and that in terms of the amounts contribution comes largely from capital gains from lands and stock shares (Fig. 5 (b)). It is expected that asymmetric behavior of price fluctuations in those risky assets and the accompanying increase in high-income persons cause the breakdown of time-reversal symmetry and/or the statistical independence, which necessarily invalidates Pareto's law.

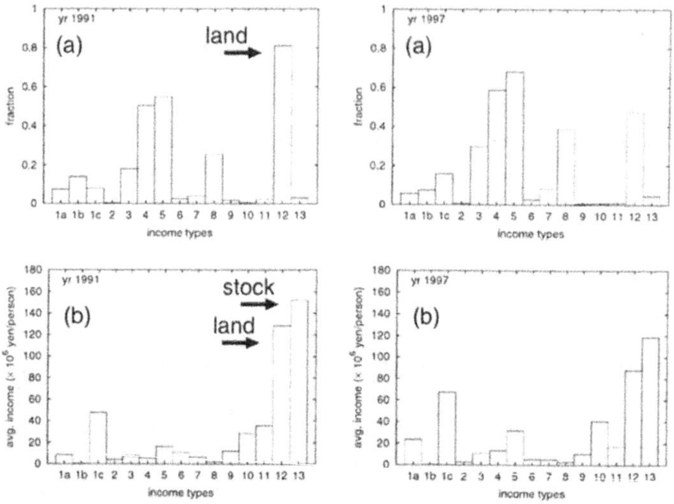

Fig. 5. See the list of income types below. *Left panels*: 1991. *Right panels*: 1992. (a) Fraction of the numbers for income earners with total income exceeding 50 million yen with a particular income type. A person can have more than a single income type. (b) Fraction of the amounts in average over all the earners.

1a	business income
1b	firm income (agricultural)
1c	other operating income (independent professionals; lawyers, doctor, entertainers, etc.)
2	interest income
3	dividends
4	rental income (mainly of real estate)
5	wages/salaries
6	comprehensive capital gains
7	sporadic income
8	miscellaneous income (including public pension, etc.)
9	forestry income
10	retirement income
11	short-term separate capital gains (selling real estate possessed in 5 years)
12	long-term separate capital gains (selling real estate possessed over 5 years)
13	capital gains from stocks, etc.

4 Summary

Exhaustive list of high-income taxpayers in Japan for two successive years provided us the first empirical study on growth of individual income. We found (i) that the dynamics has approximate time-reversal symmetry in the period, (ii) that the distribution of growth rate is independent of income-tax in the past year, and (iii) that the Pareto's law breaks down in "bubble" collapse of Japanese economy. Our findings in this work shall serve as an empirical test for models (such as [9,6,12,14,11,5,17]) of personal income and wealth. Comparison with and similar analysis in company growth, which has been studied extensively [19,1,15,13], would also be an interesting subject, where the Zipf law ($\mu = 1$) is observed.

Acknowledgement Supported in part by grants from the Telecommunications Advancement Organization in Japan. We thank M. Ojima and Teikoku Databank, Ltd. for kindly providing high-quality data, M. Aoki and T. Kaizoji for useful discussions.

References

Amaral LAN (1998) Power law scaling for a system of interacting units with complex internal structure. Phys Rev Lett 80:1385–1388
Aoyama H et al (2000) Pareto's law for income of individuals and debt of bankrupt companies. Fractals 8:293–300
Aoyama H, Fujiwara Y, Souma W (2003) Growth and Fluctuations of Personal Income II. in this volume
Badger WW (1980) An entropy-utility model for the size distribution of income. in Mathematical models as a tool for the social science (ed. West, B.J.) 87–120, Gordon and Breach, New York
Bouchaud J-P, Mézard M (2000) Wealth condensation in a simple model of economy. Physica A282:536–545
Champernowne DG (1953) A model of income distribution. Econometric Journal 63:318–351
Drăgulescu A, Yakovenko VM (2001) Exponential and power-law probability distributions of wealth and income in the United Kingdom and the United States. Physica A299:213–221
Fujiwara Y, Souma W, Aoyama H, Kaizoji T, Aoki M (2003) Growth and Fluctuations of Personal Income. Physica A in press
Gibrat R (1932) Les inégalités économiques. Sirey, Paris
Kesten H (1973) Random difference equations and renewal theory for products of random matrices. Acta Mathematica 131:207-248
Levy M, Solomon S (1996) Power laws are logarithmic Boltzmann laws. Int. J. Mod. Phys. C7:595–601
Mandelbrot BB (1961) Stable Paretian random functions and the multiplicative variation of income. Econometrica 29:517–543
Mizuno M et al (2002) Statistical laws in the income of japanese companies. in Empirical Science of Financial Fluctuations: The Advent of Econophysics (ed. Takayasu, H.) 321–330 Springer-Verlag, Tokyo
Montroll EW, Shlesinger MF (1983) Maximum entropy formalism, fractals, scaling phenomena, and 1/f noise. J. Stat. Phys. 32:209–230
Okuyama K, Takayasu M, Takayasu H (1999) Physica A269:125–131
Pareto V (1897) Le Cours d'Économie Politique. Macmillan, London
Solomon S, Richmond P (2001) in Economics with Heterogeneous Interacting Agents (ed. Kirman, A. & Zimmermann, J.-B.) Lecture notes in Economics and Mathematical Systems 503:141–145, Springer-Verlag, Berlin
Souma W (2001) Universal structure of the personal income distribution. Fractals 9:463–470
Stanley MHR et al (1996) Scaling behaviour in the growth of companies. Nature 379:804–806
Sutton J (1997) Gibrat's legacy. Journal of Economic Literature 35:40–59

Growth and Fluctuations of Personal (and Company's) Income II

Hideaki Aoyama[1], Yoshi Fujiwara[2] and Wataru Souma[2]

[1] Faculty of Integrated Human Studies, Kyoto University, Kyoto 606-8501, Japan** aoyama@phys.h.kyoto-u.ac.jp
[2] ATR Human Information Science Laboratories, Kyoto 619-0288, Japan

Abstract
We report a modern-day theory, kinematics of income dynamics. The phenomenological study done on Japanese tax payers in high income bracket for 1997 and 1998, with identifications of individuals over these two years, revealed a law we called *Universality.* We show here that this property combined with Stationarity of the income probability distribution function (pdf) leads to the Pareto's law and the relation between the growth-rate pdf in the positive-growth region and that in the negative-growth region.

1 Income distribution

In the article *"Growth and Fluctuations of Personal Income I"* [1] we reported a recent empirical study of the high personal-income. This was carried out on a database that contains names and addresses of all the people who paid income-tax over 10 million yen in years 1997 and 1998, which is ideal for tracking individuals over these two years and identify how their yearly income changed.[2, 3] There are some 80,000 individuals in each years and 52,902 individuals has been identified in both years. This kind of study was never heard of in the long history of research on personal income, and a surprising empirical law, which we call *Universality* emerged from it.[4]

To summarize, we found the following three properties,

1. *Pareto's law:*
   The probability distribution function (pdf) $P(x)$ for income x obeys the power law for large x,
   $$P(x) \propto x^{-\mu-1}, \tag{1}$$
   where the value of μ changes between 1.8 and 2.2 from year to year, depending on economical situation. The Pareto's law has been known for

** Address after April 1st, 2003: Graduate School of Science, Kyoto University, Kyoto 606-8501

more than a century.[5] Therefore, there is no surprise in this finding: This study confirmed this empirical law with very high accuracy by covering three orders of magnitude in x (and six orders of magnitude in $P(x)$) for the first time.

2. *Universality*:
Pdf of the growth rate $R \equiv T_2/T_1$, where T_1 is the income of the first year and T_2 the second year, was measured in five income brackets. The result is plotted in Fig.1. From this, it is apparent that this pdf is *Universal*, that is, it is independent from the first year's income. In terms of the joint pdf of the first year's income T_1 and the growth rate R, it means the following factorizability;

$$P_{1R}(T_1, R) = P_1(T_1) P_R(R). \tag{2}$$

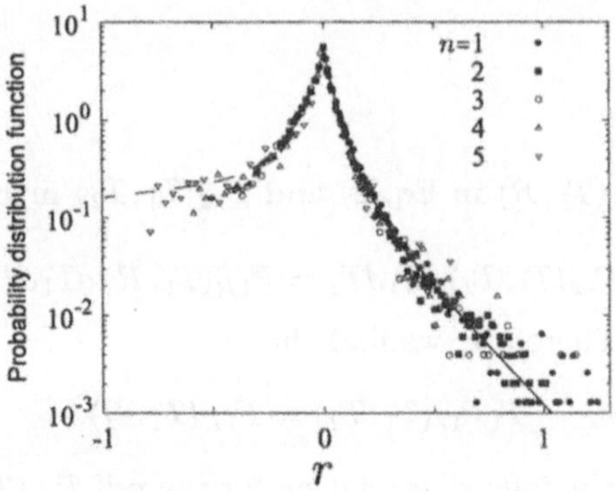

Fig. 1. Pdf of (logarithm) of growth rate $r \equiv \log_{10}(R)$ for each income bracket, $T_1 \in [10^{4+0.2(n-1)}, 10^{4+0.2n}]$ ($n = 1, \cdots, 5$).

2' *Universality in Company's income*:
We have also studied company's income, taxable profit, for the top 70,000 companies in Japan for 2000 and 2001. The income pdf is plotted in Fig.2(a) and obeys the Pareto's law with index $\mu = 1$ very closely. The growth rate pdf is plotted in Fig.2(b), where *Universality* is apparent as in the case of personal income. This is in clear contrast to company size data, which has been studied extensively in recent years.

3. *Stationarity*:
The joint pdf $P_{12}(T_1, T_2)$ for the first year's income T_1 and the second year's income T_2 is approximately symmetric:

$$P_{12}(T_1, T_2) = P_{12}(T_2, T_1), \tag{3}$$

which is apparent from the scatter plot Fig.1(a) of Ref.[1]. This implies that the economy was relatively stable during this period.

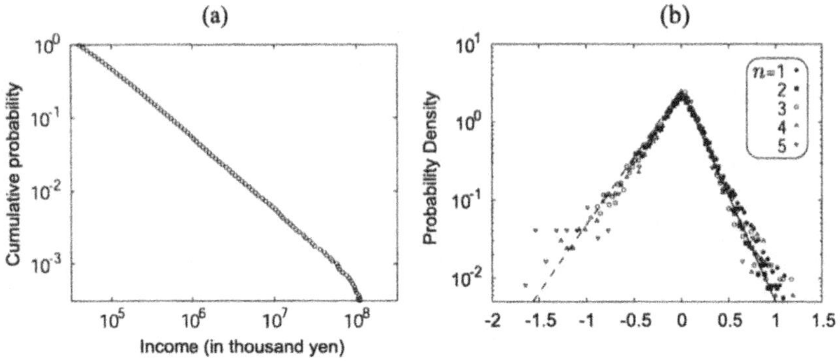

Fig. 2. (a) Cumulative pdf of company's income (profit) in 2000. The Pareto's index is very close to 1. (b) Pdf of the growth rate of company's income (profit) in 2000. Universality is apparent.

2 Implications

The two pdf's, $P_{1R}(T_1, R)$ in Eq.(2) and $P_{12}(T_1, T_2)$ in Eq.(3) are related as follows;

$$P_{12}(T_1, T_2)\, dT_1 dT_2 = P_{1R}(T_1, R)\, dT_1 dR,$$

where $R = T_2/T_1$. Therefore, we find that

$$T_1 P_{12}(T_1, T_2) = P_{1R}(T_1, R). \tag{4}$$

Using this, we find the following relation for the pdf $P_{12}(T_1, T_2)$ from Stationarity condition (3):

$$\begin{aligned} P_{1R}(T_1, R) &= T_1 P_{12}(T_1, T_2) \\ &= T_1 P_{12}(T_2, T_1) \\ &= \frac{T_1}{T_2} P_{1R}(T_2, R^{-1}). \end{aligned} \tag{5}$$

Substituting Eq.(2) into the above, we obtain the following:

$$P_R(R) = \frac{P_1(T_2)}{P_1(T_1)} \frac{1}{R} P_R\left(\frac{1}{R}\right) \tag{6}$$

This means that $P_1(T_1)/P_1(T_2)$ is a function of the ratio of their arguments, $R = T_2/T_1$. This can be satisfied only when the function $P_1(x)$ is proportional to a power of its argument:

$$P_1(x) \propto x^{-\mu-1}, \tag{7}$$

which is the Pareto's law, with μ as a free parameter. In this manner, we find that *Universality and Stationarity leads to the Pareto's law.*

There is yet another important implication: By substituting Eq.(7) into Eq.(6), we obtain the following relation between the growth-rate pdf on the positive side ($R > 1$) and the negative side ($R < 1$):

$$P_R\left(\frac{1}{R}\right) = R^{\mu+2} P_R(R). \tag{8}$$

It may be noted that this kind of relation is self-consistent, in the sense that if used twice it leads to a triviality:

$$P_R(R) = R^{-\mu-2} P_R\left(\frac{1}{R}\right) = R^{-\mu-2}\left(\frac{1}{R}\right)^{-\mu-2} P_R(R) = P_R(R). \tag{9}$$

The dashed lines for $R < 1$ ($r < 0$) in the growth-rate plot in Fig.1 and Fig.2(a) have been obtained by using the relation (8) to the solid lines obtained by fitting to the positive side $R > 1$ ($r > 0$). Agreement between the dashed lines and the real data is apparent for both cases.

We have derived Pareto's law from Universality and Stionality. This implies that deviation from Pareto's law is a sign of breaking of either Stationarity or Universality. The former case is seen in the 1991 income pdf given in Fig.4 of Ref.[1].

It should be noted that the average growth-rate pdf $\bar{P}_R(R)$ defined by the following,

$$\bar{P}_R(R) \equiv \int_0^\infty P_{1R}(T_1, R) dT_1, \tag{10}$$

is *different* from our $P_R(R)$ defined in Eq.(2). One may think that by applying Universality condition (2) into the above definition one obtains $\bar{P}_R(R) = \int_0^\infty P_1(T_1)\, P_R(R) dT_1 = P_R(R)$. This, however, is not true, since Universality holds only for the high-income. It was empirically found in that region, and furthermore, if it holds over the whole region, it leads to an inconsistency, since the resulting Pareto's law (1) is not normalizable in $x \in [0, \infty]$ for any value of μ. In contrast to Eq.(8), $\bar{P}_R(R)$ obeys a different reflection property, which can be obtained from Stationarity condition (3) or its corollary (5) as follows:

$$\begin{aligned}
\bar{P}_R(R) &= \int_0^\infty P_{1R}(T_1, R) dT_1 \\
&= \int_0^\infty \frac{1}{R} P_{1R}(T_2, R^{-1}) dT_1 \\
&= \int_0^\infty \frac{1}{R^2} P_{1R}(T_2, R^{-1}) dT_2 \\
&= \frac{1}{R^2} \bar{P}_R(R^{-1}).
\end{aligned} \tag{11}$$

Namely, we obtain the following:

$$\bar{P}_R\left(\frac{1}{R}\right) = \bar{R}^2 P_R(R), \tag{12}$$

which is quite different from Eq.(8) since $\mu \sim 2$ for personal income and $\mu \simeq 1$ for companies income. Apparent agreement between Eq.(8) and the real data means that Eq.(12) does NOT agree with our high-income data. This also shows the power of our analysis. The existence of Universality is the key to the relation (8), which differs considerably from the usual Stationarity result (12).

3 Conclusion

In conclusion, we have studied fluctuations of personal-income data and company's profit data and found *Universality* in the growth-rate pdf. We also gave a simple argument that together with Stationarity, this leads to the Pareto's law. We also gave relation between the positive side and the negative side of the growth-rate pdf, which is empirically confirmed.

The discussion presented here, we believe, opens a new modern-day view of income dynamics, which goes beyond the old and trusted Pareto's law. In our analysis we have not used any hypothesis on its dynamics, but rather, have concentrated on revealing kinematical relation between the new empirical laws. Based on this argument, we next need to see what kind of dynamics satisfies these empirical laws, especially, Universality. There are several promising theories along the line of Refs.[?] and [7], which will be presented in near future.

The authors would like to thank Dr. Masanao Aoki (UCLA) and Dr. Taisei Kaizouji (ICU) for discussions.

The research reported here was supported in part by a contract with the Telecommunications Advancement Organization of Japan entitled, 'Research on Human Communication'.

References

1. Y. Fujiwara, H. Aoyama and W. Souma, *"Growth and Fluctuations of Personal Income I"*, in this Proceedings.
2. H. Aoyama, *et. al.*, *"Pareto's law for income of individuals and debt of bankrupt companies"*, Fractals **8** (2000) 293-300.
3. W. Souma, *"Universal structure of the personal income distribution"*, Fractals **9**, (2001) 463–470.
4. Y. Fujiwara, W. Souma, H. Aoyama, T. Kaizoji and M. Aoki, *"Growth and fluctuations of personal income"*, cond-mat/0208398.

5. V. Pareto, *"Le Cours d'Économie Politique"*, (Macmillan, London, 1897).
6. J.-P. Bouchaud and M. Mézard, *"Wealth condensation in a simple model of economy"* Physica **A282** (2000) 536–545.
7. W. Souma, Y. Fujiwara and H. Aoyama, *"Small-world effects in wealth distribution"*, cond-mat/0108482.

A model of high income distribution

Keizo Yamamoto[1], Sasuke Miyazima[2], Rajindar K. Koshal[3], Majulika Koshal[4], Yuko Yamada[5]

[1]Faculty of Engineering, Setsunan University, Neyagawa, Osaka 572-8508, Japan
[2]College of Bioscience and Biotechnology, Chubu University, Kasugai, Aichi 487-8501, Japan
[3]Department of Economics, Ohio University, Athens-Ohio 45701-2979, USA
[4]Department of Management Systems, Ohio University, Athens-Ohio 45701-2979, USA
[5]Graduate School of Engineering, Nagoya Institute Technology, Nagoya 466-8555, Japan

Summary. A model discussed in this paper shows the power-law application to high-income against the ranking. This stochastic model is so simple that we can analyze it in an analytical method. The Monte Carlo simulation result of our model shows a good fit with the data for Japan and US CEO as well as the analytical results. We could also obtain the stationary solution for this model. The power index of the stationary solution also coincides satisfactorily with the estimated results.

Key words. High income rank, Power-law distribution, Fractals, High income model

1. Introduction

A log-normal distribution of a wide range of variable at the center and Pareto distribution (power-law distribution) for high-income people have been investigated since 1897 (Pareto 1897); (Champernowne 1952); (Lydall 1959); (Mandelbrot 1960) and (Shlesinger and Montroll 1983). As the data of income which were discussed in their analysis include some ambiguities, we have used tax amounts of high tax payer which we can get from tax office publication in the middle of May, every year. We plot the tax amount against its ranking in a double logarithmic diagram, where we found a power-law behavior of the tax amount against the ranking (Yamamoto and Miyazima 2001). The power is same as that that is obtained from the similar analysis for salary and bonus of US CEO. On the other hand, the powers which are obtained from the data of various prefectures in Japan are different from prefectures to prefectures. The power could become a new indicator of economical actions.

Economic activities are very intricate structure for many factors to be considered. The economic system is viewed as a prototype of complex system. Each income is given by employment incomes from companies as a result of various economic activities. These processes perform endlessly complicated nests of trades which people get incomes from unspecific random objects and give incomes other people more and more. Therefore here are intricate nests of structures and these should exist some fractality.

The fractality in the social economy exists not only in the case of high income ranking but also there are some other examples. Many analyses such as the ranking of firm size (Stanley et al 1996), incomc distribution of companies (Okuyama 1999), the fluctuations in finance (Madelbrot 1997) and so on, have been reported.

In this paper we will suggest a model for the distribution of high income. Out presented model is so simple. Then we don't understand that it can explain all properties of economics. Now we are very interested to know why the economic system shows the fractality.

2. Model and simulation

It is our purpose to explain the power-law of high income against ranking by proposing a simple mathematical model (Yamamoto et al 2002). Our high income data are derived from tax amounts of high-tax payer. The tax rate is constant for high-tax payers who are top a hundred thousand persons in Japan. Therefore the income is proportional to the tax amount for those people. We use the income amount in the following analysis.

The following two conservation conditions are assumed:

C1. Our system consists of N homogeneous members. The number N is fixed.

C2. Our system holds S=2N units as resource money. Each member keeps one unit as a minimum amount.

We used N=100, N=1,000, N=10,000, and N=100,000, and generally averaged over 100 trials of simulation.

We introduce following simple competition in the simulation of our system.

M1. Two members who are chosen randomly from whole the members take an economic action competitively. A member who wins collects whole the competing money and the other member lose whole the resource money. In order to keep the number of active members with resource money, we add one unit to the loser.

M2. We reduce one unit from a member those who are picked up randomly who has resource money more than or equal to 2 unit in order to keep whole the resource money in our system.

Here the process which consists of a step of M1 and M2 is called one Monte Carlo step (1M.C.), hereafter.

If this simulation is carried out more than N^2 Monte Carlo steps, we obtain an equivalent stationary distribution from any arbitrary initial condition (arbitrary resource money of each member). We show a log-log plot of simulation results of N=10,000 members averaged over 100 samples against the ranking in Fig. 1. Here we assume the income of each member is proportional to one's resource money which is used for investment. The slope is measured to be -0.720 ± 0.001 and the coefficient of determination-R^2 is 0.9985, which means that the income is proportional to expressed by a power-law of the ranking r. Our present model is very

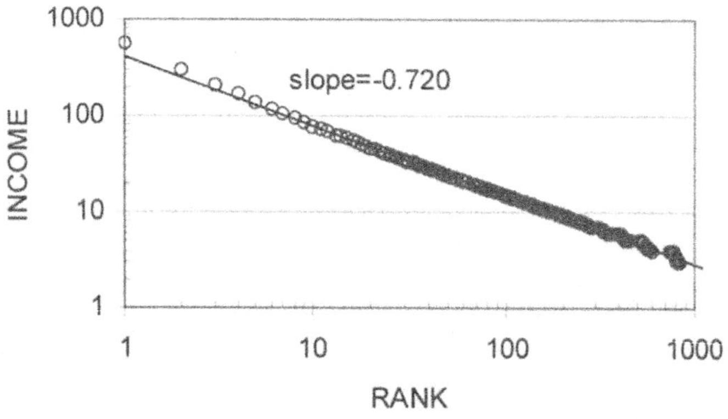

Fig. 1. The log-log plot of income against rank N=10,000 and the result is averaged over 100 samples.

simple but enough to show the power-law of the income against the ranking. Rather we can say that the simple model is easy to analyze mathematically and useful for deep understanding the phenomena of income distribution.

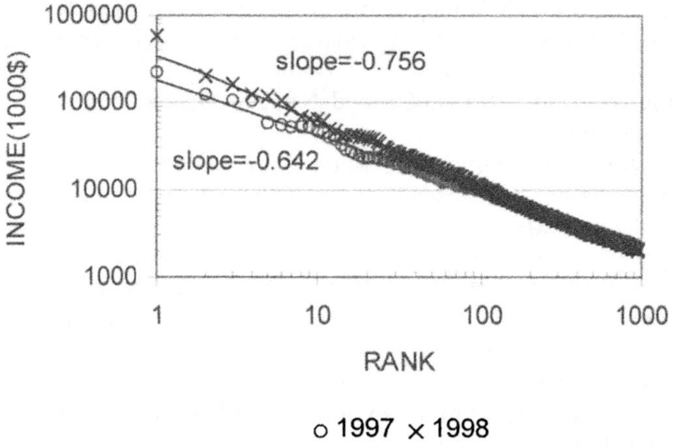

Fig. 2. The log-log plot of income of CEO in USA and Japan against rank.

The larger the ensemble size, the more straight line in log-log plot is expected. Besides the data in Japan (General Legal Security home pages) we could obtain CEO data of 1997 and 1998 in USA (Forbes home pages), which is plotted in Fig. 2. The slopes of two lines in Fig. 2 are -0.642 ± 0.001 and -0.756 ± 0.001, respectively, which show good agreement with the simulation result. (Here we used the exchange rates 1$=121yen for 1997 and 1$=131yen for 1998.)

3. Mathematical Analysis

We wish to derive the probability that an arbitrary chosen member holds X units of resource money. We denote this probability by P(X). Let us now consider the stationary state in which the probability P(X) is time-independent. If (per 1 M.C.) the same number of transitions occur in the incoming and the outgoing direction, P(X) dose not change. This requirement can be written mathematically in the following set of equations.

$$(1 - P(1))\,(1 - P(1)) + \frac{P(2)}{1 - P(1)} = (P(1))^2 \qquad (1\text{-}1)$$

for $X \geqq 2$,

$$\sum_{i+j=X} P(i)P(j) + \frac{P(X+1)}{1 - P(1)} = 2P(X) + \frac{P(X)}{1 - P(1)} \qquad (1\text{-}2)$$

Where each term on the left hand side of the above equation is the incoming rates, and one on the right hand side is the outgoing rates.

The stationary solution of Eq.(1-1) and Eq.(1-2) is obtained as follows.

$$\begin{aligned}
&P(2)= (2P(1)-1)(1-P(1)) && P(8)= (1430P(1)-1001)(1-P(1))^7 \\
&P(3)= (5P(1)-3)(1-P(1))^2 && P(9)= (4862P(1)-3432)(1-P(1))^8 \\
&P(4)= (14P(1)-9)(1-P(1))^3 && P(10)= (16796P(1)-11934)(1-P(1))^9 \\
&P(5)= (42P(1)-28)(1-P(1))^4 && P(11)= (58786P(1)-41990)(1-P(1))^{10} \\
&P(6)= (132P(1)-90)(1-P(1))^5 && P(12)= (208012P(1)-149226)(1-P(1))^{11} \\
&P(7)= (429P(1)-297)(1-P(1))^6 && P(13)= (742900P(1)-534888)(1-P(1))^{12}
\end{aligned} \qquad (2)$$

If we express $P(X)=(a_X P(1) - b_X)(1 - P(1))^{X-1}$ with coefficients a_X and b_X, the relation of these coefficients is as $a_X=a_{X-1}+b_X$. As the another relation of these coefficients has not been found, we have not expressed the general terms by a recurrence formula as yet. The first factor on the right hand side of P(X) in Eq.(2)

shows the necessary condition about P(X) > 0. It seems that the value of P(1) for the condition converges to 0.75, but the convergence is very slow. Then we decide the value of P(1) numerically. As seen from Eq.(2), P(1) dominates not only the normalization of the probability but also the relative probability.

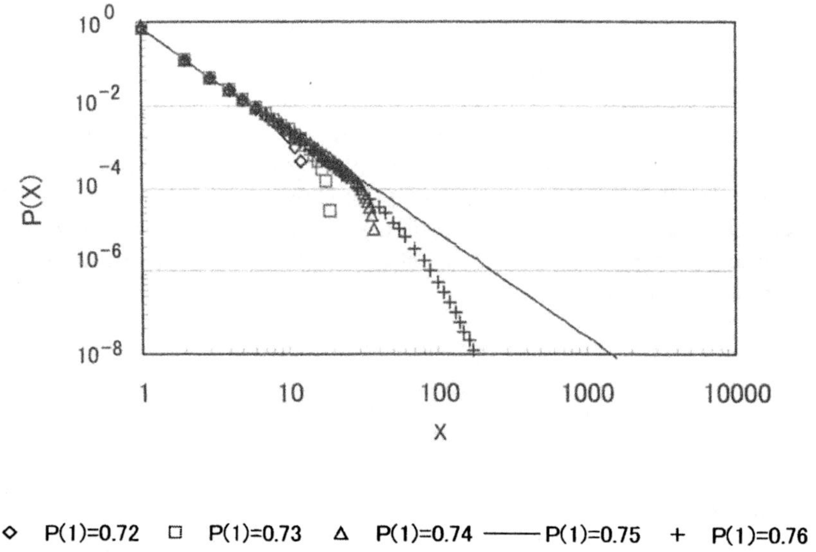

Fig. 3. The log-log plot of probability P(X) varying P(1), which is obtained from analytical equation.

Therefore, in Fig.3 we plot P(X) against X over the varying P(1) from 0.72 to 0.76. As shown in Fig.3, only when the value of P(1) is 0.75 the summation of P(X) is equal to 1, the average value <X>=2 and P(X) > 0 for all X. Then we decide P(1)=0.75 and the slope of log-log plot of P(X) against X is -2.478 ± 0.001, R^2=0.99997. If we use $X \sim r^{-\tau}$ in regard to the relation between income X and rank r, we obtain $r \sim X^{-1/\tau}$. The rank r means the number of persons who earn more than X. Therefore the probability with income more than X, $P(>X) \sim X^{-1/\tau}$, which means that the probability density

$$P(X) \sim dP(>X)/dX \sim X^{-1/\tau-1} \tag{3}$$

By equating the power of Eq.(3) to -2.478, we obtain $\tau = 0.677$. This value coincides satisfactorily with our estimates 0.642(1997) and 0.756(1998).

4. Conclusions

We examine economic activities of a large number of participants form a very complex system. It would therefore seem difficult to simulate with any regulation such a complex economic system. However, we find the proposed model shows the power-law in high-income against the ranking. This model is very simple and easy to analyze mathematically. It is also useful for further understanding of the power-law and fractal phenomena. The simulation result of our model indicates a good fit with the data in Japan and the United States' CEOs. The power index of our simulation is 0.720 and the measurement based on real world data is 0.642(1997) and 0.756(1998). We could also obtain the stationary solution for this model. The power index of the stationary solution is 0.677 and coincides satisfactorily with the estimated power index using data for two years. From this perspective, the findings of this study are not only interesting but important too.

References

Champernowne G D (1953) A model of income distribution. Economic Journal 63:318-351.

Forbes home pages, http://www.forbes.com.

General Legal Security, Inc, http://chouja.houmu.co.jp.

Lydall H F (1959) The distribution of employment incomes. Econometrica 27: 110-115.

Mandelbrot B B (1960) The Pareto-Levy law and the distribution of income. International Economic Review 1:79-106.

Mandelbrot B B (1997) Fractals and Scaling in Finance. Springer, Berlin

Okuyama K, Takayasu M and Takayasu H (1999) Zipf's law in income distribution of companies. Physica A 269:125-131

Pareto V (1897) Le Cours d'Economie Politique. Macmillan, London

Shlesinger M F and Montroll E W (1983) Fractal stochastic process: clusters and intermittancies. Lecture Note in Math 1035:138-152.

Stanley M H R, Amaral L A N, Buldyrev S V, Havlin S, Leschhron H, Maass P, Salinger M A and Stanley H E (1996) Scaling behavior in the growth of companies. Nature, 379:804-806.

Yamamoto K and Miyazima S (2001) Power-law distribution of high income taxes in Japan. Sociological Theory and Methods 16:133-138. In Japanese.

Yamamoto K, Miyazima S, Koshal K R and Yamada Y (2002) A model for distribution of high-tax payers. Physica A 314:743-748

Ideal Gas-Like Distributions in Economics: Effects of Saving Propensity

Bikas K. Chakrabarti and Arnab Chatterjee

Saha Institute of Nuclear Physics, 1/AF Bidhan Nagar, Kolkata-700064, India
bikas@cmp.saha.ernet.in

We consider the ideal-gas models of trading markets, where each agent is identified with a gas molecule and each trading as an elastic or money-conserving (two-body) collision. Unlike in the ideal gas, we introduce saving propensity λ of agents, such that each agent saves a fraction λ of its money and trades with the rest. We show the steady-state money or wealth distribution in a market is Gibbs-like for $\lambda = 0$, has got a non-vanishing most-probable value for $\lambda \neq 0$ and Pareto-like when λ is widely distributed among the agents. We compare these results with observations on wealth distributions of various countries.

1 Introduction

I should start by thanking the organisers of this Symposium for inviting me (BKC) and also perhaps more importantly for highlighting our humble efforts in Kolkata (erstwhile Calcutta), by mentioning in their (symposium) homepage that "The term econophysics was ... first used in 1995 at an international conference ... in Calcutta" [1]. Indeed Kolkata is the formal birth place of this new term. It was first used by Stanley [2] in the second of the Statphys-Kolkata series of conferences (being held over more than a decade now [3]), where several important developments in this interdisciplinary emerging field had been reported by many physicists.

In an earlier conference in Kolkata in 1994, many leading Indian economists from the Indian Statistical Institute and physicists met and discussed about the possible formulations of some economic problems and their solutions using tricks from physics [4]. In one of these papers [5], possibly the first published joint paper with both physicist and economist Indian coauthors, the possibility of a kinetic theory of (ideal) gas-like model of trading in the market was discussed. Among other things, it tried to identify, from the known effects of various fiscal policies, equivalence of the (kinetic) energy of the gas molecules (money) and the temperature (average money in the market). Such a "finite temperature" market model and the corresponding distributions were also noted by others [6, 7]. With the possibility of putting more than one agent in the same (micro) state, identified by the price or money income of the agent in the market, the likely distribution was concluded there to be Bose-Einstein like, rather than Gibbs like [5]. These studies of course had the limitation of absence of any comparison with real income distributions (in any market or country). In a recent paper by Dragulescu and Yakovenko [7] a simple (trading) market model was developed with fixed (total) money and number of agents in the market. Random two-agent exchanges (with local money conservation) lead to Gibbs-like steady income distribution. This was also confirmed by simple numerical simulations. Modifications due to savings was studied simultaneously [8]. In a very recent review [9], a popular introduction to these developments is given.

Saving propensity among the agents, a very selfish and local feature of the tradings, introduce in effect some global co-operative feature (cf. [8]). We show that a fixed and uniform saving propensity of all the agents in the market shifts the most-probable money of the distribution away from zero (as given by Gibbs for zero savings), while a random distribution of saving propensity among the population can give the Pareto (power) law [10]

$$P(m) \sim m^{-(1+\nu)} \tag{1}$$

for the wealth or money (m) distribution. We intend to discuss here in brief the effects of various kinds of savings on the ideal gas-like money distributions in the above-mentioned market models, and compare our observations with those from real markets.

2 An Ideal Gas-Like Market Model

Let us consider a simple model of a closed economic system where the total amount of money M and the total number N of agents are fixed. No development (production) or migration (death/birth of agents) occurs and the only economic activity is confined to trading. Each agent i, individual or a corporate, possess a money $m_i(t)$ at (discretised) time t. Time changes after each trading. In any trading, two randomly chosen agents i and j exchange their money such that their total money is (locally) conserved and none ends up with negative money (debt not allowed):

$$m_i(t) + m_j(t) = m_i(t+1) + m_j(t+1) \tag{2}$$

where $m_i(t) \geq 0$ for all i and t; $\sum_{i=1}^{N} m_i = M$. Since money is conserved, in the steady state ($t \to \infty$), the probability $P(m)$ of the density of people with money m will satisfy

$$P(m_1)P(m_2) = P(m_1 + m_2) \tag{3}$$

which corresponds to the Gibbs distribution [7, 8]

$$P(m) = (1/T)\exp(-m/T); \quad T = M/N. \tag{4}$$

Numerical simulations in the model also confirmed the steady state distribution of money, no matter what initial distribution of money the agents had, to be Gibbs' one: after sufficient number of tradings, most of the agents end up with very little money! This result is quite robust (and realistic too!). In fact, several variations of the trading, and of the 'lattice' (on which the agents can be put and each agent trade with its 'lattice neighbours' only), whether compact, fractal or small-world [6], leaves the distribution unchanged. Some other variations like random sharing of an amount $2m_2$ only (not of $m_1 + m_2$) when $m_1 > m_2$ (trading at the level of lower economic class), lead to even drastic situation: all the money in the market drifts to one agent and the rest of the agents all become truely pauper [9, 11]! Attempts have also been made [12] to get Pareto-like power-law distribution here with changed definition of entropy or the conservation law (cf. eqn. (3)).

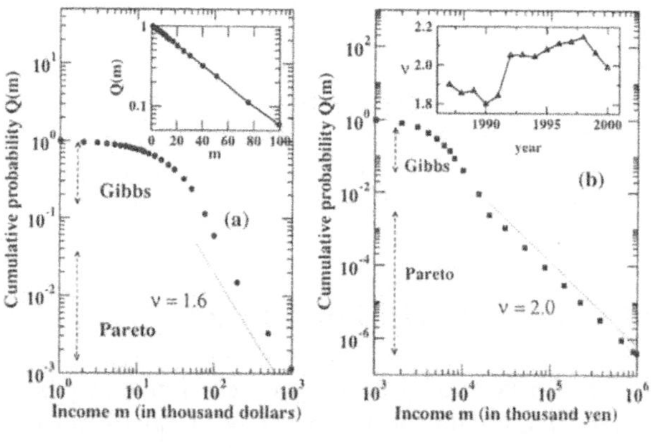

Fig. 1. Cumulative distribution $Q(m) = \int_m^\infty P(m)dm$ of individual income (a) in USA in 1997 (data taken from [13]), (b) in Japan in 2000 [14]. The inset in (a) shows the validity of Gibbs law for low income and in (b) the recently observed variations of the Pareto index ν (for higher income group) in Japan. Typically, 10% of the higher income group people possess about 40% of the wealth in any economy and follow Pareto law (1) with index ν in the range 1 to 2 [6].

Chakrabarti and Marjit [5] argued for the Bose-Einstein like distribution (rather than Gibbs) in such a market (with the temperature T similarly identified with the average money M/N per agent), as one can put more than one agent in the same economic state (specified by the income) and the maximisation of the consequent entropy. For the possibility of adding and subtracting agents into/from the market, one similarly needs (negative) "chemical potential" which becomes zero at a finite temperature or money level in the market.

when the "Bose condensed" fraction of the agents will fall out of the market distribution and might be identified as unemployed.

The real income distributions did not indicate so far anything like the Bose distribution; rather considerable evidences support the possibility of Gibbs like distribution (4) in the income (almost for 90% of the low-income range) of various countries (see e.g., [13], see also data in [14]; Fig. 1).

3 Model with fixed saving propensity of the agents

Here we assume [8] that each economic agent i saves a fraction λ of its money $m_i(t)$ before the trading at time t. We again assume that an agent's money is non-negative and no debt is permitted. Let us now consider an arbitrary pair of agents i and j, who get engaged in a trade, and their money $m_i(t)$ and $m_j(t)$ before the trade change respectively to

$$m_i(t+1) = m_i(t) + \Delta m; \quad m_j(t+1) = m_j(t) - \Delta m \tag{5}$$

$$\text{where} \quad \Delta m = (1-\lambda)[\epsilon\{m_i(t) + m_j(t)\} - m_i(t)]; \tag{6}$$

with ϵ as any random fraction. As may be checked by straight-forward substitution, this kind of trading again satisfies eqn. (2), while each agent saves a fixed fraction λ of its money before the trade and exchanges randomly (with fraction ϵ) the rest of the money.

One finds here that at $\lambda = 0$ the market becomes non-interacting and the steady state money distribution becomes the Gibbs' one. For any nonvanishing λ, the equilibrium distribution becomes asymmetric Gibbs-like (see Fig. 2a) with the most-probable money m_p per agent (corresponding to the peak in $P(m)$) shifting away from $m = 0$ (for $\lambda = 0$) to M/N as $\lambda \to 1$ [8]. This self-organising feature of the market, induced by sheer self-interest of saving by each agent without any global perspective, is very significant as the fraction of paupers decreases with saving fraction λ and most people end up with the average money in the market (socialists' dream achieved with just people's self-interest of saving)! Interestingly, self-organisation also occur in such market models when there is restriction in the commodity market [15]. Although this fixed saving propensity does not give yet the Pareto-like power-law distribution, the Markovian nature of the scattering or trading processes (eqn. (3)) is lost and the system becomes co-operative. Indirectly through λ, the agents get to know (start interacting with) each other and the system co-operatively self-organise towards a most probable distribution ($m_p \neq 0$).

4 Model with random saving propensity of the agents

We now consider a market again with fixed N and M but with random saving propensity λ_i ($0 \leq \lambda_i < 1$) fixed or "quenched" for each agent (λ_i are independent of trading or t, but vary randomly from agent to agent) [16]. One again follows the same trading rules as mentioned in the previous section (eqn. (5)), except that

$$\Delta m = \epsilon(1-\lambda_j)m_j(t) - (1-\lambda_i)(1-\epsilon)m_i(t) \tag{7}$$

here; λ_i and λ_j are the saving propensities of agents i and j. We first take a market with N agents, each having a fixed saving propensity λ distributed independently, randomly and uniformly (white) within an interval 0 to 1. Having assigned each agent i the saving propensities λ_i, and starting with an arbitrary initial (uniform or random) distribution of money among the agents, we start the tradings. At each time, two agents are randomly selected and the money exchange among them occurs, following the above mentioned scheme. We check for the steady state, by looking at the stability of the money distribution $P(m)$ in successive Monte Carlo steps t.

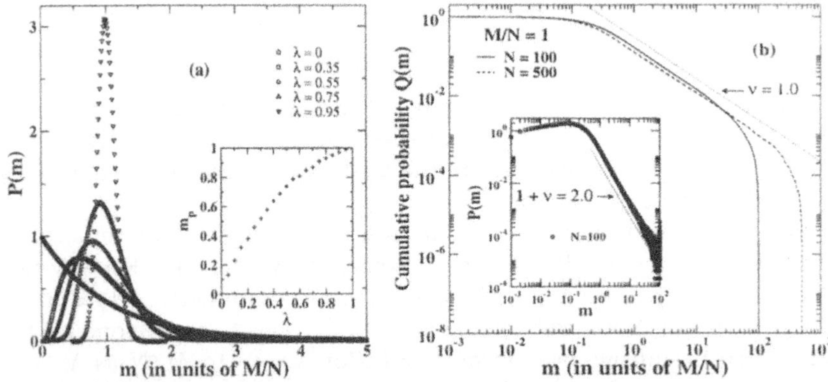

Fig. 2. Simulation results in the model with saving propensities: (a) Stable distribution of money $P(m)$ vs. m for different (but fixed for all agents) saving propensity λ of agents ($N = 100$). The inset shows that the most-probable money m_p of any agent in the market shifts with λ ($m_p = 0$ for $\lambda = 0$; Gibbs law). (b) Stable cumulative probability distribution of money ($Q(m) = \int_m^\infty P(m)dm$) in a model market with quenched distribution of λ within the range $0 \le \lambda < 1$; $N = 100, 500$. The dotted line corresponds to a power law decay with $\nu \simeq 1.0$. The inset shows the stable probability distribution of money $P(m)$ vs. m, averaged over 10^6 initial configurations.

In the inset of Fig. 2b, we show the money distribution $P(m)$ vs. m (in units of M/N) for $N = 100$, $M/N = 1$, after averaging over 10^6 initial configurations (λ_i distribution among the agents) at t/N =10,000. There is an initial growth of $P(m)$ from $m = 0$, which quickly saturates and then a long range of power-law decay in $P(m)$ for large m values (for less than 10% of the population N in the market) is observed (for more than two decades in m). This decay, when fitted to Pareto law (1), gives $\nu = 1.03 \pm 0.03$.

5 Summary and Conclusions

We have considered ideal-gas models of trading markets. In these models, we introduce saving propensity λ of agents, such that each agent saves a fraction λ of its money and trades with the rest. We show the steady-state money or wealth distribution $P(m)$ in the market is that of Gibbs (4) for $\lambda = 0$, has got a non-vanishing most-probable value for $\lambda > 0$ (but fixed for all agents), and one gets Pareto distribution (1) with $\nu \simeq 1.0$ when λ is widely distributed among the agents. These results in simple ideal-gas like market models also compare well with real market observations.

References

1. Second Nikkei Econophysics Symposium (2002) homepage: http://web.nikkei-r.co.jp/sae
2. Stanley H E et al (1996), Anomalous Fluctuations in the Dynamics of Complex Systems: From DNA and Physiology to Econophysics, Physica A 224: 302-321
3. Statphys-Kolkata I-IV: Procs. in Physica A 186: (1 & 2) (1992) I; Physica A 224: (1 & 2) (1996) II; Physica A 270:(1 & 2) (1999) III; Physica A 318: (1 & 2) (2003) IV
4. Proc. School on Complex Systems, Indian J. Phys. B 69: 483-698
5. Chakrabarti B K, Marjit S (1995), Self-organisation in Game of Life and Economics, Indian J. Phys. B 69: 681-698
6. Moss de Oliveira S, de Oliveira P M C, Stauffer D (1999), Evolution, Money, War and Computers, Tuebner, Stuttgart, pp 110-111, 127
7. Dragulescu A A, Yakovenko V M (2000), Statistical Mechanics of Money, Euro. Phys. J. B 17: 723-726
8. Chakraborti A, Chakrabarti B K (2000), Statistical Mechanics of Money: Effects of Saving Propensity, Euro. Phys. J. B 17: 167-170
9. Hayes B (2002), Follow the Money, Am. Scientist, 90: (Sept-Oct) 400-405
10. Pareto V (1897), Le Cours d'Economique Politique, Lausanne & Paris
11. Chakraborti A (2002), Distribution of Money in Model Markets of Economy, arXiv:cond-mat/0205221, to appear in Int. J. Mod. Phys. C 13 (2003)
12. Tsallis C (2003), An Unifying Concept for Discussing Statistical Physics and Economics, in this Proc. Vol.; Reiss H, Rawlings P K (2003) The Natural Role of Entropy in Equilibrium Economics, in this Proc. Vol.
13. Dragulescu A A, Yakovenko V M (2001), Evidence for the Exponential Distribution of Income in the USA, Euro. Phys. J. B 20: 585-589; Dragulescu A A, Yakovenko V M (2002), Statistical Mechanics of Money, Income and Wealth, arXiv:cond-mat/0211175
14. Fujiwara Y, Aoyama H (2003), Growth and Fluctuations of Personal Income I & II, in this Proc. Vol., arXiv:cond-mat/0208398
15. Chakraborti A, Pradhan S, Chakrabarti B K (2001), A self-organising Model Market with single Commodity, Physica A 297: 253-259
16. Chatterjee A (2002) unpublished; Chatterjee A, Chakrabarti B K, Manna S S (2003), Pareto Law in a Kinetic Model of Market with Random Saving Propensity, arXiv: cond-mat/0301289.

II-2 Company's Risks

Enterprise money system - an ultimate risk hedge

Hideki Takayasu

Sony Computer Science Laboratories, 3-14-13 Higashi-Gotanda, Shinagawa-ku, Tokyo, Japan

Summary. For the purpose of hedging an international company from foreign exchange risks, I review the novel idea of electrical enterprise money system. In this system the unit of currency is defined by a basket of ordinary currencies with the weights proportional to the ratio of amount of financial asset in each currencies. It is proved that the economical measure in this system is free from fluctuations of foreign exchange markets. It is possible to link this money system with "points" or "mileage" and there are several merits for a company not just for a new tool of hedging.

Key words. Basket currency, enterprise money, risk hedge

1. Introduction

Econophysics study has revealed the fact that any open market can never reach a steady equilibrium state, but market prices are always accompanied with unstable fluctuations whose probability density is characterized by fat-tails [1, 2]. From the viewpoint of physics such a behavior is a typical universal property of a critical phenomenon. Actually, we can regard an open market as a system whose control parameter is kept around the phase transition point between the excess-demand and excess-supply phases by continuously tuning the market prices [3].

In the first Nikkei Econophysics symposium it was demonstrated that market prices and electrical noises can show quite similar statistical properties if the electrical circuit is designed to maintain a critical state between damping and amplifying phases [4]. Extending this analogy between markets and electrical circuits we can consider application of a new noise reduction method of markets by translating technical terminologies in physics into those in markets [5, 6].

2. An example of foreign exchange risk

International companies are suffering ridiculous troubles due to foreign exchange fluctuations. For example, we encounter such a case that an international company's income is negative if we measure the total money-flow by Japanese Yen (JPY) while it becomes positive if it is measured by US dollar (USD) as shown in Fig.1.

This apparent contradiction is caused by the following way. Assume that the company has a combination of money, w_1 in JPY and w_2 in USD, the total value measured in JPY is given by the projection of this point on the JPY axis with the slope defined by the exchange rate as shown by the straight line, $w_1+w_2R(t)$, where R(t) is the exchange rate of 1 USD into JPY at time t. Similarly the value measured in USD is given by the projection to the USD axis, $w_1R(t)^{-1}+w_2$. The total value measured in JPY is decreased when $R(t)<R(t+t')$ and that measured in USD is increased. Note that even there is no money flow and the company has the identical combination of money, the total value can be changed due to the fluctuations of foreign exchange rate.

When this contradictory behavior is taken place, investors in Japan may think that this company is declining while those in USA may think that the company is growing, and even the headquarter of the company does not know which is correct. The real situation is much more complicated than this simple example as there are many kinds of currencies involved in the financial description of an international company and the foreign exchange markets are continuously fluctuating 24 hours per day.

Not only foreign exchange rate fluctuations are confusing investors, but also the fluctuation is causing troubles in financial section for payment of salary, for example, as the value of cash in each currency fluctuates all the time, and of course the exchange rate is directly responsible to the sales.

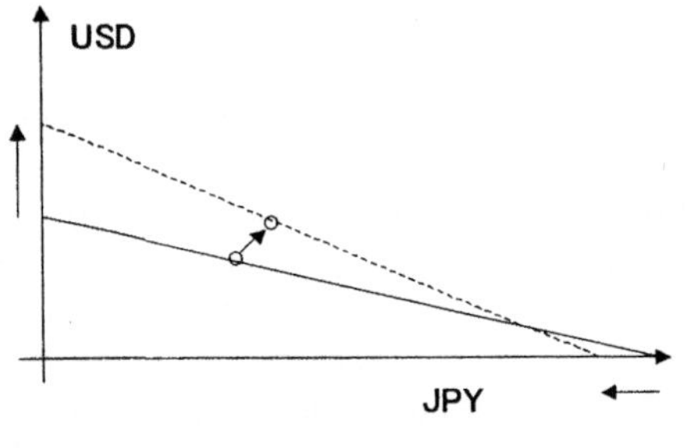

Fig.1 The effect of foreign exchange rate change for a company. Assuming that a company's asset is a composite of Yen and Dollar, and both amounts have increased. The total asset observed by Yen, however, decreased due to the exchange rate change toward stronger Yen. By measuring the same phenomenon by Dollar the company's income is evaluated as a big gain.

3. How to manage the exchange fluctuations

The example in the previous section demonstrates that the foreign exchange rate is a big problem for ordinary international companies, and the exchange rate fluctuations can have innumerable harmful effects. As the exchange rate can be viewed as a kind of ruler or a yardstick in measuring economical values, the present situation of fluctuating exchange rates may look like a situation of constructing a house using a ruler made of rubber. Remembering that the progress of material science was due to introduction of universal measuring systems, we need to establish a firm ruler in international economics to avoid confusion.

There have been four approaches for management of foreign exchange fluctuations summarized as follows:

1 Prediction either empirical or computer-aided.

2 Risk-hedge using financial technology.

3 Controls including intervention by the central banks.

4 Introduction of new currencies.

The 4-th approach includes introduction of Euro and various kinds of local currencies like "hours" in Ithaca city. In this paper we pursue this approach, namely, we introduce a new currency system to avoid the ridiculous troubles. As mentioned in the first section, market fluctuations are generally similar to electrical noises in a circuit having critical properties, and noise reduction in an electrical circuit is a routine work, so we may invent a new noise reduction method in a foreign exchange system [5, 6].

The first step of noise reduction in an electrical circuit is to remove noise generators from the circuit. In markets noise generators are those people who prefer larger fluctuations; they are called the speculators. Therefore, we need to introduce a shut-out mechanism of speculators in the new currency system. The simplest way of speculator shut-out is to allow the usage of the new currency only for non-speculators. However, it is generally not possible to categorize a person to speculator or non-speculator. The difference of speculation and investment is rather delicate but there is a significant difference in time scales, namely, the typical time scale of investment is several months or several years, while the time scale of speculation in foreign exchange is a few minutes up to several hours. Also speculative trades are generally consisted of a pair of round-trip money exchange. Therefore, if we set a rule that a short-term round-trip money exchange is either forbidden or expensive, then speculative actions can be eliminated in the new currency system.

The second step of noise reduction is to cancel positive and negative noises by summation as noises always includes random fluctuations of both signs. This noise canceling mechanism can be implemented in the money system by concentrating all orders at one place and handling them with cancellation of opposite orders. To be more precise, here, we consider an international bank system that manages an international company's finance, so orders are mainly money transfers among customers and branches in various countries. We introduce an electrical money system with a central bank as shown in Fig.2. Inside of the dotted area are filled with the new electrical currency and orders are executed immediately via the central bank. Any customer inside the dotted area has his bank account and money transfer between two customers in the dotted area are done with the new currency even they are in different countries. Money transfer order from inside of the dotted area to a customer in Japan, for example, is done in the following way: Firstly the new currency is sent to the central bank from the sender inside of the area, then the central bank exchange the currency to Yen at the present exchange rate which will be defined later, then Yen is transmitted to the customer in Japan. The inverse money flow goes in the opposite way.

Note that the central bank has buffers in various currencies. The key point is that the system is obviously in an excess-supply phase and any order of foreign exchange is done immediately using the buffer. Therefore, the system is far-from the equilibrium point and the noises are smoothed out. The central bank needs to have a trade in the foreign exchange market (in short FOREX) if one of the buffer money is going to be empty, otherwise the central bank does not need to exchange any currencies in the FOREX market. In this way the exchange of currencies as a whole is minimized.

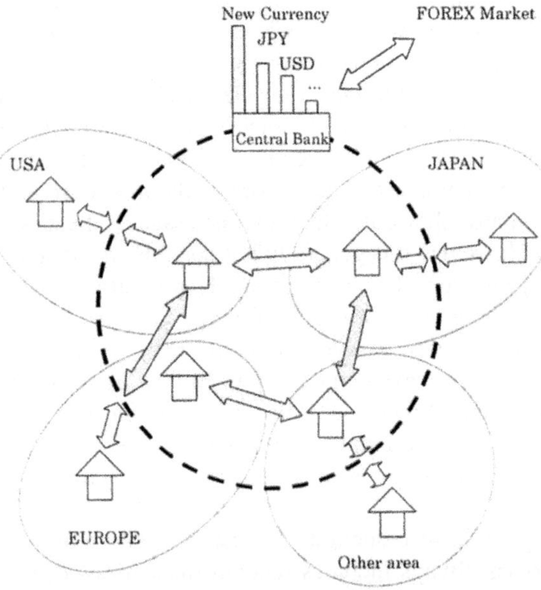

Fig.2. A schematic view of the new electrical money system. All orders are done by the central bank. Money exchanges with existing currencies are done with an appropriately given exchange rate immediately.

4. How to determine the exchange rate

The most important task of the central bank is to exchange the new currency to existing currencies, and the way of determining the exchange rates play the central role. It is shown in ref.[6] that the best way is to define the new currency by a basket currency with the weights proportional to the ratio of the total assets inside the new currency domain.

For simplicity we describe the case of two currencies, Yen and US dollar. Assume that the company possesses w_1 JPY and w_2 USD, and the exchange rate at FOREX market is given by R(0), then the weight of the new basket currency (NBC) is given by the following way:

$$1 \text{ NBC} \propto w_1 \text{ JPY} + w_2 \text{ USD} = w_1 + w_2 R(0) \text{ JPY} = w_1/R(0) + w_2 \text{ USD} \quad . \quad (1)$$

The exchange rate is automatically determined by the second and the third equations. The initial condition of the unit of NBC can be started arbitrarily, for example, with 1 NBC = 1 USD. Then the total asset of the company is given by $w_1/R(0)+w_2$ NBC.

Even started with 1 NBC = 1 USD, this relation is broken when the exchange rate of USD between JPY is changed to R(t) with the identical asset combination. As the total asset combination is identical by measuring NBC, the exchange rate between NBC and USD at the new FOREX rate of R(t) is given by solving the following relation:

$$w_1/R(0)+w_2 \text{ NBC} = w_1/R(t)+w_2 \text{ USD} \quad (2)$$

The value of w_1 and w_2 changes when there is any cash-flow, and similar results holds in the case of many currencies.

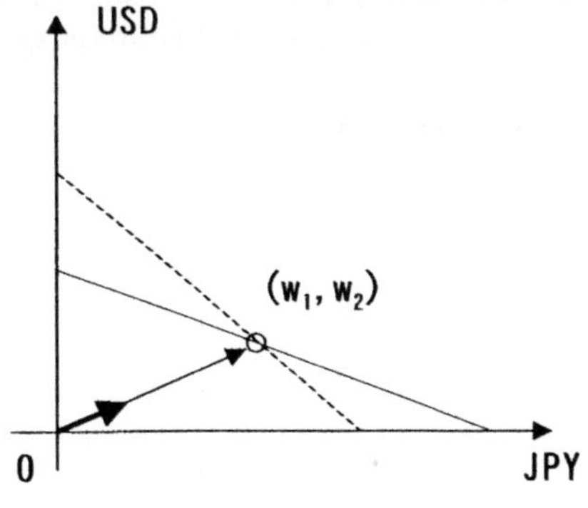

Fig.3. The axis of the basket currency. The line crosses the asset combination, w_1 JPY and w_2 USD. The measure on this line is invariant for any change of Yen-Dollar exchange rate in FOREX market.

5. Discussions

This new currency system can be applied to any international company or a union of international companies. Electronic currency might sound new, however, money transfers among companies are now done nearly 100% via bank transfer, namely, electronic money transfers. As the new currency is an electronic basket currency of existing currencies it is not difficult to implement to the existing bank systems.

By applying this new currency system we can expect to minimize the amount of foreign exchange, and also the ruler of measuring the company's finance is firmly fixed. Also the new currency can be merged with the existing point or mileage systems for lock-in customers, or the whole money flow management is similar to credit card systems.

This type of international currency can be designed rather freely by the company, so by introduction of this system the company's financial strategy will become much stronger.

References

[1] R.N.Mantegna and H.E.Stanley, 2000, "An Intorduction to Econophysics: Correlations and Complexity in Finance", Cambridge Univ. Press

[2] H.Takayasu(editor), **Empirical Science of Financial Fluctuations** – The Advent of Econophysics, (ISBN 4-431-70316-0) Springer Verlag, Tokyo, 2002,

[3] H.Takayasu and M.Takayasu, Critical fluctuations of demand and supply [**Physica** A, 269(1999), 24-29]

[4] A.-H.Sato and H.Takayasu, Market price simulator based on analog electrical circuit, pp.214-221 in Ref.[2].

[5] Hideki Takayasu, Econophysics; empirical laws, theory and application [**Evolutionary Controversies in Economics**, Ed. by Yuji Aruka as a Publication of Japan Association for Evolutionary Economics, Springer Verlag Tokyo, 2001, 191-200]

[6] Hideki Takayasu and Misako Takayasu, **Econophysics - Toward Scientific Reconstruction of Economy** (Japanese textbook), Nikkei Tokyo, 2001.

Visualization of business networks

Wataru Souma[1], Yoshi Fujiwara[1], and Hideaki Aoyama[2]

[1] ATR Human Information Science Laboratories, Kyoto 619-0288, Japan
[2] Faculty of Integrated Human Studies, Kyoto University, Kyoto 606-8501, Japan

Summary. We propose a method of visualizing business networks in order to connect the knowledge of complex networks and actual business. Although business networks are constructed with various economic agents and complex relationships, we consider only networks constructed with listed companies and shareholding relationships. We propose a center of mass method to draw a global map of business networks, and visualize that four major banks have a strong influence on almost every listed company. We also propose a visualization method based on the construction of stock portfolios, and draw a local map centered on the bank.

Key words. Business networks, Shareholding, Corporate complex, Stock portfolio

Introduction

Recent studies of complex networks (for reviews, see Watts 1999; Barabási 2002; Dorogovtsev and Mendes 2003) have clarified some topological aspects of many kinds of social and natural networks, such as WWW, power grids, movie actor collaboration networks, science collaboration networks, networks of words, metabolic networks, the neural network of nematode worm *C. elegans*, etc. In some of those studies, various quantities are proposed to characterize networks, and the calculation of those quantities for real-world networks brought about new findings, such as small-world networks (Watts and Strogatz 1998), scale-free networks (Albert et al. 1999), spectral properties (Goh et al. 2001, Farkas et al. 2001), community structure (Girvan and Newman 2002), etc.

The economy is simply regarded as the dynamics in complex networks. In economics and game theories, the characteristics of economic agents and their interactions have been extensively studied, but the topological characteristics of business networks[3] have not been studied well. Hence, from the viewpoint of

[3] In this article we use "business networks" as a generic term for complex networks in the economy.

economics and physics, it is meaningful to reveal the topological characteristics of business networks.

Recently, the topology of the world trade web (Serrano and Boguñá 2003) and Japanese business networks (Souma et al. 2003) have been studied, and the characteristics of business networks are gradually being clarified. It is extremely important to calculate the quantities characterizing business networks, but if we are to consider the application of the knowledge of complex networks to actual business, we need some ideas to bridge the gap between them. We think that the visualization of business networks will be an effective tool to fill the gap. As the first step towered this purpose, in this article, we consider the visualization of business networks constructed by listed companies and long-term shareholding relationships.

Major shareholder data and their constraints

We use data reported by Diamond Inc. (2002) and Toyo Keizai Inc. (2002). These data contain various types of information about Japanese companies, and a list of the top ten shareholders for each company and the number of shares held by each shareholder. Shareholders hold shares as an investment under sound capitalism, but this common sense is not necessarily true in Japan.

In Japan, major shareholders are banks, life/nonlife insurance companies, big companies, and trust companies. Trust companies hold shares as an investment, and life/nonlife insurance companies hold them for risk hedging purposes. However, in other big shareholders, the aim of shareholding is not for investment or risk hedging but for the solidarity of corporate complexes. The structure of corporate complexes in the Japanese economy has protected Japanese companies from foreign corporate raiders, and also brought about the bubble economy and the enormous amount of bad debts after the collapse of the bubble economy.

In this article we visualize business networks constructed with listed companies and banks, which are connected by long-term shareholding relationships. Hence we remove trust companies and life/nonlife insurance companies from the list of shareholders, because these companies do not necessarily hold shares for the solidarity of corporate complexes.

Global map of Japanese business networks

When considering the visualization of business networks, we must first decide the location of the companies. As mentioned before, we have data about the number of shares held by each shareholder. Hence if we regard the number of shares as the mass, we can define the location of the companies as the center of mass.

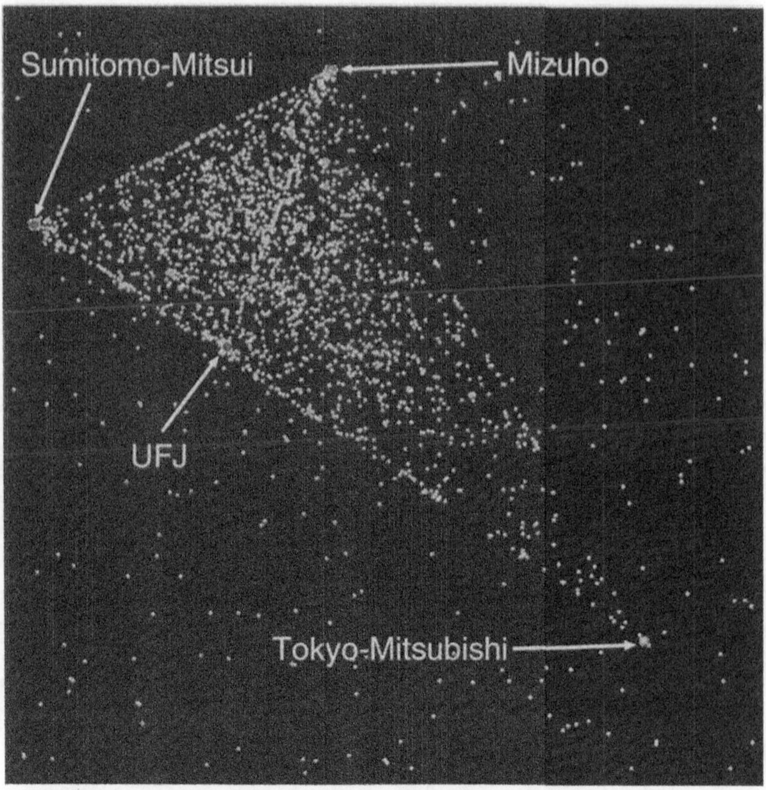

Fig. 1. A global map of the Japanese business network, which is constructed with listed companies and long-term shareholding relationships.

$$g_i \equiv \frac{\sum_j^n g_j s_j}{\sum_j^n s_j},$$

where g_i is the location of company i, s_j is the number of shares held by company j, and n is the number of shareholders of company i.

The locations of the companies are summarized by Fig. 1. We visualized this in two-dimensional space, but the x-axis and y-axis have no special meaning. The circles correspond to listed companies, however, to distinguish the four major banks (Mizuho Bank, Sumitomo-Mitsui Bank, Tokyo-Mitsubishi Bank, and UFJ Bank) from the other companies we marked them with a different size. In this figure we can draw lines between companies and their shareholders, but we abbreviated these for simplicity.

This figure suggests that almost every listed company has their shares held by the four major banks. This means that the four major banks have a strong influence on Japanese business networks. At first glance, we get the impression that UFJ Bank is directly governed by Sumitomo-Mitsui Bank and Tokyo-Mitsubishi Bank, but this is not true. As mentioned in the previous section, we removed trust companies and life/nonlife insurance companies from the top ten list of shareholders. This constraint leaves only one shareholder, Toyota Motor Corp., in the shareholder list of UFJ Bank. On the other hand, for the shareholder list of Toyota Motor Corp., this constraint leaves four shareholders, Toyota Industries Corp., UFJ Bank, Sumitomo-Mitsui Bank, and Tokyo-Mitsubishi Bank. The Toyota Motor Corp. is located on the finite line connecting Sumitomo-Mitsui Bank and Tokyo-Mitsubishi Bank. Hence UFJ Bank is located on that finite line[4].

Local map centered on the banks

In this section we consider the local map centered on the banks. In the same manner as the previous section, we visualize the map in two-dimensional space. Hence we need two parameters to decide the location of the companies, and adopt a polar coordinate system. As an example, we take up the local map centered on Sumitomo-Mitsui Bank, which is fixed on the origin of the coordinate system.

One of the most useful quantities to specify the location of companies is the securities code number, which is given for all listed companies and reflects the business category. We arrange companies in the ascending order of securities code numbers, and identify the order, i, as the newly defined company's number. We define the angle around the origin (Sumitomo-Mitsui Bank) as

$$\theta_i \equiv \frac{2\pi i}{m},$$

where m is the number of companies which have their shares held by Sumitomo-Mitsui Bank.

As mentioned before, we have data about the number of shares held by each shareholder. Hence, for each company, we can obtain data about the number of shares held by Sumitomo-Mitsui Bank. In addition, we can easily obtain data about the daily share price. We define a quantity,

$$l_i \equiv \frac{s_i p_i}{\sum_j^m s_j p_j} \leq 1,$$

where s_i is the number of shares held by Sumitomo-Mitsui Bank, and p_i is the daily share price of company i. Hence the denominator corresponds to the portfolio of Sumitomo-Mitsui Bank, and l_i represents the ratio of shares possessed by company i in that portfolio. We define the distance between the companies and the origin as

$$r_i \propto -\log l_i \geq 0.$$

[4] Strictly speaking, if a company has only one shareholder, that company and shareholder are located on the same position. To avoid this problem, we located such a company around the shareholder with a certain distance.

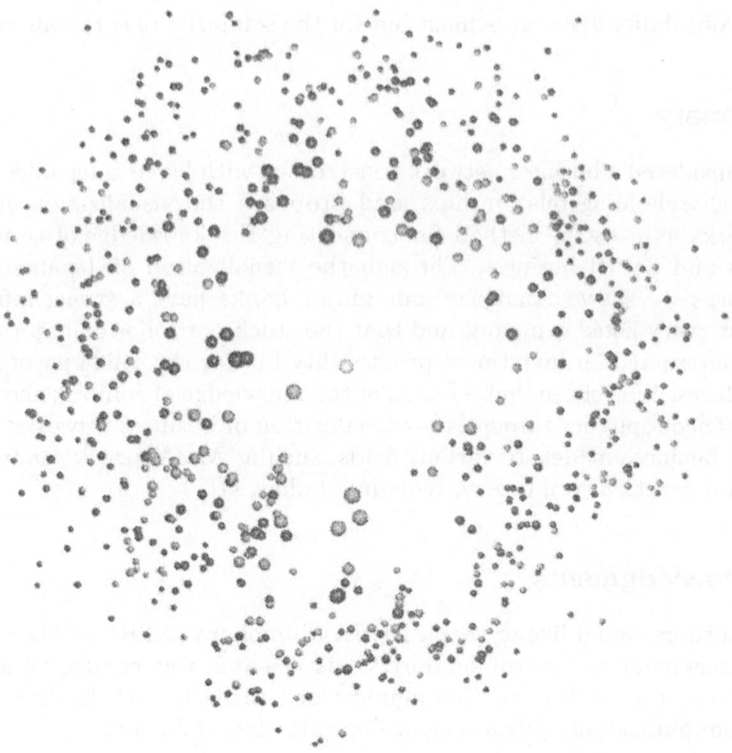

Fig. 2. A local map centered on Sumitomo-Mitsui Bank.

Hence companies with a small r_i dominate the principle part of the stock portfolio.

We can decide the location of companies with θ_i and r_i, and obtain Fig. 2. In this figure the centered circle represents Sumitomo-Mitsui Bank and the other circles represent companies whose shares are held by Sumitomo-Mitsui Bank. Excluding Sumitomo-Mitsui Bank, the size of the circle reflects the number of consolidated subsidiaries. We used different colors to paint the circles representing the companies, according to the value of consolidated return on equity (CROE). Roughly speaking, dark circles correspond to companies with low CROE, and bright circles correspond to companies with high CROE. In this figure we can draw lines between companies and their shareholders, but we abbreviated them for simplicity. As mentioned previously, the angle around the origin reflects the business category. Thus we can radially draw lines to distinguish business categories.

If you are a good portfolio manager you will set the bright, large circles close to the origin, and remove the dark circles. However this is not true in

Fig. 2. The reason why this bank constructs such a stock portfolio is not for the profitability of its investment but for the solidarity of corporate complexes.

Summary

We considered a business network constructed with listed companies and long-term shareholding relationships, and proposed the visualization of business networks as a useful method for connecting the knowledge of complex networks and actual business. Through the visualization of Japanese business networks we showed that the four major banks have a strong influence on almost every listed company, and that the stock portfolios of major banks are not constructed for investment profitability but for the solidarity of corporate complexes. We believe that a fusion of the knowledge of complex networks and that of econophysics through the visualization of business networks will bring useful business models to various fields, such as M&A, prediction/protection of chain-reaction bankruptcy, economic policy, etc.

Acknowledgments

The authors would like to thank Dr. K. Shimohara (ATR) for his continuous encouragement and warm support. This research was conducted as part of the "Research on Human Communication" project, with funding from the Telecommunications Advancement Organization of Japan.

References

Albert R, et al (1999) Diameter of the world-wide web. Nature 401:130–131

Barabási AL (2002) Linked: The new science of networks. Perseus Publishing, Cambridge Massachusetts

Diamond Inc. (2002) Japanese company file. Diamond inc., Tokyo

Dorogovtsev SN, Mendes JFF (2003) Evolution of networks: From biological nets to the internet and WWW. Oxford University Press

Farkas IJ. et al (2001) Spectra of "real-world" graphs: beyond the semicircle law. Phys Rev E64:026704

Goh HI, et al (2001) Spectra and eigenvectors of scale-free networks. Phys Rev E64: 051903

Girvan M, Newman MEJ (2002) Community structure in social and biological networks. Proc Natl Acad Sci USA 99:7821-7826

Serrano MA, Boguñá M (2003) Topology of the world trade web. arXiv:cond-mat/0301015

Souma W, et al (2002) Complex networks and economics. Proc Int Econ Conf, to be published in Physica A

Toyo Keizai Inc. (2002) Japan company handbook. Toyokeizai inc., Tokyo

Watts DJ (1999) Small worlds: The dynamics of networks between order and randomness. Princeton University Press, Princeton New Jersey

Watts DJ, Strogatz SH (1998) Collective dynamics of 'small-world' networks Nature 393:440–442

Bankruptcy Prediction Using Decision Tree

Shigeo Aoki, and Yukio Hosonuma

Graduate School of Business and Commerce, Tokyo International University, 1-13-1, Matobakita, Kawagoe, Saitama 350-1197, Japan

Summary. Many prior researches have used discriminate analysis, however, this paper uses a decision tree model to predict bankruptcy of Japanese companies.

Using 73 bankrupt and 73 non-bankrupt companies, we got the model's accuracy rate 91.2%. As a result, the interest coverage (CF) ratio is the most important measure in predicting bankrupt Japanese companies.

Key words. bankruptcy prediction, CHAID, decision tree, discriminate analysis, interest coverage ratio

1. Introduction

Bankruptcy prediction is a major challenge in Japan and this paper uses a decision tree model to predict bankruptcy of Japanese companies. In 1968 Altman used discriminate analysis as a tool to predict bankruptcy of American companies. The Altman model is widely used by financial analysts to observe the probability of a U. S. company going bankrupt. Also, other researchers have used different inputs in the Altman model, as well as employing other statistical models, such as logit and probit, to predict bankruptcy of U. S. companies. The Altman model discriminated between two categories, failed and non-failed companies. The user arbitrarily selects the independent variables to be included in the discriminate analysis model. The independent variables are selected on the basis of a single Z score. Companies whose Z score is located close to the aggregated Z score were possible candidates for misclassification, that is a failed company being classified as a non-failed company, or vice versa.

A decision tree model overcomes shortcomings of discriminate analysis. A decision tree model chooses independent variables from a large sample of variables via a t test or F test algorithm, and classifies them into a range of possible outcomes rather than using a single point as used in the discriminate analysis model.

Gentry et al. used cash flow components in a decision tree model to predict the bankruptcy of U. S. companies. Because the causes of bankruptcy in the United States and in Japan are vastly different, the proposed study will use both cash flow measures and financial ratios in a decision tree model to predict the likelihood that a Japanese company will go bankrupt.

The proposed study will use computer software *Answer Tree Ver.2.0J* that was developed by SPSS Co. The software supplies four methods, CHAID, Exhaustive CHAID, C&RT and Quest as decision tree model. We used CHAID because it has the highest accuracy rate among the four decision tree models.

2. Sample

The sample includes 73 bankrupt (failed) and 73 non-bankrupt (non-failed) companies during the period April 1, 1980 to March 31, 2000. A paired sample methodology was used in selecting the sample. A bankrupt company was selected and then matched with a non-bankrupt company within the same industry and with sales that were reasonably close in size. The 146 companies stocks were listed on the securities exchange or traded over-the-counter. A learning samples of 114 companies (80 percent of the total sample) was selected and a training sample of 32 companies (20 percent of the total) was used for each trial.

3. Results

CHAID algorithm chose five variables from a total set of 184 financial ratio . that were calculated from the non-consolidated financial statements. CHAID creates different trees for each trial. The trees were adjusted to enhance the predictive accuracy rate. The final model is presented in Figure 1, the model's accuracy rate is 91.2%. We tested the model using 23 bankrupt companies that failed during the period September 1, 2001 to March 1, 2002. This trial sample is not included in the sample used for the study. Of the 23 bankrupt companies, 21 of them were predicted correctly, that is an accuracy rate of 91.3 percent.

The independent variables selected by CHAID were the Interest Coverage Cash Flow, Interest Coverage, CB ratio, Equity ratio and RE ratio. For the pilot sample, Interest Coverage Cash Flow was the most important variable in predicting bankruptcy as shown in Figure 1. The definition of the ratios is shown in Figure 1.

4. Conclusions

The interest coverage (CF) ratio is the most important measure in predicting bankrupt Japanese companies. The numerator of Interest Coverage (CF) and Interest Coverage (OI) reflects the importance of current operating earnings and the denominator is a surrogate for interest bearing debt, a measure of financial soundness. The CB ratio is a liquidity measure. Equity and RE ratios imply the importance of past earnings power and financial soundness. The RE ratio was also significant in Altman's and Shirata's discriminate analysis model. The RE ratio implies past earnings power and financial soundness. In conclusion, earning power, liquidity and overall financial health are crucial in predicting bankruptcy.

References

- E.L. Altman. "Financial ratios, discriminate analysis and the prediction of corporate bankruptcy". *Journal of Finance*. vol.23. pp.589-609. Sept.1968.
- J.A.Gentry, M.J.Shaw, A.C.Tessmer, and D.T.Whitford. "Using Inductive Learning to Predict Bankruptcy". *Journal of Organizational Computing and Electronic Commerce*. 12(1). 39-57. 2002.
- Yoshiko Shirata. "*Kigyo Tousan Yochijoho no Keisei*". Chuokeizai-Sha. 1999.

Figure 1 Bankruptcy Prediction Model (Adjusted Tree)

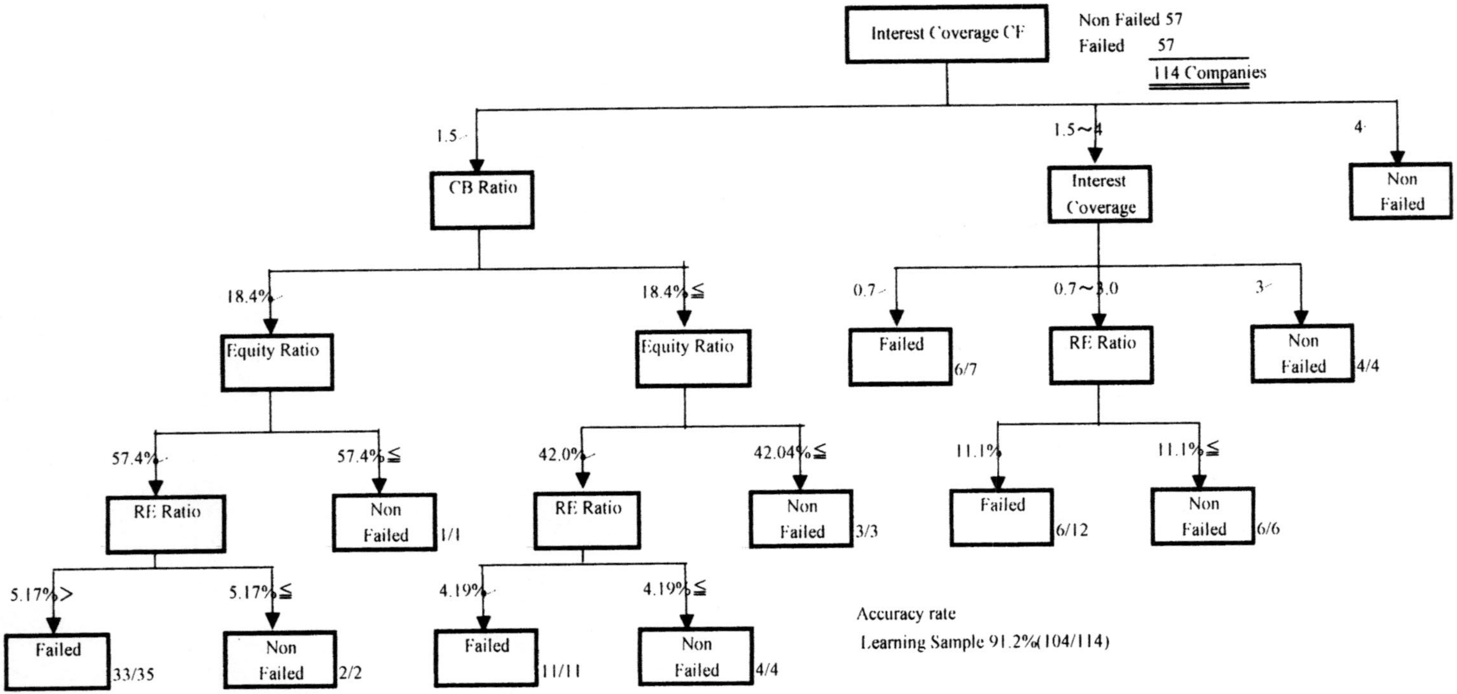

Interest Coverage CF (Net Income+Depreciation-Dividends-Officers' Bonus+Interest Expense+income Taxes)/Interest Expense
Interest Coverage(OI) (Operating Income+Interest Expense)/Interest Expense
CB Ratio (Cash and Time Deposits)/(Short & Long Term Borrowing+Notes Receivable Discounted)
Equity Ratio Stock Holders' Equity/Total Assets
RE Ratio Retained Earnings/Total Assets

Premium Forecasting of an Insurance Company

M. Ebrahim Fouladvand and Amir Darooneh

Department of Physics, Zanjan University, Zanjan, Iran.
foolad@mail,znu.ac.ir; darooneh@mail.znu.ac.ir

Summary. We present an analytical study of an insurance company. The company's performance is modeled on a statistical basis. The predicted annual income of the company is evaluated in terms of insurance parameters namely the premium, the total number of insured, average loss claims etc. We restrict ourselves to a single insurance class the so-called *automobile insurance* . We obtain the corresponding risk as well as ruin probability in terms of premium. Furthermore, we obtain the optimal premium p_{opt} which maximizes the company's profit.

1 Introduction

The rapidly expanding field of *econo-physics* has witnessed the desire of statistical physicists to apply their knowledge to some areas, as broad as they can, of financial systems. Stock markets is a prototype example on which there is a now a rich amount of results both empirically and analytically [1, 2, 3]. Another domain in finance which has, for many years, been investigated within rigorous mathematical techniques is *insurance* [4, 5, 6]. The purpose of insurance is to indemnify policy holders against the occurrence of adverse events such as illness, earthquake, death etc. Common to all insurance systems, there exists a *risk* i.e., a condition in which there is a possibility of an adverse deviation from an expected outcome. In this paper we aim to model the performance of an insurance company in order to find a better insight into the stochastic nature of the problem. we restrict ourselves to a single insurance class the so-called automobile insurance. In principle, a number of clients (policy holders) insure their cars. The company issues the corresponding insurance policies and receives an annual premium p from the policy holders. Generally speaking, the amount of premium depends on many characteristics of the generic car such as type, model, production year and driving record of the policy holders. However for simplicity, here we restrict ourselves to a category of cars for which the premium is almost a constant amount. The company's performance is modeled by the series of loss events which randomly occur in the course of time. Each loss event is incorporated with a loss amount. The loss size is itself a random variable ranging from minor to large amounts. Upon incurring a loss event, the company pays the claim amount to the policy holder which reduces the company's capital. On the other hand, the premium income which is injected to the company raises its capital. It is the competition of these two factors which determines the financial status of the company. Evidently the amount of premium plays a dominant role in insurance industry . The task of premium calculations and the related strategies have been challenging and controversial subjects for insurance companies and can drastically affect the financial status of the company [5, 6, 7]. In a competitive economy, there is an abundance of insurance companies each of which offer its own premium. If a company increases the premium, the number of clients willing to insure their cars decreases (due to competition of the other companies offering lower premiums). In other words, the company's attempt to raise its income through increasing the premium may fail due to decrease of the number of insured which obviously reduces the premium income. Therefore, increasing the premium is a highly risky action and it is of prime importance for the company's managers to have an estimation of the risk amount. To be more specific, each day a random number of clients insure their car in the company. We denote this number by $\mathcal{I}_t$ for t-th day. We assume that the number of daily policy issued obeys a simple statistic with premium-dependent characteristics such as mean, standard deviation etc. In this paper we assume that $\mathcal{I}_t$ is uniformly drawn from the interval $[a, b]$ with $b = \frac{3\bar{\mathcal{I}}}{2}$ and $a = \frac{\bar{\mathcal{I}}}{2}$ such that $\bar{\mathcal{I}}_t = \bar{\mathcal{I}}$ for all t. As a simple choice, we take an exponential form $\bar{\mathcal{I}}(p) = \mathcal{I}_0 e^{-\frac{p-p_0}{p_0 \tau}}$ for the average number of daily policy holders. p_0 is an arbitrary reference point, $\mathcal{I}_0$ is average number of insured for $p = p_0$ and the parameter τ measures the predicted fall of the number of policy holders upon increasing the premium. For instance $\tau = 0.45$ corresponds to a 20 percent fall upon a 10 percent premium increase over p_0. We now discuss the distribution of the daily number of loss events in $t-$th day which is denoted by $\mathcal{A}_t$. Evidently it implicitly depends on the previous number of policies issued $\sum_{t'<t} \mathcal{I}_{t'}$. However, for simplicity, we make a simple assumption such that $\bar{\mathcal{A}}_t = \theta \bar{\mathcal{I}}$ where constant θ measures the ratio of loss events to number of issued policies at an average level. We recall that θ can be interpreted as the probability of occurrence of loss event (per car). Analogous to $\mathcal{I}_t$, it is assumed that $\mathcal{A}_t$ is uniformly distributed in the interval $[a\theta, b\theta]$ so that $\bar{\mathcal{A}}_t = \theta \bar{\mathcal{I}}$ for all t. The loss amount itself should be considered as a stochastic variable. In this paper we adopt Erlang's model [7, 8] for statistical description of loss amounts. According to this model, the distribution of loss amount obeys an exponential function of the form $e^{-\frac{x}{\xi}}$ where ξ denotes the average loss.

2 Statistical Description

Now we present the analytical results which give the annual capital of the company in terms of insurance characteristics i.e., premium, statistics of the number of insured, statistics of loss events etc. For this purpose, during the course of each working day we randomly draw two integer number $\mathcal{I}$ and $\mathcal{A}$ as the numbers of issued policies and loss events. Correspondingly, for each loss event, we attribute a random loss amount. The sum of loss amounts gives us the daily loss which is denoted by $\mathcal{L}$. Denoting the number of annual working days by $\mathcal{N}$ and the capital at the end of t-th day by C_t, we simply have

$$C_t = C_{t-1} + p\mathcal{I}_t - \mathcal{L}_t \tag{1}$$

where t denotes the day number. Focusing our attention on the long-term profit, we now evaluate the average long-term capital of the company for a period of $\mathcal{N}$ working days. From the above recursive relation, one simply finds by iteration:

$$C_{\mathcal{N}} = C_0 + p\sum_{t=1}^{\mathcal{N}} \mathcal{I}_t - \sum_{t=1}^{\mathcal{N}} \mathcal{L}_t \tag{2}$$

which determines the long-term capital in terms of aggregate loss and premium income. In order to predict the capital, we now average the above relation over many virtual realisations of future each of which corresponds to $\mathcal{N}$ working days. Denoting such averaging by $< \cdots >$, one obtains the following relation where we have dropped the initial capital C_0 for simplicity.

$$< C_{\mathcal{N}} >= \mathcal{N}(p < \mathcal{I} > - < \mathcal{L} >) \tag{3}$$

We define the critical value of premium p_c at which the average long-term profit equals zero $< C_{\mathcal{N}} >= 0$. It is worth obtaining $< \mathcal{L} >_t$ in terms of statistics of loss events. It can be easily shown that the average loss amount for each day is simply the average loss amount, in each single loss event, multiplied by the average number of daily loss events i.e., $< \mathcal{L}_t >= \xi\bar{\mathcal{A}}$. Regarding the above considerations one obtains the following expression for the averaged long term capital :

$$< C >= \mathcal{N}\mathcal{I}_0 e^{-\frac{p-p_0}{\tau p_0}} (p - \theta\xi) \tag{4}$$

In our study, the numerical values of insurance parameters have been set from the empirical data taken from *Iranian Insurance Industry* for the year 2000. Specifically, we analysed the data taken from *Iran Insurance Company* the largest company operating in Iran. Based on the above data, we set our parameter as: $\theta = 0.19$, $\xi = 2.12$ million Rls and $\mathcal{I}_0 = 6070$. Also we take $\mathcal{N} = 300$ active days. Figure (1) shows the behaviour of $< C >$ for different values of τ.

Accordingly, two distinct regimes are identified: profit and loss. There is crossover premium $p_c = \xi\theta$ above which the company profits. The above

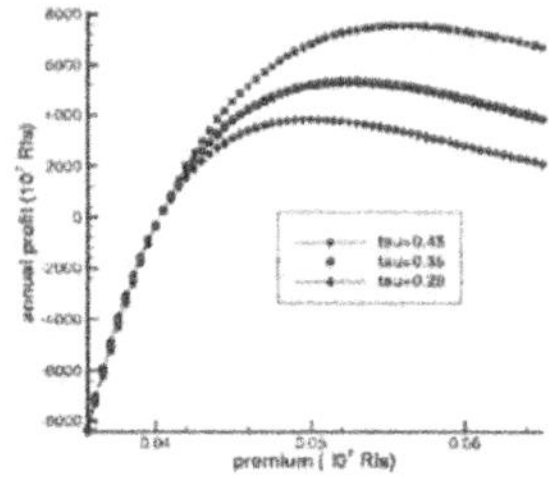

Fig. 1. Averaged long term profit of the virtual company for three values of τ: 0.45, 0.35 and 0.28 which correspondingly refer to 20, 25 and 30 percent fall in the number of issued policies upon increasing the premium by 10 percent.

graph gives us another useful fact. It determines the optimal premium p_{opt} at which the company's capital is maximized. Nevertheless, we should note that the aforementioned conclusions are based on a mean field approach and in principle the company can not rely on these average-based arguments. The simplest quantity which can give us an approximate measurement of the risk is the variance of long term profit in the predictive averaging process [4]. At this stage, we try to evaluate the variance of the long-term capital. After some straightforward mathematics we reach the following formula:

$$< (C-\bar{C})^2 >= \mathcal{N}[p^2(< \mathcal{I}^2 > -\bar{I}^2) - 2p(< \mathcal{I}\mathcal{L} > -\bar{I}^2\xi\theta) + < \mathcal{L}^2 > -\bar{I}^2\xi^2\theta^2] \tag{5}$$

The stochastic variable $\mathcal{I}$ has a uniform distribution function therefore its variance is simply obtained via the relation $\sigma^2 = \frac{1}{12}(b-a)^2$. One can simply show that the variance of daily loss is θ times the variance of $\mathcal{I}$. Recalling the definition of the covariance between two arbitrary stochastic variables X and Y:

$$Cov(X,Y) = \frac{< XY > - < X >< Y >}{\sigma_X \sigma_Y} \tag{6}$$

It can be shown that inequality $-1 \leq Cov(X,Y) \leq 1$ holds. We thus write the remaining term in the above equation as $\omega\sigma_{\mathcal{I}}\sigma_{\mathcal{L}}$ where $-1 \leq \omega \leq 1$ denotes the covariance between $\mathcal{I}$ and $\mathcal{L}$. Putting everything together we simply arrive at the following approximate equation for the variance of long term capital:

$$< (C-\bar{C})^2 >= \frac{\mathcal{N}\bar{I}^2}{12}[p^2 - 2p\omega\xi\theta + \xi^2\theta^2] \tag{7}$$

It is seen that risk value has a quadratic dependence on premium p and is proportional to the number of working days $\mathcal{N}$. Figure(2) depicts the behaviour of standard deviation as a function of premium for some values of τ.

It is observed that risk value is an increasing function of τ. This indicates that for larger values of τ we have a less-valued variance in long term capital.

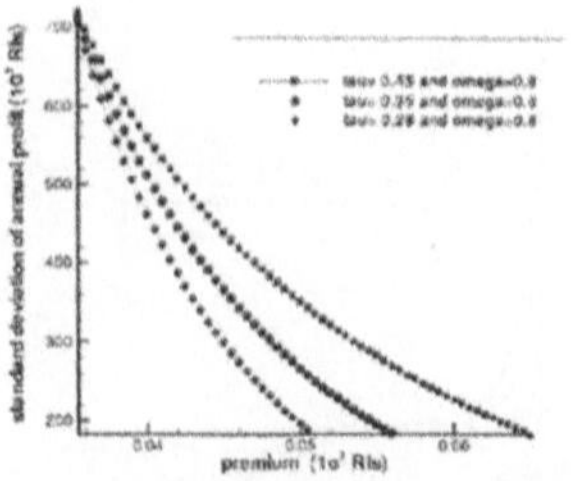

Fig. 2. mean-field standard deviation of annual capital for same three values of τ as in fig.1 . $N = 300$, $\xi = 2.12$ million Rls. and $\omega = 0.8$.

This means that the company makes less profit but with more certainty, a situation which is desirable for cautious managers who wish to avoid risky decisions. The next concept we deal with is bankruptcy. A useful quantity in risk management is the company's ruin probability. This can be expressed as the probability that the long term capital falls below zero i.e., $P(C < 0)$. To estimate the ruin probability, we use of the so-called Tchebycheff inequality. This inequality establishes a relation between the variance and the probability that a stochastic variable, with finite average and variance, can deviate by an arbitrary amount ϵ ($\epsilon > 0$) from the mean value:

$$P(|z- < Z > | \geq \epsilon) \leq \frac{\sigma_Z^2}{\epsilon^2} \tag{8}$$

where σ_Z denotes the standard deviation of the stochastic variable Z. Now let us estimate the probability that for a prescribed p, the long term profit equals $-\lambda\bar{C}(p)$ where the dimension-less parameter $\lambda > 0$ denotes the depth of ruin. According to Tchebycheff inequality if we take Z as C one simply finds:

$$P(C = -\lambda\bar{C}) \sim \frac{p^2 - 2p\omega\theta\xi + \theta^2\xi^2}{24\mathcal{N}(p - \xi\theta)^2(1+\lambda)^2} \tag{9}$$

Figure(3) exhibits the ruin probability dependence on premium for various ruin depths. λ is set to 0.1 .

In order to find a deeper insight into the problem nature, we have carried out simulations to obtain the variance of C. We have simulated the company's performance for one year and obtain the annual income. Figure(4) shows the simulated variance of C in terms of p.

The simulation results show a notable deviation from those of mean-field approach and gives $\omega \approx 0.5$ as the best fitting value for the covariance parameter ω. To conclude, we have modelled the performance of a an insurance company both analytically and numerically. Our results illustrate the dependence of long-term profit in terms of insurance parameters especially that of premium. We have shown the existence of an optimal premium which

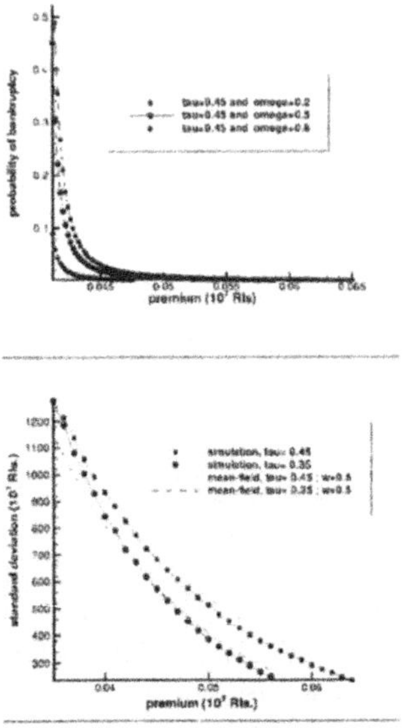

Fig. 3. Ruin probability and simulation of standard deviation. $N = 300$, $\xi = 2.12$.

maximizes the company's capital. Furthermore, we have been able to give a quantified estimation for the risk of premium increasing. We wish to express our gratitude to *Iranian Centre of Actuarial Research* for providing us with the empirical data. We highly acknowledge enlightening discussions with G.M. Schütz.

References

1. J.P. Bouchaud and M. Potters, *Theory of Finantial Risks*, Cambridge Ur:versity Press, Cambridge, (2000).
2. R. Mantegna and H.E. Stanely, *An Introduction to Econophysics*, Cambridge University Press, Cambridge, (2000).
3. D. Sornette, *Critical Market Crashes*, To appear in Phys. Rep. 2003, arXive cond-mat/0301543.
4. G. Ottaviani (Ed.) *Financial Risk in Insurance*, Springer Verlag,(1995).
5. A. Klugman, H.H. Panjer and G.E. Willmot in *Loss Models : from data to decisions*, Wiley-Interscience (1998).
6. C.D. Daykin, T. Penticainen and M. Pesonen in *Practical Risk Theory for Actuary* , CRC Press, (1994).
7. E. Straub, *Non-Life Insurance Mathematics*, Swiss Assisiation of Actuaries, Zürich, (1997).
8. S. Kotz and N.L Johonson, *Encyclopedia of Statistisc Sciences*, vol 3, 292 (1983).

II-3 Theories

A View from an Economist on Econo-physics

Koichi Hamada

Department of Economics, Yale University, New Haven, CT06520-8268, USA

Summary. Economists have learned much from physicists since the earlier days of economics. Mathematical optimization of the satisfaction of consumers is the key concept where economics find the analogy from the method of physics. It is highly welcomed that economists are now under the new stimulus from new concepts in physics such as fractals and the power distribution.
I will present some views from economists in the study of stock-price determination, a topic I worked a little with Dr. Hideki Takayasu. The advantage of econo-physics is that it can apply the idea of statistical mechanics in which each particle moves differently but the total substance can exhibit certain regularities. On the other hand, physicists may learn from economists' recognition that one has to specify precisely the information content each agent obtains. Also it will be useful to note that one has to take into account that the economic universe is closed and required to satisfy certain consistency conditions.
We first estimate the average growth of a company's annual income and its variance by using both real company data and a numerical model which we already introduced a couple of years ago. Investment strategies expecting for income growth is evaluated based on the numerical model. Our numerical simulation suggests the possibility that an investment strategy focusing on the medium-sized companies gives the best asset growth with relatively low risk.

1. Introduction

It is a pleasure to be invited to this meeting, where I feel somewhat awesome surrounded by so many brilliant physicists. In fact, economists and physicists have been in friendly relations since the beginning days of economics.

We may set aside the fact that Isaac Newton was known to be interested in alchemy. But at the turn of the 20th century, Knut Wicksell, a Swedish economist who was a forerunner of J. M. Keynes, as well as Irving Fisher, a Yale economist who laid the foundation of modern macroeconomics both studied physics first.

In the first page of his path-breaking Foundations of Economic Analysis (Samuelson,1947) Paul Samuelson cited a motto from J. Willard Gibbs, - "Mathematics is a Language". • Samuelson developed modern economic theory based on the economic optimization referring to such analogues in physics as the Hamiltonian and Le Chatelier principle.

The optimization or the maximization of an economic agent such as a consumer or a firm was formulated. Its dynamic formulation often makes use of the Hamiltonian, and the results of comparative statistics are interpreted in terms of Le Chatelier principle. More recently, the pricing of an option (the right to sell a stock at a certain specified time in the future) was elegantly formulated by the formula (Black and Sholes) by utilizing the property of a Brownian motion, somewhat similar to hyperbolic partial differential familiar in thermodynamics.

In some years, it is to be noted, some mathematical economists tried to develop suitable methods for economics rather than, or in addition to, borrowing from physicists. The emphasis of strategic interactions in human society led to the development of game theory, which was again originated by von Neumann, a physicist, and Morgenstern, in the direction of analytical tool well adapted to economics. John Nash gave foundations of the application of game theory particularly effective to economics, both for cooperative and non-cooperative human interactions.

Mandelbrot (e.g., 1997. 1999) was instrumental in uniting again economics and physical science. He noticed a long time ahead that distribution of economic variables follows the "fat-tail" power distribution and not the normal distribution. According to Professor Gene Stanley, the leading motif of econo-physics is to find that fractal structures and the power (or Pareto) distribution are effective ways of understanding economic phenomena. Mandelbrot results not only created the foundation of econo-physics approach to industrial organization, but also can be used to explain the failure of the formulae of price formation of derivatives in stock or exchange market that is based on the assumption of normal distribution.

In the mean time, economists have been benefited from conversing with other natural and social sciences, and many interdisciplinary areas have been born: political economy, law and economics, economics and sociology, economic anthropology, evolutionary game, and psychology and economics. Incidentally, the Nobel Prize in Economics in 2002 was given to Daniel Kahneman, psychologist, and Vernon Smith, an experimental economist.

2. The Stock Market and Econo-physics

The purpose of this talk is not to present a result of new research but to give a number of observations from an economist on recent trends in econo-physics. Let me focus on the price determination in stock market, a topic I worked a little in the past.

After the financial turmoil in Russia, the Long-Term-Capital-Management (LTCM) operated by a team with Nobel laureate economists failed. The reason was apparently because of the inefficacy of the financial techniques based on the normality assumption. This fact alone suffices to make us recognize necessity of econo-physics.

First, let me answer the criticism by physicists [e.g. Takayasu] that economists neglect the existence of inventories or undelivered orders. Indeed, an elementary textbook does. The analysis of unmatched orders and offers is the center of economics called disequilibrium economics. · (See Negishi, 1972, and their many follow-up papers.) At least, I have to defend our colleague economists that they are seriously considering the market without instant clearing. A part of the stock market is probably the only market where demand and supply clear instantaneously.

I had the fortune to work with Dr. Hideki Takayasu in physicists · earlier attempt to formulate the structure of the stock market from their point of view. We attempted to formulate the process of stock market trading utilizing the analogy of seismology. That is, a divergence from a previous prices will at one time be contained by inertia but at another time create a drastic movement of the stock price as a landslide. The resulting stock price distribution is not a normal distribution but the power distribution (Takayasu, Miura, Hirabayashi and Hamada, 1992).

Now, experimental methods are developed in economic analysis of stock market and econo-physicists may find it interesting to contrast physicist's results with those from the experimental approach.

I would point out several features of econo-physic approach to stock market price formation. I will begin with their advantages. First, physicists have long tradition of statistical mechanics where each particle moves differently but the total substance can exhibit certain regularities. Economists are aware of the fact that each individual moves differently, but in practice they are hard to deviate from the assumption of the representative agent in their analysis. The analysis with a representative agent goes as if the economy consists of a single individual agent or an aggregate of identical agents.

Second, physicists' attention is directed to the dynamic process of price formation rather than to the equilibrium where all the forces are working out and they are counterbalanced. The approach allows us to accommodate psychological, social and behaviorist and experimental dimensions of the stock market.

Third, physicists' approach is free of the normal distribution assumption and many restrictive assumptions economists often insist on to impose. To this aspect we will soon return.

On the other hand, there are a couple of tendencies in physicists · approach that economists find hard to understand. First, econo-physicists always concentrate their attention on market prices. By almost exclusively focusing on the market price, they tend to neglect other information that individuals obtain by interacting with each other rather than through prices. Market prices are, needless to say, the most important source of information. There are, however, many news and relevant statistics that affect the stock prices. Economists may be able to and to give the importance of related data and the proper ways of analyzing them.

Second, economists are able to present a coherent view of the economic universe and emphasize the coherence in interpreting them. Economic agents are assumed to be rational and consistent in behavior, and the universe is closed and progresses in a consistent fashion. Therefore, economists model the market in such a way that every agent knows the law of motion of the market and the motivation of other players. This is like a bridge game where all players play the optimal strategy.

Let us illustrate this by an analysis of speculative attack on a currency. A currency will collapse when many market participants consider that the currency is in danger; otherwise, it will defend its parity. There was no theory which of these two situations will be selected. An equilibrium will be chosen when sunspots appear and the other when they disappear. Morris and Shin (2000) resolve this sunspot impasse by assuming each agent receives signals about the fundamentals but with certain errors. By this trick they transformed this indeterminacy situation into a determinant situation.

The agents in the Morris-Shin model do not, on other hand, pay attention to the market price. Economic agents behave as if they were isolated and guess what should happen if the as other agents receive similar signals. By combining the analysis of the market price and the analysis of idiosyncratic information, we may hope some more results to resolve the mystery of stock market prices. In fact, physicists are well known for being skillful in analyzing the contagion of rumors. In short, economists require more consistency and coherence to the economic model; physicists are less inhibited by such restrictions. Reality must be between the two approaches.

By combining those things they develop statistically the GARCH model · Generalized Autoregressive Conditional Heteroskedasticity (GARCH) model, where errors are changing over time. There seems to be a room to develop a fruitful collaboration.

3. Intensity of Time Duration in Economic Life

Misako Takayasu emphasizes that during the trading the intensity of time changes - some people become more excited when the market activities proceed with much higher velocity and calm down at another instant. This reminds me of a lecture by Noble Laurent economist Maurice Allais (1988) who conducted the study of the quality theory of money not by changing the parameters like the discount rates or preference but by changing the scale of time from when you encounter an exciting experience to when you have a less exciting experience.

4. Concluding Remarks

Since economists are trained to model the world as if all the participants knew the rules of the game, know the fact environment as much as possible and they were to exploit as much as possible. Accordingly, the market comes close to realize the efficient allocation of resources.

By those strong restrictions economists could develop rather harmonious picture of universe.

Physicists seem to be less inhibited in drawing the picture of the universe. From economists · point of view, physicists neglect the rules that economists have kept as axioms of their discipline. Economists, of course, the unrealistic nature of some axioms and now are trying to relax them. Introducing uncertainty, frictions and the capacity to compute, all of these are the expressions of this kind of attempt. Physicists start from the land of no boundaries, so that their approach may help the revolution in economics in the direction to further reality.

On the other hand, the physicists may neglect some factors with which trained economists are much concerned. Many a priori restrictions economists are accustomed to impose on variables may be helpful to understand the results physicists find. They may help their analysis simpler and more transparent.

Probably, the reality lies between the world where there are many restrictions that are often made by economists and the world where physicists explore witho ut many inhibitions. Experiments and other input from other neighboring sciences will identify where one can obtain useful information on the nature of price formation and the progress of economic activities.

References

Allais, Maurice, 1972, "The Psychological Rate of Interest," Journal of Money, Credit and Banking, August 1974, pp. 285-331.

Allais, Maurice, 1988, "Phénomènes Alétoirs et Modèles Fréquentiels–Réalité et Théorie – Prolégomènes pour une Révision des Théories Admises (Random Phenomena and Frequential Models – Reality and Theory – Prolegomena for a Revision of Current Theories)," Lecture of October 5th, 1988. Theinisch-Westfälische Akademie der Wissenschaften.

Mandelbrot, B. B., 1997, Fractals and Scaling in Finance: Discontinuity, Concentration, Risk, New York, Springer.

Mandelbrot, B. B., 1999, "A Multifractal Walk down Wall Street," Scientific American, v. 280 no. 2 (February 1999), pp. 70-73.

Morris S. and H. S. Shin, 1998, "Unique Equilibrium in a Model ofSelf-Fulfilling Currency Attacks," American Economic Review, June 1998, Vol. 88 Issue 3, pp. 587-598.

Negishi, T., 1972, General Equilibrium Theory and International Trade, Amsterdam, North-Holland Pub.Co.

Samuelson, P. A., 1947, Foundations of Economic Analysis, Cambridge, MA.,

Harvard University Press.
Takayasu, H., H. Miura, T. Hirabayashi and K. Hamada (1992), "Statistical Property of Deterministic Threshold Elements: the Case of Market Price." Physica A, 82(182), June, pp.127-34.

A New Model of Labor Market Dynamics

Masanao Aoki[1] and Hiroshi Yoshikawa[2]

[1] Department of Economics, University of California, Los Angeles
aoki@econ.ucla.edu
[2] Facluty of Economics, University of Tokyo, yoshikawa@e.u-tokyo.ac.jp

Summary. This paper introduces a new type of models of labor market dynamics with heterogenous workers in sectors of economy in analyzing stochastic events of job creations and destructions. Differences in job experiences and human captials are measured by ultrametrics which influence the stochastic processes of posted vacancies being filled.

The phenomena of job creations and destructions are endogenized by using excess demands as proxies for profitabilities of sectors of economy.

Key words: Ultrametrics, Sojourn times, Endogenous job destructions

1 Introduction

There exists a large body of economic literature devoted to labor markets, such as Blanchard and Diamond (1989, 1992), Mortensen (1989), Pissarides (2000) and many others. In addition, empirical works by Davis and Haltiwanger (1992) on job creations and job destruction have been very influential in focussing on these as important determinant of labor market behavior.

The existing models, however, still leave rooms for improvements. In particular, existing models treat workers as identical, although some attempts exist in the literature to include the duration of unemployment. Geographical differences and differences in jobs experiences are stressed in verbal discussions, but are not modeled . In this paper workers in different sectors are different in job experiences and human capitals, and their differences are incorporated into models by using the concept of ultrametrics. See Aoki (1996, Sec.2.5) on ultrametrics. The stochastic process of filling vacancies of sector i by unemployed workers from the pool of sector j depends on the ultrametric distance $d(i,j)$ between the two sectors of the economy, in addition to vacancies and unemployment rates.

In addition to this innovation, this paper brings two other new features or elements to this literature. First, we treat from more basic levels matching processes between vacancies and unemloyed, by means of transition rate specifications in continous-time Markov processes.

We endogenize job destructions and creation differently than Pissarides. Signs of excess demands are used as proxies for profitability signals for sectors which want to change the sizes of labor forces. However, at any time only one sector out of all sectors wishing to adjust the sizes of their labor forces actually can act. The sector that acts is the sector with the shortest holding or sojourn time. To present a simple model we ignore quits and on-the-job searches, and assume that is only unemployed get jobs.

2 The Model

Aoki and Yoshikawa (2001) and Aoki (2002, Sec. 8.6) presented a new model of fluctuations and growth. This paper adds the labor dynamics to that model. To briefly summarize that model, the economy has K sectors, and that each sector employ some finite number n_i, $i = 1, \dots K$ of workers. Sector i produces $c_i n_i$ output, where c_i is exogenously given productivity coefficient, and demand for good i is given by $s_i Y$, with $Y = \sum_{i=1}^{K} c_i n_i$, where s_i is the share of the total output Y, with sum 1. Each sector has the excess demand defined by

$$f_i = s_i Y - c_i n_i. \tag{1}$$

Changes in Y due to changes in any one of the sectors affect the excess demands of all sectors. There is the aggregate externalities among all sectors. We denote the sets of sectors with positive and negative excess demands by $I_+ = \{i : f_i \geq 0\}$, and $I_- = \{i : f_i < 0\}$. Changes in the patterns of s's can affect these sets.

The time evolution of the economy is modeled as a continous-time Markov chain, as described in Aoki (1996, 2002), for example. In this model at each point in time, the sectors of economy is grouped into two subgroups; one composed of sectors with positive excess demands for their products, and the other of sectors with negative excess demands. These two groups are used as proxies for groups of profitable and unprofitable sectors, respectively. All profitable sectors wish to expand their production. All unprofitable sectors wish to contract their production. A novel feature of our model is that only one sector succeeds in adjusting their productions up or down at any given time. The sector that jumps first is the sector with the smallest waiting time or sojourn time, see Lawler (1995) or Aoki (2002, p. 28) for the notion of holding time of a Markov chain.

Since the total output of the economy is endogenous and allocated to each sectors by shares (e.g., by demand policy), each time one sector (firm) suceeds in adjusting its production, total output Y changes in general , and the sets I_+ and I_- also changes.[3]

Each sector is characterized by three variables, n_i, u_i and v_i, $i = 1, \ldots, K$, where n_i is the number of employees, u_i unemployed, and v_i vacancies, respectively of sector i.

We assume that the probability that an unemployed in sector i find a job at sector j is via vacancy filling process which depends on $d(i,j)$. In general the larger the distance $d(i,j)$, the less likely for a vacancy in sector i be filled by an unemmployed of sector j. Sector i, when it decides to fill a vacancy, draws from the pool of unemployed of sector i, which forms a queue, composed of those who are layed off from sector i some time in the past, followed by those pools of other sectors with increasing ultrametric distance. An example is given below. The set of size u_i is at the top of the queue. This is in accordance with the observation that many of the unemployed or layed-off workers return to the sector that have layed them off.[4]

3 Job Creations and Destructions: Case K=2

To simplify our explanation, we discuss the economy with two sectors. In deterministic and perfect information world, the sizes of the sectors with zero excess demands are given by solving (1)

$$n_1^e = \frac{s_1/c_1}{S} n^e, \tag{2}$$

with $S = s_1/c_1 + s_2/c_2$.

The excess demand may be written in terms of the fraction of employed in sector 1, $x = n_1/n$ and its deterministic perfect information equilibrium value, denoted by superscript e as

$$f_1 = s_1 c_2 n(1 - \frac{x}{x^e}).$$

In the two sector case we have $f_2 = -f_1$.

Sector 1 jumps first with probability x. When it jumps there are two cases to consider. With $f_1 > 0$, then sector 1 posts a vacancy sign, hence $v_1^{'} = v_1 + 1$, $n_1^{'} = n_1$, $u_1^{'} = u_1$, where we use the convention to denote the value after a jump by either sector with a prime. In a more simple version $n_1^{'} = n_1 + 1$, and $u_1^{'} = u_1$, if no waiting time is considered. Either way we have $f_1^{'} = f_1$.

[3] This gives rise to endogenous cyclical patterns and growth, when we allow new sectors are being created according to a scheme we describe in Aoki and Yoshikawa and Aoki. In this paper we keep K fixed for simpler explanation of our model.

[4] We can introduce a pool u_0 of long time unemployed, those out of labor market at the bottom of the queue, as Blanchard and Diamond find that this pool is important in affecting the vacancy creation and filling stochastic process. For this model we do not include this pool in the queue.

If $f_1 < 0$ when sector 1 jumps first, then n_1 is reduced by d_1 which is determined by

$$0 = s_1 Y - c_1(n_1 - d_1),$$

or

$$d_1 = \frac{-f_1}{c_1},$$

(actually the largest integer part of the above.) This amount of job destruction is to restore zero excess demand in perfect information world. Alternatively, job destruction is given by d_1/κ, where κ is a parameter to adjust the speed to eliminate redundant jobs only partially. With $\kappa = 1$, the gap is adjusted immediately.

We have $f_1^{'} = f_1 + c_1 d_1$, $n_1^{'} = n_1 - d_1$, $u_1^{'} = u_1 + d_1$, and $v_1^{'} = max(v_1 - d_1, 0)$, where we assume that vacancy signs are also removed by the amount of job destruction.

Sector 2 jumps first with probability $1 - x$. We have similar changes: If $f_2 > 0$, then $v_2^{'} = v_2 + 1$, $n_2^{'} = n_2, u_2^{'} = u_2$, and $f_2^{'} = f_2$. If $f_2 < 0$, then d_2 is the number of job destruction, where

$$0 = s_2 Y - c_2(n_2 - d_2),$$

or

$$d_2 = \frac{-f_2}{c_2}.$$

Alternatively, we have partial adjustment parameter κ to modify d_2 accordingly. We have $f_2^{'} = f_2 + c_2 d_2$, $n_2^{'} = n_2 - d_2$, $u_2^{'} = u_2 + d_2$, and $v_2^{'} = max(v_2 - d_2, 0)$, where we assume that vacancy signs are also removed by the amount of job destruction.

The probability that a vacancy in sector 1 is filled from its own pool of unemployed, given that sector 1 jumps first is

$$\Pr(v_1 \text{is reduced by } 1 | \text{sector 1 jumps}) = \frac{u_1}{U_1},$$

where

$$U_1 = u_1 + \frac{u_2}{1 + d(1,2)}.$$

Similarly v_1 is reduced by one from pool of unemployed of sector 2 with probability

$$\Pr(v_1 \text{is reduced by 1 from sector } 2 | \text{sector 1 jumps}) = \frac{u_2/(1 + d(1,2))}{U_1}.$$

In the case sector 2 jumps first, the probability of filling v_2 is from the queue as above with indicies 1 and 2 interchanged.

Job creation is a slower process because there is no labor hoarding, and therefore, sectors which have posted vacancies are not always able to fill the vacancies, but must wait for random times before they can fill vacancies.

With 2 sectors in economy, given that sector 1 fills its vacancy, one vacancy is filled from the own pool of unemployed with probability $u_1/(u_1 + u_2/3)$, and from u_2 with probability $u_2/3/(u_1 + u_2/3)$.

Dynamics are stated by specifying transition rates, that is, by specifying stochastic events of how vacancies are created, cancelled or filled. When transition rates are linear in state variables, one then obtain constant return to scale matching functions. With state dependent rates to incorporate uncertainties about model properties, for example by means of Gibbs distributions[5], one obtain matching functions with more complex properties. We report on this elsewhere due to lack of space.

4 The Beveridge Curve

Instead of reporting on the approximate solution of the master equation of the proposed model, we describe the Beveridge curve of this model. The number of vacancies in sector 1, v_1, increase by one with probability $p := \Pr(f_1 > 0) = \Pr(x \leq \frac{s_1/c_1}{S})$. This occurs conditional on sector 1 being the jumping sector, the probability of which is n_1/n. Similarly, v_2 increases by one with probability $q := \Pr(f_2 \leq 0) = \Pr(f_1 \geq 0) = 1 - p$, conditional on sector 2 jumping first. Its probability is given by n_2/n. In this case, the vacancies of sector 2 descrease from v_1 to $v_1 - d_1$ with probabbility q, and v_2 decreases by d_2 with probability p.

Vacancy v_1 is reduced by 1 either by u_1 is reduced by 1 with probability u_1/U_1, or by u_2 being reduced by 1 with probability $(u_2/m)/U_1$. The former is the probability that sector 1 hires from the pool of own layed-off workers, and the latter is the probability of hiring one from the pool of layed-off workers associated with sector 2. v_2 is reduced by one through hiring of one worker from its own pool of u_2 workers or from that associated with sector 1, u_1. The former occurs with probability $(u_1/m)/U_2$, the latter with probability u_2/U_2.

Economy-wide unemployment is $u = u_1 + u_2$, and economy-wide vacancies is given by $v = v_1 + v_2$.

We have described above the events of changes in u and v with associated probabilities. Collecting these we calculate the expected values to be

$$E(\Delta) = E(\Delta u_1) + E(\Delta u_2)$$

where

$$E(\Delta u_1) = d_1 q \frac{n-1}{n} - \frac{u_1}{U_1} - \frac{(u_1/m)}{U_2},$$

with similar expression for the Δu_2 by interchanging 1 and 2 and p with q.

We also have

$$E(\Delta v) = \frac{n_1 p + n_2 q}{n} - \frac{p}{m}(\frac{u_1}{U_2} + \frac{u_1/m}{U_1}) - q(\frac{u_2/m}{U_1} + \frac{u_2}{U_2}).$$

[5] See Aoki (1996, p. 140)

We give some benchmark values. If $p = q = 1/2$, that is, if both sectors are equally likely to jump, then, $E(\Delta v) = -1/2$, and

$$E(\Delta u) = \frac{1}{2}(\frac{n_1}{n}d_1 + \frac{n_2}{n}d_2) - 2.$$

Recalling the definition of p and q, $E(\Delta u$ is positive if and only if $(n_1 d_1 + n_2 d_2)/n$ is greater than 4. Therefore if $d_1 = d_2$, then this common value greater than 4 would make this expression positive, for example. In other words, we have the usual downward sloping Beveridge curve in the $u - v$ plane. There are other possibilities, of course.

5 Concluding Remarks

The master equation of this model is complicated and is not amenable for a close form solution. We will report elsewhere on an approximate soluation via Talylor series expansion, much as is done in Aoki (1996, chapt. 5), or in Aoki and Shirai (2000), where connection with the matching function approach will be discussed.

6 References

Aoki, M. (2002) *Modeling Aggregate Behavior and Fluctuations in Economics* , Cambridge Univeristy Press, New York.

—-, and Y. Shirai.(2000) " A new look at the Diamond search model: Stochstic cycles and equilibrium selection in search model", *Macroeconomic Dynamics* **4**, 487–505.

—-, and H. Yoshikawa. (2001) " A new model of growth and fluctuatons" Working paper series, center for international research on Japanese economy, CIRJE-F-136, Oct.

Blanchard, O. J. and P. Diamond. (1989) "The Beveridge Curve," Brookings Papers on Economic Activities, Issue 1,

—-, and ——-. (1992)," The flow approach to labor markets," *Amer. Econ. Rev.* **82**, 354–59.

Davis, S.'J., and J. C. Haltiwanger (1992), "Gross job creations, gross job destructions, and employment reallocation," *Qurt. J. Econ.* **107**, 819-63.

Mortensen, D.,(1989) "The persisitence and indeterminacy of unemployment in search equilibrium", *Scand. J. Econ.* **91**, 347–60.

Pissarides, C. (2000) *Equilibrium Unemployment Theory* , Blackwell, Oxford

Formulating Social Interactions in Utility Theory of Economics*

Yuji Aruka

Facutly of Commerce, Chuo University, 742-1 Higashinakano, Hachioji, Japan 192-0393 aruka@tamacc.chuo-u.ac.jp

1 Towards a New Theory for Social Interactions

1.1 Obsolescence of the Old Story in Economics

Individualistic utility theory has long been the core of economics since its appearance in the end of the 19C. This theory explicitly presumes that agents in the societies are *homogeneous* in a sense of all fulfilling a certain set of similar rational preference postulates. Another conspicuous feature of the story is to assume that every individual is based on a universally given *innate* preference as never been affected by outside environments or random shocks. Into individual's decision the cost for social interaction has never been taken account. Recently, some new innovations to overcome these limitations in utility theory are coming up to renew our old story entirely. These have some different springs. One of these comes from the idea of *random preference* in economics, which could be traced back to Hildenbrand [9]. Even if agents were homogeneous, random shocks could generate a fluctuation in a macroscopic structure of social states.

The old theory of utility in economics should be replaced with a new kind of statistical mechanics. In general, our new stories prefer to argue interaction of heterogeneous agents in the heart of theory. Our main concern here is more fundamental. We favor to criticize the foundation of utility function and free market assumptions. As Arthur [1] and others mentioned, "[e]conomics agents may be influenced by neighbouring agents' choices."

1.1 A justification to create a new theory of utility comes from what it is evident that there are many subgroups in our society, namely, we cannot regard the society as an aggregate of homogeneous agents..

* The first draft of this paper was presented in ICM2002GTA, August 14–17, 2002, Qingdao, China.

1.2 The free market assumptions should be re-examined in view of the facts that a random selection of firms as supported by an active deregulation policy can often cause a or a few firms to BE the *winner-take-most.* [2]

As for a kind of effects due to neighbouring agents in regional science, we have a simple model of local interactions like in Glaser and Scheinkman [8].[3] Arthur [1] also noted that "[s]patial mechanisms have parallels with Ising models and renormalization theory in physics and with voter models in probability theory .... "(p.129) In this paper we will only focus on conformity effects in social interaction of agents in view of the Ising Model.

1.2 Conformity Effects to Form a Cooperative System

Ising Model

We suggest an elementary model to argue an occurrence of conformity effects on agent behaviors. Durlauf [4], [5] essentially underlies this idea. Hildenbrand [9] and Föllmer [6] already suggested that Ising model in statistical mechanics was promising to deal with social interaction. In the system of quite large number of agents, statistical mechanics can work well whether particles or human agents. It is well known that entropy will be maximized as any distribution approaches to a uniform distribution. Entropy of the uniform distribution will then be the largest:

$$H[X_u] = -\int_u p_u(x) \log \frac{1}{u} dx = \log u, \tag{1}$$

where $p(x)$ is a density function.[4] By this fact, subjective probability can uniformly be attached to unobservable states to maximize entropy under a macroscopically given ***a priori*** information.

Conformity Effects

We illustrate the Ising model to depict phase transitions of magnetic material. x_i indicates a state of an element $S_i (i = 1, \cdots, N)$. It is usually supposed that each state will be represented in either

$$x_i(t) = 1 \text{ or } -1. \tag{2}$$

---

[2] "The result of deregulation over the long term has been a steady decline in large carriers, from 15 in 1981 to around 6 at present [in 1996]. Some routes have become virtual monopolies, with resulting higher fares. None of this was intended. But it should have been predicted-given increasing returns."(Arthur [2] Section: What about Service Industries?) In this context, Arthur rather suggests the use of "Polya urn process" of generalized form given by Arthur WB, Ermoliev Y, Kaniovski, Y (1983) in Arthur [1] Chapter 10. This paper does not refer to this part.

[3] As for how to measure social interactions via group selection, see Aruka [3].

[4] $p(x) = 0$ outside the bounded region u.

The ∂i indicates a set of the nearest neighbors around i. If interactions on a network configuration $(x_1, \cdots, x_N)$ only occur between nearest neighbors, local potential at time t is defined to be:

$$f_i(x_i(t)) = J \sum_{i=1}^{N} f_i(x_i, x_{\partial i})(x_i(t) - x_k(t))^2 - \mu H x_i(t). \tag{3}$$

Here the second term shows the strength of the field, the effects of the whole environment by which all the elements uniformly are to be influenced. Thus the global energy at time t may be defined to be:[5]

$$\begin{aligned} f(x(t)) &= \sum_{i=1}^{N} f_i(x_i(t), x_{\partial i}(t)) \\ &= J \sum_{i=1}^{N} \sum_{k \in \partial i} f_i(x_i, x_{\partial i})(x_i(t) - x_k(t))^2 - \mu H \sum_i x_i(t) \\ &\cong - \sum_{i=1}^{N} \{ \sum_{k \in \partial i} J x_i x_k + H x_i \}. \end{aligned} \tag{4a}$$

This tends to be minimized. It is also noted that every element *cooperates* mutually if $J > 0$, while *competes* if $J < 0$. If the number of element N is too large, it holds in view of (1) a Gibbs distribution

$$\frac{1}{Z_N} e^{-\frac{f(x)}{\theta}} \tag{5}$$

to represent some average properties under the assumption of subjective probability. J represents conformity effects. In our context, it is very important to know how the conformity effects operate in grouping of human actions. Equally important how diverse the conformity can be designed.

2 Durlauf's Random Preferences Influenced by Social Interaction

2.1 Individualistic Utility

Durlauf [4] in principle never part with the *methodological individualism* in orthodox economics but depends essentially on the above formulation of interactive network of agents when he seriously is concerned with the effects of *heterogeneous subgroups* on individualistic choice. His case may be regarded as an application of the above minimization problem in view of social costs. In particular, he utilized the *logistic distribution* as a specific form of *random*

[5] The binary values (2) are used to derive the last-hand side.

preferences to derive a probability distribution of choice as resulting some similar idea of Gibbs distribution.

In Durlauf's idea the utility V of agent has been construed as a union of three different types of utility for individual $i \in I$: private utility ω as private and deterministic utility; social interaction S as social and deterministic utility; random shock ϵ as private and stochastic utility.

$$V_i = \omega_i + S + \epsilon_i. \tag{6}$$

2.2 Private Utility

Binary Choice. We indicate the outcome of binary choice whether an individual $i \in I$ can be a painter or a musician by ω_i, and its support by $\Omega_i = \{-1, 1\}$. $\omega_i = -1$ means painter, $\omega_i = 1$ a musician, for example.

Heterogeneous Agent. Agents playing in the system are heterogeneous, whose property is expressed as X_i.

Random Shocks. We assume independent stochastic disturbances among heterogeneous agents. In the case of $\omega_i = 1$ which suggests to be a musician, if utility of random shock $\epsilon(1)$ is *relevant*, utility of choice for i to be a musician will increase. $\epsilon(1)$ exhibits unobservable musical talent, while $\epsilon(-1)$ unobservable artistic talent.

There may be several possibilities for each individual. A particular one will be realized by a probabilistic distribution of random shocks. We for convenience assume a *logistic distribution* of the probability which the difference of the random shocks on a binary choice $\epsilon_i(-1) - \epsilon_i(1)$ can be within a particular z.[6]

$$\text{Prob}(\epsilon(-1) - \epsilon(1) \leq z) = \frac{1}{1 + e^{-\beta_i z}}. \tag{7}$$

The density function of logistic distribution is symmetrically bell-shaped. β_i indexes *unobservable heterogeneity*. The larger β is, the lower a probability which the difference of utility falls within a certain value of z may be.

2.3 Social Utility

Agent i has his own belief on choices of the other members of the group I based on a certain information: $\mu_i^e(\omega_{-i}|F_i)$.[7] Individuals are inclined to conform to or deviate from others' choices of the group. When all the members have the same choice, social utility for individual is of zero. When each deviates from each other, social utility for agent i is negative.

$$S(\omega_i, X_i, \mu_i^e) = -E_i\left(\sum_{j \neq i} \frac{J_{i,j}}{2}(\omega_i - \omega_j)^2\right). \tag{8}$$

[6] Here $\beta(X_i) \geq 0$.

[7] The suffix $-i$ shows the remaining members except for i.

Weight on interaction $J_j(X_i)$ simply is denoted by $J_{i,j}$, as relating the choice for i to one for j. Thus the square $(\omega_i - \omega_j)^2$ will give *conformity effects.* Hence social utility S is just a subjectively expected value as based on agent i's belief of the distribution of social interaction $[J_j(X_i)]$.

2.4 Derivation of the Probability on Individual Choice

The decision process of individual i is as follows:

$$\omega_i = \arg \max_{\omega \in \{-1,1\}} V(\omega, X_i, \mu_i^e, \epsilon_i). \tag{9}$$

Linearization of Utility Function

A linearized function of utility is given where a slope is h_i, an intercept k_i:[8]

$$u(\omega_i, X_i) = h_i\omega_i + k_i. \tag{10}$$

Under a certain set of information, a choice ω_i is generated only if the gain of the choice is greater than the opposite. A conditional choice $\mu(\omega_i)$ is the probability when the gain of ω_i is greater than the gain of $-\omega_i$.

$$\mu(\omega_i|X_i, \mu_i^e) = \mu(V(\omega_i, \epsilon_i(\omega_i)) > V(-\omega_i, \epsilon_i(-\omega_i)). \tag{11}$$

Taking account of the above (6) and (8) into (11), it follows:

$$\begin{aligned} &\mu(h_i\omega_i - E_i \sum_{j\neq i} \frac{J_{i,j}}{2}(\omega_i - \omega_j)^2 + \epsilon_i(\omega_i) \\ &> -h_i\omega_i - E_i \sum_{j\neq i} \frac{J_{i,j}}{2}(-\omega_i - \omega_j)^2 + \epsilon_i(-\omega_i)) \\ &= \mu(h_i\omega_i + \sum_{j\neq i} J_{i,j}\omega_i E_i(\omega_i) + \epsilon(\omega_i) > -h_i\omega_i - \sum_{j\neq i} J_{i,j}\omega_i E_i(\omega_j) + \epsilon_i(-\omega_i)) \\ &= \mu(\epsilon_i(\omega_i) - \epsilon_i(-\omega_i) > -2h_i\omega_i - \sum_{j\neq i} 2J_{i,j}\omega_i E_i(\omega_j)). \end{aligned} \tag{12}$$

Since the random shocks are assumed to be subject to the *logistic distribution* (11), the probability distribution on the choice of individual i, i.e., ω_i can be solved:

$$\mu(\omega_i|X_i, \mu_i^e) \propto e^{\beta_i h_i \omega_i + \sum_{j\neq i} \beta_i J_{i,j} \omega_i E_i(\omega_j)}. \tag{13}$$

The joint probability distribution of the whole population on the choice of ω then is:

$$\mu(\omega|X_1, \ldots, X_n, \mu_1^e, \ldots, \mu_n^e) = \Pi_i \mu(\omega_i|X_i, \mu_i^e) \propto \Pi_i e^{\beta_i h_i \omega_i + \sum_{j\neq i} \beta_i J_{i,j} \omega_i E_i(\omega_j)}. \tag{14}$$

---

[8] Choose such h_i and k_i that $h_i + k_i = u(1, X_i)$ and $-h_i + k_i = u(-1, X_i)$.

3 Helbing's Alternative Probabilistic Decision Theory

3.1 The Law of Relative Effect

In this section, we shall abandon the supposition of *methodological individualism* in economics. We rather move to a new setting as Helbing [10] cleverly designed.

Helbing's Assumptions

3.1 Decision making will require a finite time to reach a *final* judgment among a series of imagined consequences.
3.2 New agents will always arrive at or immigrate into a system, given an open society.

3.2 Decision Process of Social Interactions

Recently, Helbing [10](pp.132–137) skillfully in this context was successful to define an analytical form of decision of social interaction. Suppose there are behavioral alternatives:

$$x_1, \cdots, x_n, \cdots, x_s$$

We have *imagined consequences*, i.e., a sequence of consequence anticipation, of the possible decisions:

$$i_0, i_1, \cdots, i_n, \cdots$$

When i_n $f$avors x_n, we describe

$$f(i_n) = x_n.$$

We call *decision* on x has been made after k successive consequence anticipation on the same x, if

$$\left.\begin{array}{r} i_{n+1} \rightarrow x \\ \vdots \\ i_{n+k} \rightarrow x \end{array}\right\}$$

3.3 A Psychologically Complex Value

Helbing has an insight that utility has a past history of decision sequence and also cannot be independent from it. His utility function should rather be called *a psychologically complex value*. Consequently, utility will fluctuate like in human mind. Then he defines *utility* or *preference* of decision on alternative x for agent α who has finally made decision y:

$$U_\alpha(x \mid y; t).$$

The probability for occurrence of i, due to $f(i) = x$, will be

$$p_\alpha(x \mid y; t).$$

These considerations may rather be concisely summarized by Helbing [10], [11] when he argued transition rates of *master equation* for social process. Helbing assumes a kind of such *learning* that agent employs his strategy i as a function of successful times $N(i)$. [9] Underlying the Helbing assumptions, even private utility can never be independent of *the field of choice.* His definition of utility at time t when strategy i is employed, i.e., $U(i,t)$, has two parts of *personal preferences* $S(i,t)$ and *social reinforcement* $R(i,t)$. In particular, personal satisfaction $S(i',t)$ on decision i' always are supposed to be *relative*, i.e., *conditional* as a selection *after* a decision i has been employed. Two different ways may be defined:

A symmetric part:

$$S(i'|i,t) = \frac{1}{2}\{U(i'|i,t) + U(i|i',t)\} = S(i|i',t). \tag{15a}$$

An asymmetric part:

$$A(i'|i,t) = \frac{1}{2}\{U(i'|i,t) - U(i|i',t)\} = -A(i|i',t). \tag{15b}$$

$S(i'|i,t)$ can be interpreted as *transaction costs* for a transition of strategy. Helbing also deals with a representative utility of members within a subgroup a of homogeneous agents with almost the same θ_a, U_a.

3.4 A Multinomial Logit Model of Utility

Helbing [11] declares to use the *law of relative effects* in view of entropy maximization.

$$p_a(i,t) = \frac{N_a(i)}{\sum_{i'} N_a(i')}. \tag{16}$$

Here a denotes a subgroup of homogeneous agents. i, i' are strategies utilized by a. $p_a(i,t)$ therefore means a probability at time t by which a subgroup a employs strategy i. Helbing then, for some parameter θ, i.e., *social temperature*, introduced *utility function* $U(i)$ of form:

$$U_a(i,t) = \theta \log N(i) \tag{17a}$$

$$\text{such that} \quad p_a(i,t) = \frac{e^{U_a(i,t)/\theta_a(t)}}{\sum_{i'} e^{U_a(i',t)/\theta_a(t)}}. \tag{17b}$$

The (17a) is called the *multinomial logit model.* In view of the classical derivation of the multinomial model, θ is regarded as a measure for the uncertainty

[9] The similar approach may be guaranteed by *a two-armed bandit problem* which means a choice of gambler facing two slot machines. In this problem only a finite series of trials is admissible for him. See Holland [12] Chapter 5.

of information. Helbing thus defined the utility at time t of subgroup a, i.e., $V_a(i,t)$ as follows:

$$V_a(i,t) = U_a(i,t) + \epsilon_a(i,t). \tag{18}$$

Here ϵ represents a stochastic part.

3.5 The Effects of Interaction between Subgroups

Taking account of *symmetrical/asymmetrical relations* of personal preferences S_a and A_a, the conditional or transition probability will be

$$p_a(i'|i,t) = \frac{\frac{e^{A_a(i'|i,t)/\theta_a(t)}}{e^{-S_a(i'|i,t)/\theta_a(t)}}}{\sum_{i"} \frac{e^{A_a(i"|i,t)/\theta_a(t)}}{e^{-S_a(i"|i,t)/\theta_a(t)}}}. \tag{19}$$

According to Helbing, we can formulate a practical form of p_a for empirical analysis by reducing the number of parameters, for example, t.

References

1. Arthur WB (1994) Increasing returns and path dependence in the economy. University of Michigan Press, Ann Arbor
2. Arthur WB (1996) Increasing returns and the new world of business. Harvard Business Review 74(July-Aug):100–109
3. Aruka Y (2003) How to measure social interactions via group selection? Journal of Economic Behavior, special issue, forthcoming
4. Durlauf SN (1997) Statistical mechanics approaches to socioeconomic behavior. In Arthur WB, Durlauf SN, Lane, DA(eds) The economy as an evolving complex system II, Addison-Wesley, Mass., 81–104
5. Durlauf SN (2000) A framework for the study of individual behavior and social interaction. SSRI Working Paper Series(Social Systems Research Institute of the Economic Department), the University of Wisconsin, Madison
6. Föllmer H (1974) Random economies with many interacting agents. Journal of Mathematical Economics 1:51–62
7. Fukao, T (1987) The disperse system: the thermodynamic system. Shokodo, tokyo(in Japanese)
8. Glaeser EL, Scheinkman JA (2001) Measuring social interactions. In Durlauf SN, Young, HP(eds) Social dynamics, Brooking Institute, Washington DC, 83–131
9. Hildenbrand W (1971) Random preferences and equilibrium analysis. Jounral of Economic Theory 3:414–429
10. Helbing D (1995) Quantitative sociodynamics: stochastic methods and models of social interaction processes. Kluwer Academic Press, Dordrecht.
11. Helbing D (2003) Dynamic decision behavior and optimal guidance through information services: models and experiments. In Schreckenberg M, Selten R(eds.) Human behaviour and traffic networks. Springer, Berlin (in print
12. Holland J (1992) Adaptation in natural and artificial systems, MIT press, Mass: Cambridge.

Collective Behaviour and Diversity in Economic Communities: Some Insights from an Evolutionary Game

Vivek S. Borkar[1], Sanjay Jain[2], and Govindan Rangarajan[3]

[1] School of Technology and Computer Science, Tata Institute of Fundamental Research, Mumbai 400 005, India borkar@tifr.res.in
[2] Centre for Theoretical Studies, Indian Institute of Science, Bangalore 560 012, India jain@cts.iisc.ernet.in
[3] Department of Mathematics and Centre for Theoretical Studies, Indian Institute of Science, Bangalore 560 012, India rangaraj@math.iisc.ernet.in

Summary. Many complex adaptive systems contain a large diversity of specialized components. The specialization at the level of the microscopic degrees of freedom, and diversity at the level of the system as a whole are phenomena that appear during the course of evolution of the system. We present a mathematical model to describe these evolutionary phenomena in economic communities. The model is a generalization of the replicator equation. The economic motivation for the model and its relationship with some other game theoretic models applied to ecology and sociobiology is discussed. Some results about the attractors of this dynamical system are described. We argue that while the microscopic variables – the agents comprising the community – act locally and independently, time evolution produces a collective behaviour in the system characterized by individual specialization of the agents as well as global diversity in the community. This occurs for generic values of the parameters and initial conditions provided the community is sufficiently large, and can be viewed as a kind of self-organization in the system. The context dependence of acceptable innovations in the community appears naturally in this framework.

Key words: Generalized replicator dynamics, Diversity, Specialization

1 Introduction

Several complex adaptive systems in the course of their evolution exhibit the phenomenon that the individual components comprising the system evolve to perform highly specialized tasks whereas the system as a whole evolves towards greater diversity in terms of the kinds of components it contains or the tasks that are performed in it. Here are some examples:

1. Living systems are made of cells, which in turn are made of molecules. Among the various types of molecules are the proteins. Each type of protein molecule has evolved to perform a very specific task, e.g., catalyse a specific reaction in the cell. At the same time, during the course of evolution, diverse kinds of protein molecules have appeared – the range of specialized tasks being performed by protein molecules has increased.
2. In an ecology, species with highly specialized traits appear (e.g., butterflies with a specific pattern of spots on their wings). Simultaneously, the ecology evolves to support a diverse variety of specialized species.
3. Many early human societies (such as hunter-gatherer societies) were perhaps characterized by the fact that there were relatively few chores (e.g., hunting, gathering, defending, raising shelter) to be performed, and everyone in the community performed almost all the chores. These societies evolved to have specialist hunters, tool makers, farmers, carpenters, etc. Individuals specialized, and simultaneously a diverse set of specialists appeared.
4. In an economic web, we find firms exploring and occupying increasingly specialized niches, while the web as a whole supports an increasingly diverse set of specialists.

In the examples above the systems and their underlying dynamics are quite different. But they all share the twin evolutionary phenomena of individual specialization and global diversification. In all these systems, the nonlinear interaction among the components seems to play a crucial role in the manifestation of this type of behaviour. For example, in an ecology, the development of highly specialized traits in a species is a result of its interaction with and feedback from other species. In an economic community, each agent's choices depend upon feedback from exchanges (of goods, money, etc.) with other agents. Moreover, there is no purposeful global organizing agency which directs the behaviour of individual components and ordains them to be specialists. The phenomenon happens 'spontaneously', arising from the local moves made at the level of individual components. Similarly diversity also arises as individuals capitalize on innovations – mutations, technological innovations, etc. – which suit them in the existing context.

In this article, we describe a mathematical model which seems to exhibit the above twin evolutionary phenomena. The (idealized) behaviour of agents in economic communities provides the basic motivation of the model. The model consists of a set of coupled nonlinear ordinary differential equations describing the time evolution of the activities of individual agents.

2 The model

The system is a community of N agents labeled by the index $\alpha = 1, 2, \dots, N$. Each agent can perform s strategies or activities labelled by $i \in S =$

$\{1, 2, \ldots, s\}$. At time t, agent α performs strategy i with a probability $p_i^\alpha(t)$, $\sum_{i=1}^s p_i^\alpha(t) = 1$. The vector $\mathbf{p}^\alpha(t) = (p_1^\alpha(t), p_2^\alpha(t), \ldots, p_s^\alpha(t))$ is the mixed strategy profile of agent α at time t. In particular, if $p_i^\alpha(t) = \delta_{ij}$, then the agent α is said to pursue the pure strategy j or to have specialized in strategy j.

The vectors $\mathbf{p}^\alpha(t)$ constitute the basic dynamical variables of the model. The equation governing their evolution is taken to be

$$\dot{p}_i^\alpha(t) = p_i^\alpha(t)[\sum_{\beta \neq \alpha} \sum_{j=1}^{s} a_{ij} p_j^\beta(t) - \sum_{\beta \neq \alpha} \sum_{i,j=1}^{s} p_i^\alpha(t) a_{ij} p_j^\beta(t)]. \tag{1}$$

Here a_{ij} denotes the ijth element of the s-dimensional payoff matrix A.

Consider

$$J = \{\mathbf{x} = (x_1, x_2, \ldots, x_s) \in R^s | \sum_{i=1}^{s} x_i = 1, x_i \geq 0\},$$

which is the simplex of s-dimensional probability vectors. J is the full configuration space of PRD dynamics and is invariant under it. The configuration space for GRD is J^N, the N-fold product. A generic point of J^N is $p = (\mathbf{p}^1, \mathbf{p}^2, \ldots, \mathbf{p}^N)$, each $\mathbf{p}^\alpha$ being an s-dimensional probability vector $\mathbf{p}^\alpha(t) = (p_1^\alpha(t), p_2^\alpha(t), \ldots, p_s^\alpha(t))$ belonging to $J^{(\alpha)}$ (the latter being a copy of the simplex J corresponding to agent α).

A point $\mathbf{x} \in J$ such that $x_i = \delta_{ij}$ for some j is called the jth corner of J. If an agent α has specialized to the pure strategy j, then $p_i^\alpha = \delta_{ij}$, i.e., $\mathbf{p}^\alpha$ has gone to the jth corner of $J^{(\alpha)}$. If every agent has specialized to some strategy or the other, the corresponding point in J^N will be called a *corner* of J^N and we say that the community is *fully specialized.* Note that every corner of J^N is an equilibrium point of GRD, since the r.h.s. of (1) vanishes. Hence we refer to corners as corner equilibrium points (CEPs).

A CEP can be characterized by an s-vector of non-negative integers $\mathbf{n} = (n_1, n_2, \ldots, n_s)$ where n_i denotes the number of agents pursuing the pure strategy i at CEP. There can be many CEPs with the same $\mathbf{n}$ vector. These would differ only in the identity of the agents at various corners. In this article we will ignore the differences between such corners and characterize a CEP by its $\mathbf{n}$-vector alone, since the agents are identical and differ only in their strategy profile. As long as the system is not in some 'face' F_k (where all agents have opted out of strategy k), we say that the community exhibits the *"full diversity"* of strategies.

If a CEP is such that $n_i \neq 0$ for all i, i.e., each strategy is played by at least one agent at the CEP, we will refer to it as a *fully diversified* CEP or FDCEP. If one or more n_i is zero, the full diversity of strategies is lost and such CEPs are called non-FDCEPs.

We are interested in studying the circumstances under which FDCEPs are the preferred attractors of the dynamics, for then, individual specialization and global diversity will arise dynamically in the community. If it happens that the FDCEPs are attractors and their basins of attraction cover most of J^N (all of J^N except a set of lower dimension), then for generic initial conditions the community is bound to end up in an FDCEP, which means that it will exhibit individual specialization and as well as global diversity.

3 Results

In this section we discuss some results concerning attractors of GRD. The proofs of the theorems are omitted here; these and further results can be found in Borkar, Jain and Rangarajan (1998 a,b). We will discuss the significance of these results for specialization and diversity in GRD.

3.1 Interior Equilibrium Points

An equilibrium point of GRD is called an interior equilibrium point (IEP) if none of the p_i^α is zero. We have the following theorem:

Theorem 1: There is at most one isolated IEP. If there is one, it is homogeneous and is given by $p_i^\alpha = x_i \;\; \forall \;\; \alpha, i$ where $x_i \equiv u_i/\det B$, u_i is the cofactor of B_{0i}, and B is the $(s+1)\times(s+1)$ matrix (whose rows and columns are labelled by the indices $0, 1, 2, \ldots, s$)

$$B \equiv \begin{pmatrix} 0 & 1\ 1 \cdots 1 \\ -1 & \\ -1 & A \\ \vdots & \\ -1 & \end{pmatrix}. \tag{2}$$

A necessary and sufficient condition for an isolated IEP to exist is given by

A1: $u_i \neq 0 \;\; \forall \;\; i$, and all u_i have the same sign.

PRD also has an isolated IEP if condition **A1** holds, which is then unique and given by the same x_i as given above for GRD. Further, note that at the IEP in GRD, the system exhibits full diversity since no strategy is opted out of by any agent. However, there is no specialization.

3.2 Specialization

For generic payoff matrices A, generic initial conditions, and sufficiently large N, we find that the system flows into a corner of J^N. Thus specialization is a generic outcome of the dynamics. This observation is based on the following facts:

1. **Theorem 2:** (Hofbauer and Sigmund 1988, Weibull 1995) Any compact set in the interior of J^N or the relative interior of any face cannot be asymptotically stable. An equilibrium point is asymptotically stable if and only if it is a strict Nash equilibrium.
2. **Theorem 3:** Every asymptotically stable attractor must contain at least one corner equilibrium.
3. **Numerical Work:** The GRD equation for $s = 3$ was numerically integrated using Runge-Kutta method of fourth order. We randomly generated ten 3×3 payoff matrices and numerically integrated the GRD equations for long times for each payoff matrix with ten randomly chosen initial conditions. When this was done with $N = 5$, in 90 out of the 100 cases the dynamics converged to a corner. The remaining 10 cases (all corresponding to a single payoff matrix) converged to a heteroclinic cycle. When N was increased to 10 for the same ten payoff matrices studied above, all 100 cases converged to a corner. This suggests that the typically, the stable attractors are corners or heteroclinic cycles, with corners becoming overwhelmingly more likely at larger N.

The above theorems imply that the IEP is always unstable in GRD.

3.3 Collective behaviour

We have seen above that the GRD flows to corners generically. The next step is to determine *which* corners the dynamics flows to. For the moment we restrict ourselves to FDCEPs. Characterizing an FDCEP by its $\mathbf{n}$ vector (described in the previous section), we have the following result:

Theorem 4: Let $\mathbf{n}$ and $\mathbf{n}'$ be any two asymptotically stable FDCEPs with $N \geq s$. If condition **A1** holds then all components of $\mathbf{n} - \mathbf{n}'$ are bounded by a function of the payoff matrix A alone (and not of N). Further,

$$\lim_{N \to \infty} \frac{n_i}{N} = x_i.$$

Thus, out of a large number (of order N^{s-1}) of FDCEPs all of which are equilibria for GRD, only a few are stable. Further, even though the agents are acting individually and selfishly and going to corners (specializing), the system as a whole retains a memory of the unique interior equilibrium point (which is guaranteed to exist under the conditions of the above theorem) and tunes the ratios n_i/N such that they are close to the IEP x_i values.

One can give a physical or "economic" interpretation of this collective behaviour. A stable equilibrium is by theorem 2 a strict Nash equilibrium. Thus it cannot be advantageous for any agent to switch her pure strategy unilaterally. This means that all agents must receive more or less the same payoff. More precisely, since a switch of strategy by a single agent causes changes of $O(1)$ in the payoffs to other agents, at a strict Nash equilibrium it must be the case that differences of payoffs among agents could not be larger than $O(1)$, since otherwise it would be possible for some agent to make an advantageous switch without affecting others. Thus stability is achieved only at those equilibria at which differences in payoff among agents are a very small fraction of the total payoff to any agent, which is $O(N)$. This requirement of "near-equality" of payoffs narrows down the set of stable equilibria considerably. As to why the ratio n_i/N gets tuned to be close to the IEP, we remark that in both PRD and GRD, the IEP is characterized by exactly equal payoffs to all strategies.

The above remarks also help explain some of our numerical results. When we increased the number of agents from 5 to 10, we found that all cases converged to corners. This is because as N increases, the ratios n_i/N can reproduce the x_i values corresponding to IEP more accurately and thereby achieve the near-equality of payoffs required for the existence of a strict Nash equilibrium.

3.4 Diversity and self-organization

We have seen above that among the FDCEPs, only a very small subset can be asymptotically stable. Now we consider non-FDCEPs. It could happen that along with an FDCEP, some non-FDCEPs are also stable. In that case, if the community starts in the basin of attraction of a non-FDCEP, it would eventually lose its diversity. We would like to eliminate such attractors of the dynamics. It turns out that this can be achieved by imposing certain inequalities on the payoff matrix elements. Consider the following condition:

A2: The payoff matrix is diagonally subdominant, i.e., $a_{ii} < a_{ji} \forall j \neq i$.

Theorem 5: For $s = 2$, if **A2** holds, then all non-FDCEPs are unstable for $N \geq 2$. For $s = 3$, if **A1**, **A2** hold, then there exists a positive number N_0 depending on A, such that all non-FDCEPs are unstable for $N > N_0$.

The condition **A2** means that each pure strategy gives more payoff to other strategies than itself, clearly a tendency that would support diversity. It is interesting that the model also has the desirable feature that larger communities favour diversity.

It will be useful to have conditions for higher s also which make all the non-FDCEPs unstable. Partial results in this direction are contained in Borkar, Jain and Rangarajan (1998b). We expect that for higher s the condition of sufficiently large N and further inequalities on the payoff matrix elements would ensure the instability of non-FDCEP.

We now discuss the behaviour of the system when such conditions hold. Notice that since these conditions are inequalities on the payoff matrix elements (and not equalities), the behaviour of the system is structurally stable or generic, i.e., is not destroyed by a small perturbation of the parameters. From the evidence presented in section 5.2, the system is expected to go to a corner with generic initial conditions. By theorem 5 (and its generalizations to higher s), this corner cannot be a non-FDCEP. Hence it must be an FDCEP. But then theorem 4 applies and tells us that it must be a very specific corner.

At this corner the number of agents n_i pursuing the pure strategy i is fine tuned to a value close to Nx_i with x_i determined by the payoff matrix via Theorem 1, and all agents receive the same $O(N)$ payoff upto differences of $O(1)$. The final state is fine-tuned but robust in that it arises without fine-tuning the parameters or the initial state. In this sense (of spontaneous dynamical fine-tuning) the system exhibits self-organization, albeit without any obvious critical behaviour. Furthermore, this self organization is of the kind that we were originally seeking, namely, in which the community exhibits individual specialization and global diversity.

3.5 Innovations

So far we have considered strategy spaces of a fixed size s. However, the growth of diversity in the systems mentioned in the introduction has to do with the appearance of new strategies and disappearance of some old strategies. We now discuss how the above considerations of the instability of non-FDCEPs are relevant for the generation of innovations. As an example consider a community of N agents which has initially only two strategies and a payoff matrix A satisfying condition **A2**. Then all non-FDCEPs are unstable and only those FDCEPs with $\mathbf{n} = (n_1, n_2)$ such that the ratios n_i/N close to x_i of the IEP are stable. Let us assume that the system has settled into such a state. Now, assume that a new, third strategy arises which enlarges the 2×2 matrix A into a 3×3 matrix A' that contains A as a submatrix. The third row and column of A' represent the relationship of the new strategy with respect to the old – how much payoff it gives to them and receives from them. At the time this strategy arises, the system is in the state $(n_1, n_2, 0)$ since the third strategy (being new) is as yet unpopulated. Note that this state is a non-FDCEP for $s = 3$. Now if A' is such that it satisfies the conditions **A1, A2** in Theorem 5, and N is sufficiently large, then the above state is unstable. Therefore, any small perturbation in which the agents start exploring the new strategy ever so slightly will destroy the old state and take the system to a new stable state which must be an FDCEP with 3 strategies. By Theorem 4 this state will be (n'_1, n'_2, n'_3) with

$$\frac{1}{N}(n'_1, n'_2, n'_3) \approx (x'_1, x'_2, x'_3)$$

where x'_i are the components of the IEP corrsponding to the payoff matrix A' and are all non-zero since A' satisfies **A1**. Thus the innovation has destabilized the previous state of the system and brought it to a new state where a finite fraction of the population has adopted strategy 3. In such a case, we say that the innovation has been "accepted" by the community. Note that the only requirements for this to happen is that the elements of the new row and column in the payoff matrix satisfy certain inequalities with respect to the existing matrix elements (contained in the conditions **A1, A2** to be satified by A') and that the community be sufficiently large. This tells us what properties a new strategy should have *in the context of already existing activities* in order for it to be "accepted" by the community. Thus the model suggests a natural mechanism for the emergence of context dependent innovations in the community.

4 Conclusions

To summarize, Generalized Replicator Dynamics, eq. (1), is a nonlinear dynamical model of learning for a community of N mutually interacting agents with the following features:

1. Each agent is selfish and exhibits bounded rationality.
2. This is a non-cooperative game and there is no global organizing agency at work. It is in general a non-hamiltonian system.
3. Specialization of individual agents to pure strategies is a generic outcome of the dynamics.
4. Under certain generic conditions on the payoff matrix parameters the agents exhibit a collective behaviour, and for sufficiently large N, the community exhibits diversity and self-organization.
5. A 'good' innovation (one that satisfies conditions **A1, A2**, etc., with respect to the exisiting strategies) makes the society unstable and evolve until the innovation is accepted.

It is noteworthy that in this dynamical system, order is generated at large N (unlike the systems where increasing the number of degrees of freedom makes the system less orderly, in some sense). This order is not the usual statistical mechanical kind of order, the order of appropriately defined macroscopic variables, but an order in the original dynamical variables themselves. However, in another sense, this order is also statistical since we do not know which pure strategy an individual agent follows. We only know about the fraction of agents pursuing a given strategy.

Acknowledgements

SJ and GR are also associated with the Jawaharlal Nehru Centre for Advanced Scientific Research, Bangalore. This work was supported by a grant from Department of Science and Technology, India.

References

1. Borkar VS, Jain S, Rangarajan G (1998a) Dynamics of individual specialization and global diversification in communities. Complexity 3:50-56
2. Borkar VS, Jain S, Rangarajan G (1998b) Generalized replicator dynamics as a model of diversifying societies. Adv. Complex Sys. 1:325-359
3. Hofbauer J, Sigmund K (1988) The Theory of Evolution and Dynamical Systems. Cambridge University Press, Cambridge
4. Weibull JW (1995) Evolutionary Game Theory. MIT Press, Cambridge

A Complex Adaptive Model of Economic Production Networks

A.Ponzi[1], A.Yasutomi[2], and K.Kaneko[1]

[1] Department of Pure and Applied Sciences, University of Tokyo, Komaba, Meguro-ku, Tokyo 153, Japan.
adam@complex.c.u-tokyo.ac.jp

[2] Department of Area Studies, Graduate School of Arts and Sciences, University of Tokyo, Komaba, Meguro-ku, Tokyo 153, Japan.

Summary. We describe a new model of a complex adaptive economy of evolving production processes. We present numerical simulations which illustrate the emergence of price levels and multiple timescales. We briefly explain the origin of this behaviour.

1 Introduction

We desribe a new complex adaptive dynamical system model of a evolving production economy[1]. Our motivation in producing this model is to take a step towards understanding the complex dynamics of real economies, including the self-organization of complex production networks, the appearance of periodic recessions, unemployment, price inflation and deflation from the viewpoint of complex dynamical systems theory. The model is the simplest we can envisage which also produces behaviour of comparable complexity to a real economy. We imagine that this model will be useful in understanding how to make economic production processes more efficient as well as managing non-equilibrium macroeconomic behaviour. It is based on a neoclassical model of economic production processes due to von-Neumann[2, 3] and catalytic chemical reaction network dynamics[5].

The original von Neumann model (VNM) of balanced economic growth assumes that each good is produced jointly with certain others, in an analogous way to a chemical reaction. In this model capital goods at different stages of wear and tear are different goods and a production process is defined as operation which converts one bundle of goods, including capital equipment, into another bundle of goods, including the capital goods (at appropriate wear and tear) which are carried forward for use in further production. Capital goods therefore function something like *catalysts* in chemical reactions, reformed at the end of the reaction in an amount conserved in the reaction.

In the VNM it is assumed that consumption of goods takes place only through the processes of production, including necessities of life consumed by workers, and all income is reinvested in production. In the VNM there are a fixed number of processes and a fixed number of products in the system. Therefore the VNM is defined by a fixed input matrix and a fixed output matrix, representing, respectively, the fixed ratios of input products and output products for each process. An example of such a network in shown in Fig.1. Each process is assumed to have unit time duration, longer processes may be broken down into several processes with intermediate products each with unit time duration. Furthermore each process has an intensity, and each product has a price. For example a process may be given by: inputs: baker, dough, oven, coal and lunch; outputs: baker, oven, bread and waste (including heat). The ratios these products 'react' in are of course fixed and the baker and oven are catalysts (except for some 'wear and tear'). The intensity would correspond to the amount of bakers etc. employed (the quantity of baker here is considered a continuous variable like the dough.) An 'auto-catalytic' economic process may be for example: inputs: untrained workers, company training department; output: larger company training department.

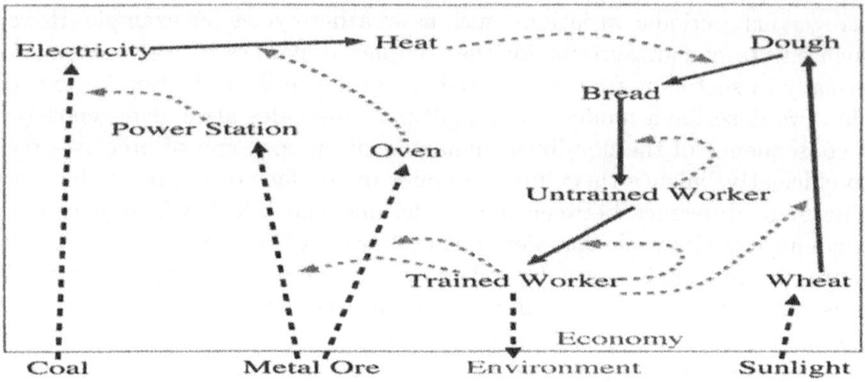

Fig. 1. An Economy. Solid lines are production flows, long dashed lines flows to and from the environment (including other countries) and curly dotted lines are catalytic effects. Some processes are autocatalytic.

The VNM is defined as a static equilibrium model describing relationships between the variables which must hold at equilibrium. Equilibrium is a state of 'balanced growth' where prices are constant and intensities of production grow or decay at constant geometric rates. There are no dynamics defined by the model which might describe out of equilibrium or approach to equilibrium behaviour. However it is rarely the case the economic processes are in equilibrium. For example consider a simple oil refinery with inputs: crude oil, trained workforce, machinery and outputs: workforce, machinery, petroleum.

Now suppose for example the supply of crude increases in the short term (another refinery breaks down!), this decreases the cost of the crude oil and the refinery may try to move to process more crude. However such action may be limited by the fact that extra trained workforce may not be available. Even if the extra crude supply continues for long enough to train sufficient extra workforce, the even longer term process of building more refineries may limit the processing capacity. At some later date the supply of crude may decrease again leaving unemployed workforce and idle machinery which cannot be sold off rapidly. Here one observes that economic processes are confounded by the multiple timescale nature of economic supplies and the inherent unpredicatibilty of future supplies and demands. The different timescales are clearly related to the very different costs of inputs. For example the machinery needed to process a barrel of oil is much more expensive than the workforce necessary to operate the machinery which is much more expensive than the barrel of oil. This is due to the different supplies of these quantities available in the economy as a whole. Process supplies varying on different timescales mean actual economic processes, and indeed entire economies, are often 'inefficient' and rarely in equilibrium with periods of idle machinery and high unemployment and other periods when the workforce cannot cope with the available work.

Multiple economic timescales are sometimes considered to be caused by 'exogenous' periodic variations, such as weather cycles for example. However such effects cannot account for the complex dynamics of real economies especially in sectors not closely related to agricultural production for example. Here we describe a model where multiple timescales arise 'endogenously' as a consequence of the non-linear dynamics of an economy of processes trying to efficiently balance their input supplies in the face of unpredictable prices. The main difference between our model and the VNM, which gives rise to this multiple timescale complexity is that we explicity take into account how a process' production rate depends on its available input supplies. Indeed as illustrated above the rate of an economic process depends only on quantity of the *minimum* of its input supplies, and not on the quantities of its other supplies. For example a baker working at full pace can only fill a certain amount of ovens, increasing the amount of ovens further will not increase the rate of production of bread. Similarly employing more bakers will not increase your bread production if the ovens are already full. This is in direct contrast with a chemical reaction obeying the law of mass action[4], where an increase in concentration of any of its input species will increase the rate of reaction.

2 Model

This model has two dynamical behaviours which are considered to be occurring simultaneously. (i) Processing, where all production processes simultaneously produce goods and consume supplies, according to their defining fixed input and ouput stochiometric ratios, a_{ij} and b_{ij} respectively, where i labels the processes and j the products. (ii) Marketing, where prices are set, processes revalued and goods exchanged. (i) Processing: Each process i at time t is considered to have access to supplies of its input products $S_{ij}(t)$. How the processes share supplies of inputs from the total supply $S_j(t) = \sum_i S_{ij}(t)$ of product j available in the economy is defined by the marketing dynamic explained below. However as in reality, each process may not be able to obtain these supplies in their exact proportions a_{ij} for the production reaction to proceed without any unused inputs. In this case a process' rate of production is given by, $R_i(t) = \alpha Min_j(S_{ij}(t)/a_{ij})$ where Min_j denotes the minimum over its input supplies and $0 < \alpha < 1$ is a parameter which describes the relative timescale of processing and marketing. If we denote a process' supplies after processing with '*', this means in general, $S^*_{ij}(t) = S_{ij}(t) + (\beta b_{ij} - a_{ij})R_i(t)$ where b_{ij} are the output stochiometric ratios for process j. If we consider the stochiometric ratios normalized so that $\sum_j a_{ij} = \sum_j b_{ij}$ for each process i the parameter $0 < \beta < 1$ is a simple way to represent waste (including heat) production, that is the process *efficiency*.

(ii) Marketing: In this model all goods j also have a dynamically changing *market price*, $p_j(t)$, defined collectively throughout the economy. They are set by continuously equalizing total *demand* for product j to total *supply* of product j. This is the simplest, yet plausible, way to allocate prices. In order to define prices we must include *money* in the model and we do this by considering each process to have a *funds* $F_i(t)$ measured in dollars. Each process uses its current funds to compete for supplies of goods available in the economy which it then uses for subsequent processing. To do this each process i divides its funds $F_i(t)$ into demands $D_{ij}(t)$ for products j appropriately so that $D_{ij}(t) = F_i(t)\frac{a_{ij}p_j(t-1)}{\sum_j a_{ij}p_j(t-1)}$. The process cannot use $p_j(t)$ to allocate its funds into demands since the price $p_j(t)$ itself depends on the demands! The new price $p_j(t)$ is then defined by $p_j(t)(\sum_i S^*_{ij}(t)+S^e_j) = (\sum_i D_{ij}(t)+D^e_j)$. Here S^e_j and D^e_j are possible external supplies and demands coming from, say, another country. Trading and process revaluation are accomplished in a simple mean field way according to this aggregate price so that $S_{ij}(t+1) = D_{ij}(t)/p_j(t)$ and $F_i(t+1) = \sum_j S^*_{ij}(t)p_j(t)$. This means that if $p_j(t) = p_j(t-1)\ \forall j$ the process will obtain input supplies in exactly the stochiometric ratios it needs.

The self-referential demand-price optimization puzzle encountered above will often be treated using a repeated Walrasian tatonnement or by iterated game theoretic analysis. Here we suggest that a more realistic 'mean-field' behaviour will be as described here, where processes with larger funds will obtain proportionally larger shares of the input supplies and processes with larger shares of the output supplies will obtain proportionally larger shares of the available funds. This means however that not all processes will obtain inputs in the exact stochiometric ratios they desire. This however is unavoidable since at any time t the supplies necessary for all processes to be perfectly satisfied may not be present in the economy, given the non-linear processing dynamics, and furthermore this is precisely what occurs in reality and is the basic cause of non-equilibrium in this model and in the real economy.

It is important to realize that in this trading process there may be no net exchange of goods. Indeed the process both sells its current supply and buys new supply at the same price, $p_j(t)$. The above prescription should be thought of as a simplified continuous *revaluation* according to the entire market demand and the subsquent expansion or contraction of the process occurs accordingly. In this way when processes are coupled together the system becomes an evolving network. Each process' current funds $F_i(t)$ representing its general *fitness value.*

3 Results & Discussion

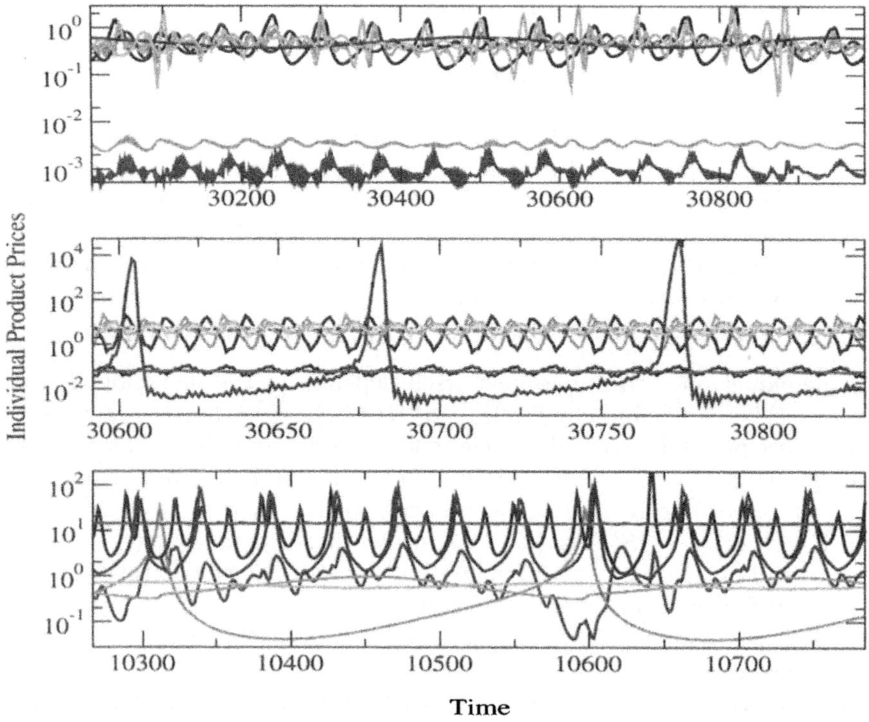

Fig. 2. Three examples of product price $p_j(t)$ time series produced from 15 process, 10 product open random networks.

Shown in Fig.2 are three examples of snapshots from product prices $p_j(t)$ time series, produced from random networks with 15 processes and 10 products, where each process is the simplest possible and has one input product, one output product and one catalyst product, each chosen randomly from the 10 products. The economy is open, where half of the ten products have non-zero fixed random external supplies or demands and where the processes generate 10% waste so that $\beta = 0.9$. Furthermore the demands prescription has been generalised to $D_{ij}(t) = F_i(t)(\gamma \frac{a_{ij} p_j(t-1)}{\sum_j a_{ij} p_j(t-1)} + (1-\gamma) a_{ij})$. This is because a process need not necessarily base its demands completely on the previous prices $p_j(t-1)$. In the results in Fig.2, $\gamma = 0.1$.

The details are to be published, however the time series have two characteristics. 1) Emergent sets of products with similar average price levels, 2) Emergent multiple timescales. These two characteristics are closely connected. In fact, allthough we start with a network of 15 processes, some of these processes usually go bankrupt, due to competition from other processes. The resultant emergent network usually has several tree-like structured chains of connected processes, in a similar way to that shown in Fig.1. Each set of processes is a chain of input and output products, where the prices of the inputs and outputs move around the same average levels. Different chains however, may cross-catalyse each other, so the chains do interact, in a way similar way to that shown in Fig.1. Infact the multiple timescales appear when different chains with very different price levels cross catalyse each other. This is because high prices in one chain, for example, usually mean restricted supplies, which through the minimum condition control the processing timescales in other chains which use the first chains products as catalysts. This is the origin of the complexity of economic processes.

We describe a new model of an emergent economy with dynamical behaviour as rich as a real economy. We expect further analysis of this model will be of great value in understanding real economic complexity and will have applications in efficiency maximization within companies where many processes are coupled.

4 Acknowledgement

A.Ponzi wishes to thank the Japan Society for the Promotion of Science for support of this research.

References

1. A.Ponzi,A.Yasutomi,K.Kaneko, *A Non-Linear Model of Economic Production Processes*, to appear in Physica A.
2. J.von Neumann, *Review of Economic Studies* **31** 1-9 (1945).
3. M.Morishima, *Theory of Economic Growth*, (Oxford University Press 1970).
4. P.W.Atkins, *Physical Chemistry*, (Oxford University Press 1994).
5. K.Kaneko & T.Yomo, *Bull.Math.Bio*, **59** 139-196 (1997).

GPSR Compliance

The European Union's (EU) General Product Safety Regulation (GPSR) is a set of rules that requires consumer products to be safe and our obligations to ensure this.

If you have any concerns about our products, you can contact us on ProductSafety@springernature.com

In case Publisher is established outside the EU, the EU authorized representative is:

Springer Nature Customer Service Center GmbH
Europaplatz 3
69115 Heidelberg, Germany

Batch number: 10755284

Printed by Printforce, the Netherlands